Genomic Perl

This introduction to computational molecular biology will help programmers and biologists learn the skills they need to start work in this important, expanding field. The author explains many of the basic computational problems and gives concise, working programs to solve them in the Perl programming language. With minimal prerequisites, the author explains the biological background for each problem, develops a model for the solution, and then introduces the Perl concepts needed to implement the solution.

The book covers pairwise and multiple sequence alignment, fast database searches for homologous sequences, protein motif identification, genome rearrangement, physical mapping, phylogeny reconstruction, satellite identification, sequence assembly, gene finding, and RNA secondary structure. The author focuses on one or two practical approaches for each problem rather than an exhaustive catalog of ideas. His concrete examples and step-by-step approach make it easy to grasp the computational and statistical methods, including dynamic programming, branch-and-bound optimization, greedy methods, maximum likelihood methods, substitution matrices, BLAST searching, and Karlin–Altschul statistics.

Rex A. Dwyer founded Genomic Perl Consultancy, Inc. in July 2001. He was formerly an Associate Professor of Computer Science at North Carolina State University, where he taught data structures, algorithms, and formal language theory and demonstrated his skill as both theoretician and practitioner. He has published more than a dozen papers in academic journals such as *Advances in Applied Probability, Algorithmica,* and *Discrete and Computational Geometry.* His accomplishments as a Perl software developer include a proprietary gene-finding system for the Novartis Agribusiness Biotechnology Research Institute (now part of Syngenta, Inc.) and a web-accessible student records database for faculty colleagues at NCSU.

Genomic Perl

From Bioinformatics Basics to Working Code

REX A. DWYER
Genomic Perl Consultancy, Inc.

CAMBRIDGE
UNIVERSITY PRESS

PUBLISHED BY THE PRESS SYNDICATE OF THE UNIVERSITY OF CAMBRIDGE
The Pitt Building, Trumpington Street, Cambridge, United Kingdom

CAMBRIDGE UNIVERSITY PRESS
The Edinburgh Building, Cambridge CB2 2RU, UK
40 West 20th Street, New York, NY 10011-4211, USA
477 Williamstown Road, Port Melbourne, VIC 3207, Australia
Ruiz de Alarcón 13, 28014 Madrid, Spain
Dock House, The Waterfront, Cape Town 8001, South Africa

http://www.cambridge.org

First published 2003

Printed in the United States of America

Typeface Times 10/13 pt.   *System* AMS-T$_{\text{E}}$X   [FH]

A catalog record for this book is available from the British Library.

Library of Congress Cataloging in Publication data available

ISBN 0 521 80177 X hardback

A mi chapinita y a mi chapinito.

Contents

Preface

This book is designed to be a concrete, digestible introduction to the area that has come to be known as "bioinformatics" or "computational molecular biology". My own teaching in this area has been directed toward a mixture of graduate and advanced undergraduate students in computer science and graduate students from the biological sciences, including biomathematics, genetics, forestry, and entomology. Although a number of books on this subject have appeared in the recent past – and one or two are quite well written – I have found none to be especially suitable for the widely varying backgrounds of this audience.

My experience with this audience has led me to conclude that its needs can be met effectively by a book with the following features.

- To meet the needs of computer scientists, the book must teach basic aspects of the structure of DNA, RNA, and proteins, and it must also explain the salient features of the laboratory procedures that give rise to the sorts of data processed by the algorithms selected for the book.
- To meet the needs of biologists, the book must (to some degree) teach programming and include *working programs* rather than abstract, high-level descriptions of algorithms – yet computer scientists must not become bored with material more appropriate for a basic course in computer programming.
- Justice to the field demands that its statistical aspects be addressed, but the background of the audience demands that these aspects be addressed in a concrete and relatively elementary fashion.

To meet these criteria, a typical chapter of this book focuses on a single problem that arises in the processing of biological sequence data: pairwise sequence alignment, multiple alignment, sequence database searching, phylogeny reconstruction, genome rearrangement, and so on. I outline both the biological origins of the input and the interpretation of the desired output; then I develop a single algorithmic approach to the problem. Finally, I show how to implement the algorithm as a working Perl program. Variations on the problem and/or improvements to the program are presented in exercises at the end of each chapter. In a few cases, I develop a straight-forward but inefficient algorithm in one chapter and devote the following chapter to a more sophisticated approach. Bibliographic notes in each chapter are limited to a half dozen of the most accessible references; I assume that a serious student can use these hooks into the literature to track down other relevant resources.

The choice of the Perl language as the medium for presenting algorithms might surprise some, but it has many advantages.

- Perl's built-in strings, lists, and hash tables make it possible to express some algorithms very concisely. (Chapter 12's 80-line program for constructing suffix trees is an outstanding example.)
- Perl is already widely used for server-side scripting (CGI) in web-based applications, and a large library of code (the bioPerl effort described at `www.bioperl.org`) is freely available to assist bioinformatic programmers.
- Perl falls outside the standard computer science curriculum. This means that attention to language details will be of some interest even to students with strong computer science backgrounds.
- Perl is portable; it runs on all major operating systems.
- Perl is available without charge as part of Linux distributions and from the World Wide Web at `http://www.perl.org`.
- Rare but legal Perl constructs like @{$_[$#_]||[ ]} inspire awe in the uninitiated; this awe sometimes rubs off on Perl programmers.

Perl also has a few disadvantages.

- Perl programs are not compiled to native-mode code, so they often run many times slower than equivalent C or C++ programs in computationally intensive applications. (In CGI applications dominated by disk-access or network-response times, this disadvantage evaporates.)
- Perl's built-in strings, lists, and hash tables sometimes hide potential performance problems that can be overcome only with nonintuitive tricks.
- Perl is profligate in its use of memory. This means that the input size that a Perl program can handle may be many times smaller than for an equivalent C or C++ program.

All in all, the advantages prevail for the purposes of this book, and – although using Perl makes many of the programs in this book teaching toys rather than production-grade tools – they do work. Unlike pseudocode descriptions, they can be modified to print traces of their operation or to experiment with possible improvements. They can also serve as prototypes for more efficient implementations in other languages.

In the interest of clarity, here are a few words about what this book is *not*.

- This book is not a consumer's guide to the many software packages available to assist the biologist with sequence assembly, database searching, phylogeny reconstruction, or other tasks. Such packages will be mentioned in passing from time to time, but the focus of this book is on how these packages solve (or might solve) the problems presented to them and not on how problems must be presented to these packages.

- This book is not an encyclopedic compendium of computational aspects of molecular biology. The problems included are presumed to be significant to biologists, but other equally significant problems have doubtless been omitted.

- Likewise, this book is not a tutorial or reference manual on the Perl language. Features of Perl will be explained and used as needed, and diligent readers should come away with a good knowledge of Perl, but algorithms are the focus, and many widely used Perl features will be left unmentioned.

I embarked upon this writing project not to convey to others my own research contributions in the area (since I have made none) but rather to study the contributions of others and make them more accessible to students. It is certain that my own lack of expertise will show through now and again. Despite this, I hope that my concrete approach will permit a much larger audience to understand and to benefit from the contributions of the real experts.

Acknowledgments

Special thanks to Cambridge University Press and my editor, Lauren Cowles, for accepting a book proposal from a first-time author with no publications dealing directly with the subject matter of the book. Ms. Cowles's careful comments and other assistance in the preparation of the manuscript have been invaluable.

Phil Green of the University of Washington was kind enough to allow me to use the source of his PHRAP program to develop Chapter 13. Alejandro Schäffer and Steven Altshul of NCBI offered helpful feedback on chapters dealing with multiple alignment and BLAST.

While I was a faculty member at North Carolina State University, the students of CSC 695 in Spring 1997 and Spring 1999 and CSC 630 in Spring 2001 accepted and proofread an ever-growing collection of notes and programs upon which this book was eventually based. Among these, Barry Kesner deserves special mention for giving very helpful feedback on several chapters. Tim Lowman, systems administrator extraordinaire, was most helpful in answering Perl questions in the early weeks of this project.

Financially, this writing project was partially supported by NC State University in the form of sabbatical support. The laboratory now known as Syngenta Biotechnology, Inc. (in Research Triangle Park, NC) complemented this support on several occasions while permitting me to experiment with proprietary data.

Last but not least, Pam and Kevin Bobol, proprietors of the New World Cafe and Larry's Beans, provided the chairs in which a large fraction of this book was written. Staff members Amy, D.J., Heather, James, Jeramy, Kathy, Margaret, Marwa, Patty, Sam, Sue, and others served the hot liquids that kept my fingers moving.

As expected, any remaining errors are rightfully counted against my own account alone and not against that of my editor, colleagues, students, or barristas.

## CHAPTER ONE

The Central Dogma

1.1 DNA and RNA

Each of us has observed physical and other similarities among members of human families. While some of these similarities are due to the common environment these families share, others are *inherited,* that is, passed on from parent to child as part of the reproductive process. Traits such as eye color and blood type and certain diseases such as red–green color blindness and Huntington's disease are among those known to be heritable. In humans and all other nonviral organisms, heritable traits are encoded and passed on in the form of *deoxyribonucleic acid,* or DNA for short. The DNA encoding a single trait is often referred to as a *gene.*[1] Most human DNA encodes not traits that distinguish one human from another but rather traits we have in common with all other members of the human family. Although I do not share my adopted children's beautiful brown eyes and black hair, we do share more than 99.9% of our DNA. Speaking less sentimentally, all three of us share perhaps 95% of our DNA with the chimpanzees.

DNA consists of long chains of molecules of the modified sugar deoxyribose, to which are joined the *nucleotides* adenine, cytosine, guanine, and thymine. The scientific significance of these names is minimal – guanine, for example, is named after the bird guano from which it was first isolated – and we will normally refer to these nucleotides or *bases* by the letters A, C, G, and T. For computational purposes, a strand of DNA can be represented by a string of As, Cs, Gs, and Ts.

Adenine and guanine are *purines* and share a similar double-ring molecular structure. Cytosine and thymine are *pyrimidines* with a smaller single-ring structure. Deoxyribose has five carbons. The conventions of organic chemistry assign numbers to the carbon atoms of organic molecules. In DNA, the carbon atoms of the nucleotides are numbered 1–9 or 1–6, while those of the sugar are numbered $1'$ ("one prime"), $2'$, $3'$, $4'$, and $5'$. As it happens, the long chains of sugar molecules in DNA are formed

---

[1] We will refine this definition later.

by joining the $3'$ carbon of one sugar to the $5'$ carbon of the next by a *phosphodiester bond*. The end of the DNA chain with the unbound $5'$ carbon is referred to as the *$5'$ end*; the other end is the *$3'$ end*. For our purposes, it is enough to know two things: that single DNA strands have an orientation, since two $3'$ ends (or two $5'$ ends) cannot be joined by a phosphodiester bond; and that strings representing DNA are almost always written beginning at the $5'$ end.

Ribonucleic acid, or RNA, is similar to DNA, but the sugar "backbone" consists of ribose rather than deoxyribose, and uracil (U) appears instead of thymine. In a few simple organisms, such as HIV,[2] RNA substitutes for DNA as the medium for transmitting genetic information to new generations. In most, however, the main function of RNA is to mediate the production of proteins according to the instructions stored in DNA.

As its name suggests, deoxyribose can be formed by removing an oxygen atom from ribose. Although RNA is itself an accomplished molecular contortionist, a chain or *polymer* made of deoxyribose can assume a peculiar coiled shape. Furthermore, pairs composed of adenine and thymine joined by *hydrogen bonds* and similarly joined pairs of cytosine and guanine have similar shapes; (A,T) and (C,G) are said to be *complementary* base pairs (see Figure 1.1). Taken together, these two characteristics allow DNA to assume the famous *double helix* form, in which two arbitrarily long strands of complementary DNA base pairs entwine to form a very stable molecular spiral staircase.[3] Each end of a double helix has the $3'$ end of one strand and the $5'$ end of the other. This means that two strands are *complementary* if one strand can be formed from the other by substituting A for T, T for A, C for G, and G for C – and then reversing the result. For example, ATTCCTCCA[4] and TGGAGGAAT are complementary:

```
5'-ATTCCTCCA-3'
3'-TAAGGAGGT-5'
```

The double helix was first revealed by the efforts of Watson and Crick; for this reason, complementary base pairs are sometimes referred to *Watson–Crick pairs*. In fact, the names "Watson" and "Crick" are sometimes used to refer to a strand of DNA and its complement.

1.2 Chromosomes

Each cell's DNA is organized into *chromosomes,* though that organization differs tremendously from species to species.

---

[2] Human immunodeficiency virus, the cause of AIDS.

[3] A spiral staircase is, in fact, no spiral at all. A spiral is defined in cylindrical coordinates by variations of the equations $z = 0$; $r = \theta$. The equations $z = \theta$; $r = 1$ define a *helix*.

[4] This sequence, known as the *Shine–Dalgarno sequence,* plays an important role in the initiation of translation in the bacterium *E. coli.*

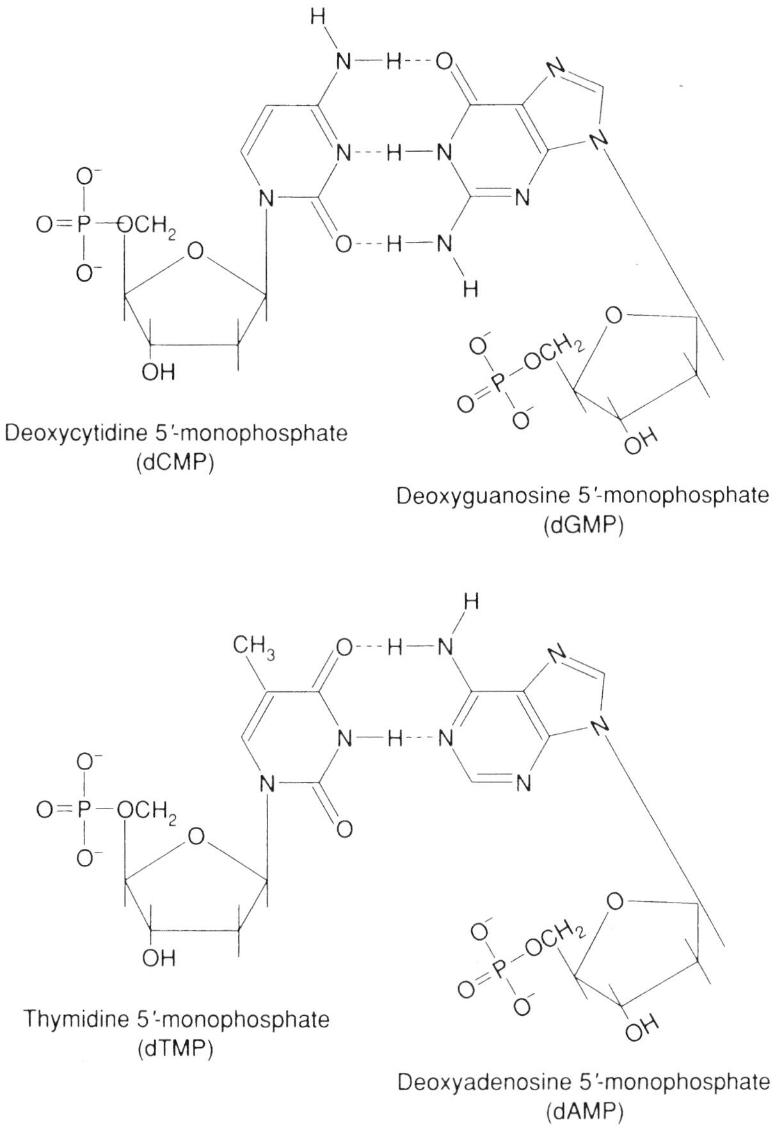

Figure 1.1: The nucleotides C and G (above) and A and T (below), showing how they can form hydrogen bonds (dotted lines). (Reproduced from Hawkins 1996.)

Human cells have 24 distinct types of chromosomes, with a total of about three billion (3×10^9) base pairs of DNA.[5] Among these, the *autosomes* are numbered 1–22 from largest to smallest, and the *sex chromosomes* are named X and Y. Each cell contains a pair of each autosome and either two X chromosomes (females) or one

---

[5] If denatured and stretched out, the DNA in each cell's nucleus would be about one yard (94 cm) long.

X and one Y chromosome (males). Egg and sperm cells, collectively known as *germ cells,* are exceptions to this rule; each contains only one of each autosome and one sex chromosome. Taken together, the 24 types of human chromosome constitute the human *genome.*

The pairs of autosomes in a cell should not be confused with the double-stranded nature of DNA. Each Chromosome 1 is double-stranded. Furthermore, the two Chromosomes 1 are nearly identical but not completely so. Wherever one contains a gene received from the mother, the other contains a gene from the father. This state of affairs is called *diploidy* and is characteristic of species that can reproduce sexually.[6]

Multiple, linear chromosomes are characteristic of the cells of *eukaryotes,* organisms whose chromosomes are sequestered in the cell's *nucleus.*[7] However, not all eukaryotes are diploid. The bread mold *Neurospora crassa* is *haploid,* meaning that each cell has only a single copy of each of its seven types of chromosomee. Mold cells reproduce asexually by dividing.

Simpler organisms called *prokaryotes* lack a cell nucleus. The bacterium *Escherichia coli,* a well-studied inhabitant of the human bowel, has a single, circular chromosome with about 4.5 million base pairs. *Viruses* are simplest of all, consisting only of genetic material – RNA, or either single- or double-stranded DNA – in a container. Viruses cannot reproduce on their own. Instead, like molecular cuckoos, they co-opt the genetic machinery of other organisms to reproduce their kind by inserting their genetic material into their host's cells. The genetic material of the virus ΦX174, which infects *E. coli,* consists of only 5386 bases in a single-stranded ring of DNA.[8]

1.3 Proteins

Like DNA and RNA, proteins are polymers constructed of a small number of distinct kinds of "beads" known as *peptides, amino acids, residues,* or – most accurately but least commonly – *amino acid residues.* Proteins, too, are oriented, and they are normally written down from the *N-terminus* to the *C-terminus.* The names of the 20 "natural"[9] amino acids, together with common three- and one-letter abbreviations, are noted in Figure 1.2.

Some proteins give organisms their physical structure; good examples are the keratins forming hair and feathers and the collagen and elastin of ligaments and tendons. Others, much greater in variety if lesser in mass, catalyze the many chemical reactions required to sustain life. Protein catalysts are called *enzymes,* and their names can be recognized by the suffix *-ase.* Proteins do not assume a predictable, uniform

---

[6] Not all diploid species have distinct sex chromosomes, however.

[7] Greek *karyos* is equivalent to Latin *nucleus; eu-* means "good, complete".

[8] Viruses that infect bacteria are also called *bacteriophages,* or simply *phages.*

[9] Selenocysteine, abbreviated by U, has been recently recognized as a rare 21st naturally occurring amino acid. When occurring, it is encoded in RNA by UGA, which is normally a stop codon.

Alanine	A	Ala	Leucine	L	Leu
Arginine	R	Arg	Lysine	K	Lys
Asparagine	N	Asn	Methionine	M	Met
Aspartic acid	D	Asp	Phenylalanine	F	Phe
Cysteine	C	Cys	Proline	P	Pro
Glutamine	Q	Gln	Serine	S	Ser
Glutamic acid	E	Glu	Threonine	T	Thr
Glycine	G	Gly	Tryptophan	W	Trp
Histidine	H	His	Tyrosine	Y	Tyr
Isoleucine	I	Ile	Valine	V	Val

Figure 1.2: Amino acids and their abbreviations.

shape analogous to DNA's double helix. Instead, protein shapes are determined by complicated interactions among the various residues in the chain. A protein's shape and electrical charge distribution, in turn, determine its function.

Predicting the shape a given amino acid sequence will assume *in vivo*[10] – the *protein-folding problem* – is one of the most important and most difficult tasks of computational molecular biology. Unfortunately, its study lies beyond the scope of this book, owing to the extensive knowledge of chemistry it presupposes.

1.4 The Central Dogma

The Central Dogma of molecular biology relates DNA, RNA, and proteins. Briefly put, the Central Dogma makes the following claims.

- The amino acid sequence of a protein provides an adequate "blueprint" for the protein's production.
- Protein blueprints are encoded in DNA in the chromosomes. The encoded blueprint for a single protein is called a *gene*.
- A dividing cell passes on the blueprints to its daughter cells by making copies of its DNA in a process called *replication*.
- The blueprints are transmitted from the chromosomes to the protein factories in the cell in the form of RNA. The process of copying the DNA into RNA is called *transcription*.
- The RNA blueprints are read and used to assemble proteins from amino acids in a process known as *translation*.

We will look at each of these steps in a little more detail.

The Genetic Code. A series of experiments in the 1960s cracked the genetic code by synthesizing chains of amino acids from artificially constructed RNAs.

---

[10] "In life" – as opposed to *in vitro* or "in glass" (in the laboratory). The process of predicting the shape computationally is sometimes called protein folding *in silico*.

Amino acids are encoded by blocks of three nucleotides known as *codons*. There are $4 \times 4 \times 4 = 64$ possible codons, and (except for methionine and tryptophan) each amino acid is encoded by more than one codon, although each codon encodes only one amino acid. The end of a protein is signaled by any of three *stop codons*.

DNA comprises more than just codons for protein production. Along with *coding regions,* DNA contains *regulatory regions* such as *promoters, enhancers,* and *silencers* that help the machinery of protein production find its way to the coding regions often enough – but not too often – to keep the cell adequately supplied with the proteins it needs. DNA also encodes some RNAs that catalyze reactions rather than encoding proteins. Most mammalian DNA, however, has no known function and is often referred to as *junk DNA*. The computational process of sifting out probable coding and regulatory regions from what we presently call "junk" is called *gene prediction*.

Replication. The hydrogen bonds that join the complementtary pairs in DNA's double helix are much weaker than the covalent bonds between the atoms within each of its two strands. Under the right conditions, the two strands can be untwisted and separated without destroying the individual strands. A new complementary strand can be constructed on each of the old strands, giving two new double strands identical to the original.

Replication is accomplished with the assistance of two types of enzymes. *DNA helicases* untwist and separate the double helix, and *DNA polymerases* catalyze the addition of free nucleotides to the growing complementary strands. These enzymes work together at the *replication fork,* where the original double strand parts into two single strands.

Transcription. *RNA polymerase* catalyzes the production of RNA from DNA during transcription. The two strands of DNA are separated, and an RNA strand complementary to one of the DNA strands is constructed.

Transcription begins at a site determined by certain *promoter elements* located in the noncoding region at the 5′ end of the coding region, the best-known of which is the "TATA box". How transcription terminates is not well understood in all cases.

In prokaryotes, translation is begun at the free end of the RNA while the other end is still being transcribed from the DNA. But in eukaryotes, the RNA (referred to as a *primary transcript*) is first subjected to a process in the nucleus called *splicing*. Splicing removes certain untranslatable parts of the RNA called *introns*. After splicing, the final *messenger RNA* (mRNA) passes from the nucleus to the cell's *cytoplasm*. Here mRNAs are translated, and many of the resulting proteins remain here to perform their functions.

Translation. Proteins are assembled by *ribosomes* that attach to RNA and advance three bases at a time, adding an amino acid to the protein chain at each step. Ribosomes consist of several proteins as well as small ribosomal RNA molecules (rRNAs) that fold like proteins and act as catalysts. Ribosomes are also assisted by so-called tRNAs.

Translation is initiated when one of the rRNAs in the ribosome binds to a particular sequence of about ten bases in the mRNA.[11] The first codon translated is always AUG (methionine). However, not every AUG codon marks the beginning of a coding region. Translation ends at the first stop codon encountered. A single mRNA molecule can be translated many times to make many identical protein molecules. However, mRNAs eventually degrade until translation is no longer possible.

1.5   Transcription and Translation in Perl

Now we develop a Perl program that can tell us what proteins a given DNA sequence can encode. There are two main steps.

1. Read in a table of codons and amino acids in text form, and construct a Perl data structure that allows codons to be translated quickly.
2. Read strands of DNA from input and write out the corresponding RNA and protein sequences.

We begin every program with

```
#!/usr/bin/perl
use strict;
```

The first line tells the operating system where to find the Perl interpreter.[12] You should consult your system administrator to learn its precise location on your system. The second line tells the Perl system to do as much error checking as possible as it compiles.

Our first programming task is to create and fill a data structure that can be used to look up the amino acid residue corresponding to a single codon. The most convenient structure in Perl is the *hash table,* or just *hash* for short.[13] An empty hash is declared (i.e., brought into existence) by the statement

```
my %codonMap;
```

Once the hash is filled, we will be able, for example, to print Arg, the residue encoded by CGA, with the statement

---

[11] This is the Shine–Dalgarno sequence given previously. The exact sequence varies among species.

[12] Many sources suggest that this line should always be #!/usr/bin/perl -w, which asks the Perl system to issue warnings about constructs in the program that are not strictly errors but that appear suspect. Unfortunately, the -w "switch" is a little too suspicious, and I recommend it only as a debugging tool. Perl also offers a way to turn the warning feature on and off as the program progresses.

[13] The terms "associative array", "look-up table", and "dictionary" are roughly synonymous with "hash".

---

print $codonMap{"CGA"};

Here CGA is the *hash key* and Arg is the corresponding *value.* When we refer to the whole hash, the hash's name is preceded by a percent sign. When referring to an individual element, we precede the name by a dollar sign and follow it by braces.

Although we could fill the hash by listing every codon–residue pair explicitly in our program, we will use Perl's DATA feature to save some typing. This feature allows free-form text to be included at the end of the program file for reading and processing like any other text file during program execution. Our text will have one line for each of the 20 residues plus one for "Stop". Each line will include a three-letter residue abbreviation followed by each of the 1–6 corresponding codons in the genetic code. The first three of these 21 lines are:

```
Ala GCU GCC GCA GCG
Arg CGU CGC CGA CGG AGA AGG
Asn AAU AAC
```

Next we must write code to read in these lines and use them to fill the hash:

```
my $in;                              ## 1
while ($in=<DATA>) {                 ## 2
    chomp($in);                      ## 3
    my @codons = split " ",$in;      ## 4
    my $residue = shift @codons;     ## 5
    foreach my $nnn (@codons) {      ## 6
        $codonMap{$nnn} = $residue;  ## 7
    }
}
```

Line 1 declares a *scalar* variable named $in. A scalar variable can hold a *single* value – an integer, a real number, a character string, Perl's special "undefined" value, or a *reference.* (We will discuss references in greater detail in later chapters.) The names of scalar variables always begin with the dollar sign. Unlike many familiar programming languages, Perl is not strongly typed. During the execution of a program, the *same* scalar variable can hold first the undefined value, later a string, and later a real number. The **my** keyword appears again in Lines 4, 5, and 6; these lines illustrate that a variable can be declared by adding **my** to its first use rather than in a separate statement.

The < > notation on Line 2 causes one line to be read from input. Here, we use the *filehandle* DATA to tell the program to read our residue–codon table from the end of the program file. Each execution of Line 2 reads the next text line, turns it into a

Perl string, and stores it in $in; then, the execution of Lines 3–8 proceeds. With the data given above, the first execution of Line 2 has the same effect as the assignment

$in = "Ala GCU GCC GCA GCG\n";

(The two symbols \n appearing together in a string represent in visible form the single end-of-line or *newline* character that separates lines in a text file.)

If no more lines of text remain, then the "undefined" value (**undef**) is assigned to $in and the **while**-loop terminates. In general, Perl's **while**-loops terminate when the value of the parenthesized condition is undefined, the number 0, the string "0", or the empty string ""; any other value causes the loop to continue. (The nonempty strings "00" and " " do *not* terminate loops!) Since the value of an assignment is the same as the value on its right-hand side, this loop will terminate when the input at the end of the program is exhausted.

Line 3 uses Perl's built-in **chomp** operator to remove the newline character from the string stored in $in.

Line 4 uses the **split** operator to break the string in $in into a *list* of strings and then assigns the resulting list to the list variable @codons. The names of list variables begin with the "at" sign (@) when referring to the whole list. When processing the first line of our data, the effect is the same as

@codons = ("Ala","GCU","GCC","GCA","GCG");

The first operand of **split** is a *pattern* or *regular expression* describing the positions at which the second operand is to be split. The pattern here, " ", is just about the simplest possible; it matches at positions containing a blank-space character. A slightly more sophisticated pattern is /[AU]/, which matches positions containing either A or U. If we had written **split** /[AU]/, $in; then the result would have been the list of four strings ("", "la GC", " GCC GC", " GCG"). We will see other more complicated patterns in future chapters.

The **shift** operator in Line 5 removes the first item from the list @codons and assigns it to $residue, and it has the same effect as

$residue = "Ala";
@codons = ("GCU","GCC","GCA","GCG");

Lines 6 through 8 use the **foreach** construct to repeat the same group of statements on each item in the list @codons. To provide uniform access to the list items, the **foreach**-loop assigns each element of the list to the loop-variable $nnn before executing the body of the loop. The ultimate effect is the same as

```
$codonMap{"GCU"} = "Ala";
$codonMap{"GCC"} = "Ala";
$codonMap{"GCA"} = "Ala";
$codonMap{"GCG"} = "Ala";
```

If we repeat this process for every line of input, then it is clear we will fill the hash %codonMap as desired.

Having filled our hash, we can now read and process DNA strings entered by our user at the terminal:

```
while (my $dna=<STDIN>) {              ## 1
    chomp($dna);                      ## 2
    print "DNA: ", $dna, "\n";        ## 3
    my $rna = transcribe($dna);       ## 4
    print "RNA: ", $rna, "\n";        ## 5
    my $protein = translate($rna);    ## 6
    print "RF1: ", $protein, "\n";    ## 7
    $rna =~ s/.//;                    ## 8
    $protein = translate($rna);       ## 9
    print "RF2: ", $protein, "\n";    ## 10
    $rna =~ s/.//;                    ## 11
    $protein = translate($rna);       ## 12
    print "RF3: ", $protein, "\n\n";  ## 13
}                                     ## 14
```

Lines 1 and 2 are similar to Lines 3 and 4 in the previous code fragment; they read a line from the terminal (STDIN), assign it to $dna, and remove the newline. Line 3 echoes the user's input. Line 4 calls the *subroutine* transcribe. This subroutine takes a string representing DNA as an *argument* and returns a string representing RNA as its *result*. In this case, the result is stored in the variable $rna. Of course, Perl doesn't have a built-in operator for transcribing DNA to RNA; we will write this subroutine ourselves shortly. Line 5 prints the RNA.

Line 6 calls the subroutine translate, which takes an RNA string as an argument and returns a string representing an amino acid sequence. Line 7 prints the sequence.

RNA is translated in blocks of three, and it is possible that the user's input begins or ends in the middle of one of these blocks. Therefore, we would like to print the amino acid sequence encoded by each of the three possible *reading frames*. To print the second reading frame, we delete the first base from the RNA and translate again. Line 8 accomplishes the deletion using Perl's pattern-matching operator =~. The string to be operated on is the one stored in $rna, and the operation will change the value of $rna. In this case, the operation is a substitution, signaled by **s**. Between the first two slashes is the pattern to be replaced, and between the second and third slashes is the replacement. The dot . is a special pattern that matches any single character. The replacement is empty. Since substitution finds and replaces only the *first* occurrence of the pattern, this substitution deletes the first character of $rna. Translating this new string gives the second reading frame, and repeating the process once more gives the third.

Now we must write the subroutines transcribe and translate. We begin with transcribe:

```
sub transcribe {                          ## 1
    my ($dna) = @_;                       ## 2
    my $rna = scalar reverse $dna;        ## 3
    $rna =~ tr/ACGT/UGCA/;                ## 4
    return $rna;                          ## 5
}                                         ## 6
```

Line 1 begins the definition of the subroutine. Line 2 is Perl's peculiar way of relating formal parameters (the variable names used inside the subroutine definition) to actual arguments (the values passed to the subroutine when it is called). For the moment, it will suffice to say that (a) transcribe expects a single value and (b) that value is assigned to the local variable $dna when the subroutine is called.

Since transcription produces RNA that is complementary to the DNA, we must both reverse the string and change bases to their complements. Perl's **reverse** operator can be used both to reverse the order of the characters in a string and to reverse the order of the items in a list, depending on context. To avoid potential confusion, we use the keyword **scalar** to specify the first option explicitly. Line 3 does not change the value of $dna; instead, it creates a new string and assigns it to $rna.

Line 4 uses the pattern-matching operator. This time, the letters **tr** indicate a *transliteration*. Transliteration replaces *every* occurrence of a character between the first two slashes by the corresponding character between the second and third, so every A is replaced by U, every C by G, and so on.

Line 5 returns the RNA string to the main program as the value of the subroutine.

Finally, we come to our RNA-to-protein translation subroutine:

```
sub translate {                           ## 1
    my ($mrna) = @_;                      ## 2
    my $pro = "";                         ## 3
    while ($mrna =~ s/(...)//) {          ## 4
        $pro = $pro . $codonMap{$1};      ## 5
    }                                     ## 6
    return $pro;                          ## 7
}                                         ## 8
```

To translate RNA to protein, we break nucleotides off the RNA string three at a time and look these codons up in %codonMap. The results are collected in the variable $pro using ., the concatenation operator. Our substitution pattern has *three* dots, meaning that the pattern will match the first three bases of the RNA. The empty replacement will delete the three bases. However, since we have wrapped parentheses around the pattern, the portion of the string matched by the pattern will be saved in the special variable $1. We use the saved codon for look-up in %codonMap in Line 5.

The **while**-loop terminates as soon as the pattern match fails – that is, when fewer than three characters remain in $mrna.

If this program is saved in the file cendog.pl, we can execute it on a Unix system by moving to the directory containing the file and typing:[14]

```
./cendog.pl
```

Under either Unix or MS/DOS, the following command line works:

```
perl cendog.pl
```

Here is output from a sample execution of the program on the input ACGGTC CTACCTTTA, including original DNA, RNA transcript, and translations in reading frames 1, 2, and 3:

```
DNA: ACGGTCCTACCTTTA
RNA: UAAAGGUAGGACCGU
RF1: ...Arg...AspArg
RF2:  LysGlyArgThr
RF3:  LysValGlyPro
```

1.6 Exercise

1. One of Perl's mottos is "There's always more than one way to do it." Find at least three other ways to write Lines 4–6 of translate. *Hint:* Check out $&, .=, and **substr** in your favorite Perl reference.

1.7 Complete Program Listings

To illustrate the overall layout of a Perl program, we include a complete listing here as well as in file chap1/cendog.pl in the software distribution.

```
#!/usr/bin/perl
use strict;              ## request conservative error checking
```

---

[14] We must also be sure that the file is "executable", using the chmod command if not.

```perl
my %codonMap;       ## declare a hash table

## transcribe translates DNA strings to RNA
sub transcribe {
    my ($dna) = @_;
    my $rna = scalar reverse $dna;
    $rna =~ tr/ACGT/UGCA/;
    return $rna;
}

## translate translates mRNA strings to proteins
sub translate {
    my ($mrna) = @_;
    my $pro = "";
    while ($mrna =~ s/(...)//) {
        $pro = $pro . $codonMap{$1};
    }
    return $pro;
}

## construct hash that maps codons to amino acids by reading table
## from DATA at the end of the program, which have the following form:
## Residue Codon1 Codon2 ...

while (my $in=<DATA>) {       ## assigns next line of DATA to $in; fails if none
    chomp($in);                    ## remove line feed at end of $in
    my @codons = split " ",$in;
    my $residue
        = shift @codons;      ## remove first item from @codons and assign
    foreach my $nnn (@codons) {
        $codonMap{$nnn} = $residue;
    }
}

## now read DNA strands from input <STDIN> and print translations in all six
## possible reading frames

while (my $dna=<STDIN>) {
    chomp($dna);
    print "DNA: ", $dna, "\n";
    my $rna = transcribe($dna);
    print "RNA: ", $rna, "\n";
    my $protein = translate($rna);
    print "RF1: ", $protein, "\n";
```

```
    $rna =~ s/.//;
    $protein = translate($rna);
    print "RF2: ", $protein, "\n";
    $rna =~ s/.//;
    $protein = translate($rna);
    print "RF3: ", $protein, "\n\n";
}
```

The lines below are not Perl statements and are not executed as part of the
program. Instead, they are available to be read as input by the program
using the filehandle DATA.

```
__END__
Ala GCU GCC GCA GCG
Arg CGU CGC CGA CGG AGA AGG
Asn AAU AAC
Asp GAU GAC
Cys UGU UGC
Gln CAA CAG
Glu GAA GAG
Gly GGU GGC GGA GGG
His CAU CAC
Ile AUU AUC AUA
Leu UUA UUG CUU CUC CUA CUG
Lys AAA AAG
Met AUG
Phe UUU UUC
Pro CCU CCC CCA CCG
Ser UCU UCC UCA UCG AGU AGC
Thr ACU ACC ACA ACG
Trp UGG
Tyr UAU UAC
Val GUU GUC GUA GUG
... UAA UAG UGA
```

1.8   Bibliographic Notes

Two widely used but weighty introductory genetics textbooks are Lewin's (1999) and
Snustad and Simmons's (1999). A lighter but still informative approach is taken in
Gonick and Wheelis's *Cartoon Guide* (1991). Goodsell's (1996) pencil-drawn protein portraits are no less artistic and extremely effective in conveying both the wide
range of functions performed by proteins and the relationship of their forms and
functions.

A visit to the local chain book store will turn up a book on Perl for each of the Arabian Nights. Those in O'Reilly's series are among the most frequently seen on the bookshelves of Perl programmers. Vroman's (2000) slim pocket reference may provide supplement enough to readers who already program well but just not in Perl.[15] The "camel book" (Wall, Christiansen, and Orwant 1997) will be more appropriate for those less comfortable at the keyboard.[16] Srinivasan's (1997) more advanced book may be useful to both types of reader as an accompaniment to the more complex data structures of our later chapters.

A small sense of the history of genetic research can be gained from the following articles. Pelagra once ravaged the poor of the American South; its history (Akst 2000) provides one of many possible illustrations of the unfortunate social consequences of ignorance of the mechanisms of heredity and roles of heredity and environment in the propagation of disease. The discovery of the structure of DNA is famously documented by one of its most active participants in Watson's *Double Helix* (1998). A decade later, his senior partner wrote about the cracking of the genetic code (Crick 1966) for the educated general reader. Frenkel (1991) described the promise of bioinformatics around the time of the field's christening. The cataloging of the three billion nucleotides of the human genome (the collection of all 24 chromosomes) was recently completed; the announcement (Pennisi 2001) was accompanied by widespread discussion about the future importance of the data.

---

[15] It is an elaboration of a free version downloadable from `www.perl.com`.
[16] One of its authors, Larry Wall, is the inventor of Perl.

## CHAPTER TWO

RNA Secondary Structure

In the previous chapter, we concentrated on the flow of information from DNA to RNA to protein sequence and emphasized RNA's role as a "messenger" carrying copies of DNA's protein recipes to the ribosome for production. In this function, RNA's role is similar to the paper in a photocopy machine: though perhaps not always flat, it must be flattened out and dealt with in a linear fashion when used. RNA also plays catalytic roles in which it more resembles the paper used in origami: the shape into which it is folded is what matters most.

2.1 Messenger and Catalytic RNA

RNA is sufficiently versatile as a catalyst that some scientists postulate that the origins of life lie in an "RNA world" preceding both DNA and proteins. Other scientists have invented new RNA catalysts in the laboratory by experimentation with random RNA sequences. Important categories of catalytic RNAs in higher organisms include the following.

transfer RNA (tRNA). These RNAs, typically around 85 bases long, are crucial in the translation process. tRNAs assume a shape approximating the letter "L". In the presence of a ribosome, the three-base *anticodon* at the top of the L is able to bind to a single codon in the messenger RNA. The shape of the base of the L allows it to bind to a single molecule of a particular one of the 20 amino acids. During translation, tRNAs collect their respective amino acids from the cytoplasm with the assistance of enzymes called aminoacyl-tRNA synthetases. When a tRNA approaches a ribosome poised over the appropriate codon on the mRNA, the tRNA binds to the codon and the ribosome. Next, the amino acid is released from the tRNA and attaches to the growing protein. Finally, the tRNA is released from the mRNA and departs in search of another amino acid molecule.

Clearly, at least one form of tRNA must exist for each of the 20 amino acids. In fact, most organisms have at least one tRNA for each of the 61 possible non-stop codons, and even several distinct tRNAs for some codons. Conversely, some anticodons are capable of binding to two distinct codons.

    ribosomal RNA (rRNA). The ribosome itself is a complex of several RNAs and proteins. The size and structure of the ribosome differ from organism to organism. The mammalian ribosome is largest, and it is roughly equal parts RNA and protein by weight. Its four types of RNA are known as the 18S, 28S, 5.8S, and 5S rRNAs,[1] which are roughly 1900, 4700, 156, and 120 bases long (respectively). During translation, the ribosome binds simultaneously to an mRNA and two tRNAs.

    small nuclear RNA (snRNA). This is a catch-all category, but snRNA most often refers to the six RNAs involved in mRNA *splicing*. Splicing is an editorial process in which noncoding regions of mRNAs called *introns* are removed. Splicing occurs between transcription and translation. The portions of the mRNA that are eventually translated are called *exons*. The complex of snRNAs and proteins involved in splicing is called the *spliceosome*. The six major human spliceosome snRNAs are named U1–U6; in size they are 165, 186, 217, 145, 116, and 107 bases long.

    signal recognition particle RNA (SRP RNA). The signal recognition particle is a complex of six proteins and one RNA molecule of about 300 nucleotides. In glandular cells, the SRP helps ribosomes producing protein secretions to attach to the rough endoplasmic reticulum, an early step in the protein's passage through the plasma membrane to the cell's exterior.

2.2   Levels of RNA Structure

Like the bases of DNA, those of RNA are able to form hydrogen bonds with their "opposites": A with U (2 bonds) and G with C (3 bonds). However, RNA is single-stranded, and RNA nucleotides form bonds with other nucleotides on the *same* strand. These bonds are important determinants of the RNA's three-dimensional structure and of its chemical properties. The structure of catalytic RNAs is normally described at four distinct levels.

1. *Primary structure* refers to the linear sequence of an RNA's nucleotides, conventionally given in $5'$-to-$3'$ order. Primary structure can be obtained using the polymerase chain reaction and automated sequencing machines, as described in the next chapter. Thus, the 79-base-long sequence

   ```
   CUCUCGGUAGCCAAGUUGGUUUUAAGGCGCAAGACUGUAAAUCUUGAG
   AUCGGGCGUUCGACUCGCCCCCGGGAGACCA
   ```

   is the primary structure of a tyrosine tRNA in yeast.

---

[1] Here "S" stands for *Svedberg units,* which measure sedimentation rates in a process known as *sucrose gradient centrifugation*. Although sedimentation rates are often referred to as "sizes", shape also influences sedimentation rate.

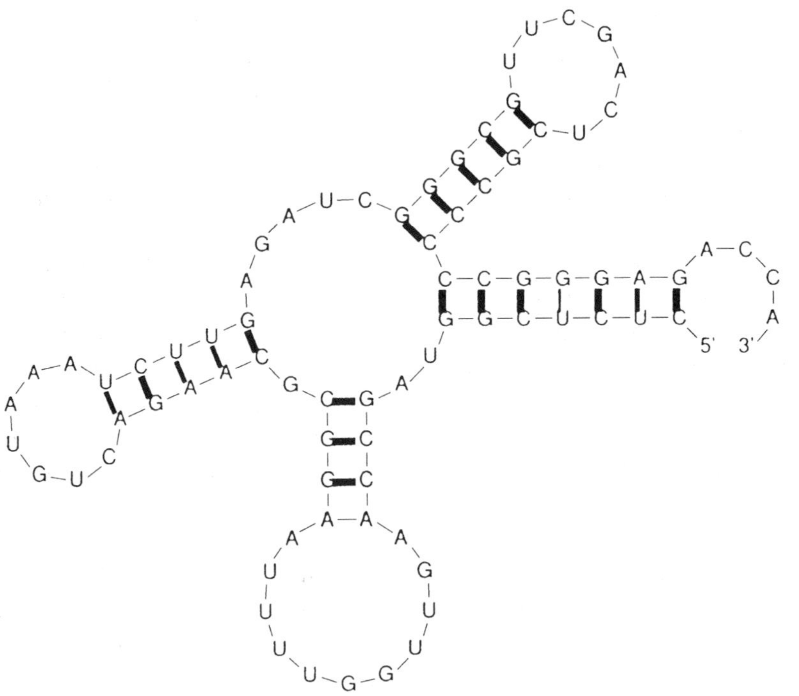

Figure 2.1: Secondary structure of a typical tRNA. The anticodon AUG is on the loop at far left.

2. *Secondary structure* refers to the set of pairs of nonadjacent nucleotides that form hydrogen bonds in the RNA. In addition to the "Watson–Crick pairs" AU and GC, guanine and uracil can pair to form a single hydrogen bond. These GU pairs are sometimes called *wobble pairs*. The secondary structure of the yeast tRNA is depicted in Figure 2.1 by a *cloverleaf diagram*. (Appendix A describes a program for drawing these diagrams.)

3. *Tertiary structure* refers to the precise location of every atom of the RNA molecule in three-dimensional space. A two-dimensional picture of a three-dimensional model of a tRNA is shown in Figure 2.2.

4. *Quaternary structure* refers to the relative location in three-dimensional space of the various RNAs and proteins that form complexes like the spliceosome or the ribosome.

2.3   Constraints on Secondary Structure

It is the tertiary structure of an RNA molecule that tells us the most about its chemical properties. The three-dimensional structure of small RNAs like tRNAs can be determined by X-ray crystallography. Larger RNAs are not amenable to this sort

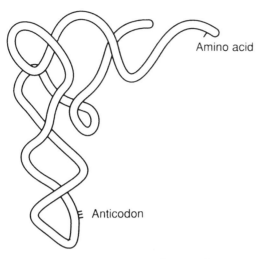

Amino acid

Anticodon

Figure 2.2: Tertiary structure of a typical tRNA (sugar backbone).

of analysis, since they are too difficult to crystallize. Although tertiary structure is not determined solely by secondary structure, secondary structure does provide very strong hints about tertiary structure. Therefore, it is natural to attempt to predict secondary structure from an RNA's nucleotide sequence.

Like all physical systems, RNA tends to move from less stable high-energy states to more stable low-energy states. Since the hydrogen bonds formed by AU, CG, and GU pairs are a key component of RNA stability, a promising simplification of the problem is to seek the configuration of the RNA that maximizes the total number of hydrogen bonds. Of course, the strong covalent bonds holding RNA chains together are not infinitely flexible, and this imposes certain structural constraints on any pairing. For example, the RNA chain cannot twist tightly enough to allow a base to form hydrogen bonds with any of its six nearest neighbors in the chain (three on each side). This can be expressed more formally as follows.

> ***Primary Proximity Constraint.*** If (i, j) is a pair, then $|i - j| > 3$.

From the computational point of view, the following structural constraint is even more significant.

> ***Nesting Constraint.*** If (i, j) and (p, q) are pairs with $i < p < j$
> and $p < q$, then $q < j$.

For example, if the 23rd and 37th bases form a bonded pair, then the 29th base cannot pair with any base preceding the 23rd or following the 37th. The Nesting Constraint prohibits pairings like the ones in Figure 2.3.

In reality, such crossed pairings, called *pseudoknots,* do indeed exist in some RNA molecules. However, they are estimated to constitute only 1–3% of all pairings. The

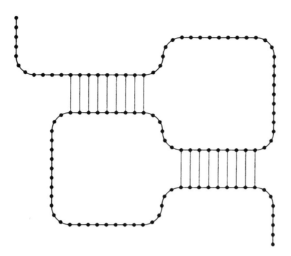

Figure 2.3: Example of a pseudoknot in an RNA secondary structure.

cost in computational efficiency so greatly outweighs the gain in accuracy of secondary structure that researchers generally agree to accept the Nesting Constraint as a valid assumption for the prediction of secondary structure; they relegate pseudoknots to the realm of tertiary structure.

Acceptance of the Nesting Constraint allows secondary structures to be represented by strings of balanced parentheses and dots. For example, the string

$$((\ldots)\ldots(((\ldots))\,.\,))\,.\,.$$

for RNA

AGAAGCCCUUCCCCCGGUAUUU

represents the structure depicted in Figure 2.4 with base pairs $(1, 20) = $ AU, $(2, 6) = $ GC, $(9, 19) = $ UA, $(10, 17) = $ UG, and $(11, 16) = $ CG and with a total of $2 + 3 + 2 + 1 + 3 = 11$ hydrogen bonds.[2]

2.4   RNA Secondary Structures in Perl

In this section, we will develop two programs. The first reads an RNA sequence and proposed folding and evaluates the number of hydrogen bonds in the folding. This program will give us a chance to learn a new Perl feature or two and to become familiar with *recursion*. The second program will read just a sequence and print out

---

[2] Of course, this short example violates the Primary Proximity Constraint.

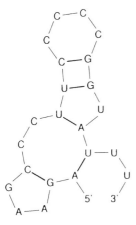

Figure 2.4: An RNA secondary structure satisfying the nesting constraint.

the folding that gives the most hydrogen bonds possible. In it, the powerful *dynamic programming* technique makes the first of its many appearances in this book.

2.4.1   Counting Hydrogen Bonds

Our program to count the H-bonds in a given structure has three components: two subroutines named evalRnaStructure and evalRna, and a short main program. We will write the three components in this order.

The proposed RNA secondary structure is passed to subroutine evalRnaStructure as two strings: $basestring, the nucleotide sequence; and $structurestring, a string of balanced parentheses and dots. evalRnaStructure wraps a new artificial pair of 5 and 3 around the structure to mark the 5′ and 3′ ends of the sequence, splits the strings into a list of single characters stored globally in @bases and @structure, and then calls a recursive subroutine named evalRna to do the bulk of the work.

---

my @structure; **my** @bases;

sub evalRnaStructure {
    my ($basestring,$structurestring) = @_;   *## 1*
    @bases = **split**(//,5.$basestring.3);   *## 2*
    @structure = **split**(//,"($structurestring)");   *## 3*
    return evalRna(0,$#structure);   *## 4*
}

The first Perl feature of interest is the use of the empty pattern // with **split** in Lines 2 and 3. The empty pattern matches at *every* position in the string, so an *n*-character string is split into a list of *n* one-character strings.

Variable Interpolation. The second notable Perl feature used is *variable interpolation* in Line 3. When double quotes are used around a string, any dollar sign within the string is assumed to begin the name of a variable whose *value* is to "interpolated" into the string. The following prints Pal belongs to Arthur..

```
my $owner = "Arthur";
my $pet = "Pal";
print "$pet belongs to $owner.\n";
```

"($structurestring)" is exactly equivalent to the string "(" . $structurestring . ")" formed by explicit use of the concatenation (dot) operator. Double quotes also cause processing of control sequences like \n for newline. Since interpolation is so frequently useful, we will make a habit of using double quotes for all strings unless we have a specific need to avoid variable interpolation – for example, if we must print a dollar sign. No interpolation is performed on strings surrounded by single quotes like 'only $99.95'.[3]

The real work of counting hydrogen bonds is carried out by evalRna. The arguments of evalRna are the starting and ending points in @structure of a balanced string of parentheses. The first and last characters in the list must be a left and a right parenthesis. The corresponding nucleotides in @bases must form a hydrogen-bonded pair, and the number of bonds can be looked up in a hash table (Lines 2 and 3).[4] The rest of the string can be decomposed into zero or more shorter strings of balanced parentheses whose bonds can be counted by calling evalRna again.

In order to identify the shorter balanced strings, we will assign a *nesting level* to each symbol in the structure string. Here is an example of one structure and its nesting levels:

$$((\ldots)\ldots(((\ldots))\ldots))$$
$$01111100123333332210$$

On a left-to-right scan, nesting level increases by 1 when a left parenthesis is encountered and decreases by 1 when a right parenthesis is encountered. The balanced substrings that must be evaluated by recursive calls are the ones that begin and end with (left and right) parentheses at level 1: in our example, $(\ldots)$ and $(((\ldots))\ldots)$.

Lines 6 through 11 perform this left-to-right scan. The current nesting level is maintained in the variable $level. When a right parenthesis at level 1 is found, $ii holds the index of the most recently encountered left parenthesis at level 1. (In general, $ii holds the index of the most recently seen symbol whose predecessor was at level 0.)

---

[3] But we could also write "only \$99.95" for the same result.
[4] Line 2 illustrates one way to initialize a hash. Note that quotes on the strings GU, UG, etc. are optional in this context.

```
sub evalRna {
    my ($l,$r) = @_;                                        ## 1
    my %bonds
        = (GU=>1,UG=>1,AU=>2,UA=>2,CG=>3,GC=>3);           ## 2
    my $numBonds = $bonds{$bases[$l].$bases[$r]};           ## 3
    my $level = 0;                                          ## 4
    my $ii = $l;                                            ## 5

    for (my $i=$l+1; $i<=$r; $i++) {                        ## 6
        $level-- if ($structure[$i] eq ")");               ## 7
        if ($level==0) {                                   ## 8
            $numBonds += evalRna($ii,$i)
                if ($structure[$i] eq ")");                ## 9
            $ii = $i;                                      ## 10
        }
        $level++ if ($structure[$i] eq "(");               ## 11
    }
    return $numBonds;                                      ## 12
}
```

Subroutines like evalRna that call themselves in this fashion are said to be *recursive*. Although it may seem to the uninitiated that a subroutine that calls itself is doomed to do so ad infinitum, this method is in fact safe so long as certain *well-foundedness conditions* are met. In our cases, we can convince ourselves of eventual termination by considering the quantity $r - $l: It always decreases by at least 2 with every recursive call, and no recursive calls are made if it is less than 2. It is very important to remember that every recursive invocation of the subroutine has *its own* local variables like $l, $r, $level, and $numBonds; there is not just a single $numBonds being shared and updated by all the different invocations.

To illustrate the operation of evalRna, we will trace through the recursive calls made to evaluate an example:

```
call evalRna(0,19) for GAGGGUCCUUUCAGUAGCAC, ((...)..(((....)).))
    call evalRna(1,5) for AGGGU, (...)
    evalRna(1,5) returns 2
    call evalRna(8,18) for UUUCAGUAGCA, (((....)).)
        call evalRna(9,16) for UUCAGUAG, ((....))
            call evalRna(10,15) UCAGUA, (....)
            evalRna(10,15) returns 2
        evalRna(9,16) returns 1+2=3
    evalRna(8,18) returns 2+3=5
evalRna(0,19) returns 3+2+5=10
```

The main program does little more than read the input strings and pass them to the subroutines.

```perl
my $basestring = <STDIN>;
chomp($basestring);
my $parenstring = <STDIN>;
chomp($parenstring);
print evalRnaStructure($basestring,$parenstring),
    " hydrogen bonds in this structure.\n";
```

2.4.2   Folding RNA

Our final task is to find the RNA secondary structure that maximizes the number of hydrogen bonds formed, subject to the constraints outlined in Section 2.3.

To begin, let's assume we have the RNA sequence UUUCAGUAGCA. For our purposes, there are really only two kinds of foldings for this sequence: those in which the U in the first position and the A in the last position hydrogen-bond, and those in which they don't. We can construct any folding of the first type by first constructing a folding of the substring UUCAGUAGC and then adding the UA pair:

$$(U, A) + \left(\begin{array}{c} \text{some folding of} \\ \text{UUCAGUAGC} \end{array} \right).$$

We can construct any folding of the second type by first dividing the string UUUCAGUAGCA into two substrings at one of ten possible positions and then folding the two substrings independently. Every such folding falls into one or more of the following ten categories:

$$\left(\begin{array}{c} \text{some folding of} \\ U \end{array} \right) + \left(\begin{array}{c} \text{some folding of} \\ \text{UUCAGUAGCA} \end{array} \right),$$

$$\left(\begin{array}{c} \text{some folding of} \\ UU \end{array} \right) + \left(\begin{array}{c} \text{some folding of} \\ \text{UCAGUAGCA} \end{array} \right),$$

$$\left(\begin{array}{c} \text{some folding of} \\ UUU \end{array} \right) + \left(\begin{array}{c} \text{some folding of} \\ \text{CAGUAGCA} \end{array} \right),$$

$$\left(\begin{array}{c} \text{some folding of} \\ UUUC \end{array} \right) + \left(\begin{array}{c} \text{some folding of} \\ \text{AGUAGCA} \end{array} \right),$$

$$\left(\begin{array}{c} \text{some folding of} \\ UUUCA \end{array} \right) + \left(\begin{array}{c} \text{some folding of} \\ \text{GUAGCA} \end{array} \right),$$

$$\left(\begin{array}{c} \text{some folding of} \\ UUUCAG \end{array} \right) + \left(\begin{array}{c} \text{some folding of} \\ \text{UAGCA} \end{array} \right),$$

$$\left(\begin{array}{c} \text{some folding of} \\ UUUCAGU \end{array} \right) + \left(\begin{array}{c} \text{some folding of} \\ \text{AGCA} \end{array} \right),$$

$$\left(\begin{array}{c} \text{some folding of} \\ \text{UUUCAGUA} \end{array} \right) + \left(\begin{array}{c} \text{some folding of} \\ \text{GCA} \end{array} \right),$$

$$\left(\begin{array}{c} \text{some folding of} \\ \text{UUUCAGUAG} \end{array} \right) + \left(\begin{array}{c} \text{some folding of} \\ \text{CA} \end{array} \right),$$

$$\left(\begin{array}{c} \text{some folding of} \\ \text{UUUCAGUAGC} \end{array} \right) + \left(\begin{array}{c} \text{some folding of} \\ \text{A} \end{array} \right).$$

We say "one or more" because some foldings can be subdivided in more than one way without separating bonded pairs. For example, the 30-base folding `..((.....))....(((....)))(...)` can be divided in eight different positions without separating bonded pairs:

```
          . | .((.....))....(((....)))(...)
         .. | ((.....))....(((....)))(...)
      ..((.....)) | ....(((....)))(...)
      ..((.....)). | ...(((....)))(...)
      ..((.....)).. | ..(((....)))(...)
      ..((.....))... | .(((....)))(...)
      ..((.....)).... | (((....)))(...)
..((.....))....(((....))) | (...)
```

If we needed to know precisely *how many* distinct foldings there were, it would be important that every folding fall into *exactly* one category. But because our goal is simply to find one best folding, it suffices that every folding be included in *at least* one category.

It should be clear that, if we are looking for the single best possible folding of UUUCAGUAGCA and we decide to break this string into UUUCAGU and AGCA, then we need to concern ourselves only with the *best* possible foldings of these two substrings – poor foldings of the substrings will not lead to the best folding of the whole string. More specifically, the best folding for our string is one of the eleven possibilities

$$(\text{U}, \text{A}) + \left(\begin{array}{c} \textit{best } \text{folding of} \\ \text{UUCAGUAGC} \end{array} \right),$$

$$\left(\begin{array}{c} \textit{best } \text{folding of} \\ \text{U} \end{array} \right) + \left(\begin{array}{c} \textit{best } \text{folding of} \\ \text{UUCAGUAGCA} \end{array} \right),$$

$$\left(\begin{array}{c} \textit{best } \text{folding of} \\ \text{UU} \end{array} \right) + \left(\begin{array}{c} \textit{best } \text{folding of} \\ \text{UCAGUAGCA} \end{array} \right),$$

$$\vdots$$

$$\left(\begin{array}{c} \textit{best } \text{folding of} \\ \text{UUUCAGUAGC} \end{array} \right) + \left(\begin{array}{c} \textit{best } \text{folding of} \\ \text{A} \end{array} \right).$$

This suggests that we adopt a strategy of finding the best foldings of shorter substrings first and building on these to find the best foldings of longer substrings. Our problem is typical of those that succumb to *dynamic programming*: Solutions to large problems can be decomposed cleanly into combinations of solutions of similar but smaller subproblems directly derived from the original, and suboptimal solutions of the subproblems need not be considered in order to find the optimal solution of the larger problem.

In our case, we will keep track of the solutions of subproblems in a table c in which the entry c_{ij} in the ith row and jth column is the number of hydrogen bonds in the best folding of the ith through jth bases of our string. We know that no bonds are possible with fewer than five bases, so we will begin with all substrings of length 5, then proceed to all substrings of length 6, then 7, and so forth, until we have found c_{1n}, the number of H-bonds in the best folding of the entire string.

We will implement this strategy in Perl using (a) a two-dimensional array named @c to hold the scores of the best foldings of the substrings and (b) a one-dimensional array @s to hold the bases of the RNA. We fill @s in Line 3 of subroutine foldRna (to follow). We squeeze an X into position 0 of the list @s so that the first character of the input string $s is at list position 1 rather than position 0, since 1-based indexing is customary in biology.

Line 4 controls the loop that advances through all possible sizes of substrings one by one. Controls for **for**-loops have three elements:[5] an *initializer,* a *continuation test,* and an *update,* each separated from its neighbors by semicolons. The behavior of this loop is regulated by manipulating the variable $len. The initializer $len = 5 sets it to 5, and the update $len++ increases it by 1 after each execution of the loop body. The loop continues only as long as $len is less than or equal to $slen, the length of the entire input string. In general, the initializer is executed only once, before the first execution of the loop body; the test is evaluated just prior to every execution of the loop body, including the first; and the update is performed just after every execution of the body. Line 5 controls an inner loop that iterates similarly through all possible substrings of a given length.

Lines 7–10 consider all possible ways that a folding of bases $i through $j can be constructed from foldings of shorter substrings.

- Line 7's $c[$i+1,$j] term accounts for all cases in which base $i is unbonded.
- Line 8's $bonds{$s[$i].$s[$j]}+$c[$i+1][$j−1] term accounts for all cases in which base $i is H-bonded to base $j.
- Line 10's $c[$i][$k]+$c[$k+1][$j] term accounts for all cases in which base $i is H-bonded to base $k. Line 9 causes this term to be evaluated for every base lying between base $i and base $j.

---

[5] Just as in the C and C++ languages.

An oft-exploited Perl feature is its handling of uninitialized variables. The very first execution of Line 7 will be for $i = 1 and $j = 5, causing the program to make an immediate reference to $c[2][5], which has not been defined. In a numerical context such as this one, Perl treats an undefined value as 0. This is precisely what we desire of table entries for substrings shorter than four bases. Hence we are spared the trouble of explicitly initializing these entries.

```
my (@c,@s);
my %bonds = (GU=>1,UG=>1,AU=>2,UA=>2,CG=>3,GC=>3);

sub foldRna {
    my ($s) = @_;                                                    ## 1
    my $slen = length $s;                                            ## 2
    @s = ('X',split(//,$s));                                         ## 3
    for (my $len=5; $len<=$slen; $len++) {                           ## 4
        for (my $i=1; $i<=$slen-$len+1; $i++) {                      ## 5
            my $j = $i+$len-1;                                       ## 6
            $c[$i][$j] = max($c[$i+1][$j],                           ## 7
                        $bonds{$s[$i].$s[$j]}+$c[$i+1][$j-1]);       ## 8
            for (my $k=$i+1; $k<$j; $k++) {                          ## 9
                $c[$i][$j] = max($c[$i][$j],
                            $c[$i][$k]+$c[$k+1][$j]);                ## 10
            }
        }
    }
    return $c[1][$slen];
}
```

At the end of the execution of foldRna, we know the number of hydrogen bonds in the best folding but not the folding itself. However, the bonded pairs can be inferred quickly from the information in @c by a recursive subroutine we call traceBack. This subroutine returns a string of parentheses and dots indicating the best folding. Again, there are several cases corresponding to different types of constructions.

- The line marked ## easy! handles the case of no bonds. Perl's **x** operator is used to produce a string of dots of the required length. The right operand of **x** is always an integer copy count; its left operand can be either a string or a list. Its result is a string or a list containing the specified number of repetitions of the left operand.
- If $c[$i][$j] is the same as Line 7's $c[$i+1][$j] term, then base $i is unbonded in the best possible folding. We construct a string with a dot for position $i and make a recursive call to find the string for positions $i+1..$j.

- If $c[\$i][\$j]$ is the same as Line 8's $bonds\{\$s[\$i].\$s[\$j]\}+\$c[\$i+1][\$j-1]$ term, then bases i and j are H-bonded in the best possible folding. We construct a string with parentheses for positions i and j and make a recursive call to find the string for positions $i+1..\$j-1$.
- Similarly, if we find a position k between i and j for which $c[\$i][\$j]$ is the same as Line 10's $c[\$i][\$k]+\$c[\$k+1][\$j]$ term, then we make *two* recursive calls to find strings for positions $i+1..\$k$ and positions $k+1..\$j$. We concatenate and return these strings. (Execution of **return** terminates both the **for**-loop and the subroutine immediately; some values of k may never be considered.)

```
sub traceBack {
    my ($i,$j) = @_;
    my $cij = $c[$i][$j];
    return ("." x ($j-$i+1)) if ($cij==0);                    ## easy!
    return "." . traceBack($i+1,$j)                           ## foldRna 7
        if ($cij==$c[$i+1][$j]);
    return "(" . traceBack($i+1,$j-1) . ")"                   ## foldRna 8
        if ($cij==$bonds{$s[$i].$s[$j]}+$c[$i+1][$j-1]);
    for (my $k=$i+1; $k<$j; $k++) {
        return traceBack($i,$k) . traceBack($k+1,$j)
            if ($cij==($c[$i][$k]+$c[$k+1][$j]));             ## foldRna 10
    }
}
```

The main program invokes subroutines foldRna and traceBack like this:

```
my $basestring = <STDIN>;
chomp($basestring);
foldRna($basestring);
print "$basestring\n", traceBack(1, length $basestring), "\n";
```

2.5 Exercises

1. Why wasn't Line 2 of evalRnaStructure written using variable interpolation like this?

```
@bases = split(//,"5$basestring3");
```

2. Subroutines evaluateRnaStructure and evalRna were coded recursively partly as an exercise in recursion. In fact, they could be coded more efficiently and perhaps more easily without recursion. Use Perl's **push** and **pop** operators to write

a nonrecursive bond-counting subroutine. Scan the structure left to right, pushing a nucleotide when a left parenthesis is encountered and popping it when the corresponding right parenthesis is encountered.

3. We traced the operation of evalRna for the folding $((\ldots)..(((\ldots))\ldotp))$ of GAGGGUCCUUUCAGUAGCAC. Is this folding the one that maximizes the number of H-bonds?

2.6 Complete Program Listings

The programs of this chapter are included in the software distribution as chap2/evalrna.pl and chap2/foldrna.pl.

2.7 Bibliographic Notes

Dynamic programming was first applied to the prediction of RNA secondary structure by Waterman and Smith (1986). Although hydrogen bonding between pairs of bases is an important determinant of RNA structure, it is far from the only one. The program given in this chapter is, in fact, a very poor predictor of RNA secondary structure. A much better but much more complex program – still based on the dynamic programming approach – is described by Zuker (1989) and Zuker, Mathews, and Turner (1999).

Interesting speculation about an "RNA world" predating the emergence of "life as we know it" can be found in Gesteland, Cech, and Atkins (2000).

## CHAPTER THREE

Comparing DNA Sequences

The human genome comprises approximately three billion (3×10^9) base pairs distributed among 23 pairs of chromosomes. The Human Genome Project commenced in the 1990s with the primary goal of determining the[1] sequence of this DNA. The task was completed ahead of schedule a few months before this book was completed.

Scientists have been anxious to remind the public that the completion of the Human Genome Project's massive DNA sequencing effort will hardly mark the end of the Project itself; we can expect analysis, interpretation, and medical application of the human DNA sequence data to provide opportunities for human intellectual endeavor for the foreseeable future.

It is equally true, though less well understood, that this milestone in the Human Genome Project will not mark the end of massive sequencing. *Homo sapiens* is just one species of interest to *Homo economicus,* and major advances in agricultural productivity will result from ongoing and new sequencing projects for rice, corn, pine, and other crops and their pests. Because of the large differences in size, shape, and personality of different breeds, the Dog Genome Project promises to bring many insights into the relative influences of nature and nurture. Even recreational sequencing by amateur plant breeders may not lie far in the future.

3.1 DNA Sequencing and Sequence Assembly

For many years, DNA sequencing was a time-consuming and labor-intensive procedure. More recently, DNA sequencing has been automated to the extent that strands

---

[1] Of course, it is wrong to speak of "the" sequence, since each of us, as a unique individual, has a different sequence. In fact, due to mutations occurring during development, most of us are *mosaics,* carrying slightly different sequences in different cells. Among the most easily recognized mosaics are persons with one brown and one blue eye. However, it has been estimated that two randomly chosen individuals differ in only about one million base pairs, a number far less than 1% of the total and likewise far less than the number of errors expected to be introduced by the laboratory procedures we are about to discuss. For the moment, we will continue to focus on similarities and speak of *the* sequence of human DNA.

of 500 or more bases are routinely sequenced in large sequencing laboratories in a matter of hours. Here are the main steps of the process.

1. The DNA molecule to be sequenced, or *template,* must be available in many copies. One way to do this is to replicate it repeatedly by subjecting it to several rounds of the *polymerase chain reaction* (PCR): A soup of the template, free nucleotides, DNA polymerase, and short *oligonucleotide primers* is prepared. The primers are bits of DNA complementary to a short segment at one end of the template. The soup is heated until the two strands of the template separate. As the soup cools, the primers attach themselves to the separated strands, and the DNA polymerase and free nucleotides extend the primers to form new strands that are complementary to the template. This cycle of *heat denaturation, primer annealing,* and *primer extension* can be repeated every few minutes, with a doubling in the number of DNA strands every time.[2]

2. A last round of heating and cooling is carried out with four new ingredients added to the soup: free nucleotides with "defective" 3′ ends that cannot be extended. To each of the four types of defective nucleotides (A, C, T, G) is attached one of four types of small dye molecules. These special nucleotides are randomly inserted during construction of the complementary strands, causing them to be terminated prematurely. As a result, the final soup will contain various *prefixes* of the DNA to be sequenced.

3. The next step is to perform *electrophoresis* on the DNA soup. Samples of the soup are placed at the end of a long, flat agarose or polyacrylimide gel. (A typical gel is about 1 foot wide and 2 feet long.) The gel is subjected to an electric current, and the DNA strands migrate toward the positive pole. Shorter strands move more quickly through the gel than longer strands; thus, when the current is stopped, the strands are spread out along the gel according to their length.

4. The gel is inserted into a sequencing machine. The sequencer moves the gel past a laser and an array of light sensors. The sequence of bases in the DNA is "called" by observing the colors emitted by the dye-labeled bases as they pass the laser. (Typically, the gel has a few dozen "lanes" for different DNAs, and the sequencer can read them all concurrently. A few lanes are devoted to calibration sequences of known lengths.)

This whole process takes eight or more hours.

Newer laboratory instruments replace the gel with a set of fine capillaries through which short sequences travel more quickly than long sequences. The process goes roughly twice as quickly, and these instruments also have many more lanes.

---

[2] A major innovation in PCR was the introduction of enzymes from the bacterium *Thermus aquaticus.* This organism's DNA polymerase is not damaged by PCR's repeated heatings to temperatures near the boiling point – predictably, perhaps, since *T. aquaticus* inhabits the hot springs and geysers of Yellowstone National Park.

Although future increases in the length of DNA amenable to automatic sequencing are predictable, even a hundredfold increase would not approach the 130 million base pairs of an "average" human chromosome. One way to sequence longer strands is called *shotgun sequencing,* in which the long strand is replicated and the copies are broken into pieces at random points. This is typically accomplished by subjecting the DNA to *nonspecific DNAses, restriction enzymes,* or other chemicals. The important point is that different copies are broken in different places so that the pieces from the various copies overlap. Once the short pieces have been sequenced, the remaining task is entirely computational: to determine which pairs of short sequences overlap and to use this information to recreate the original long sequence – the *target sequence* – from the overlap information. This task is referred to as the *sequence assembly problem.*

As a first step in solving the sequence assembly problem, we will investigate how to determine whether two sequences overlap. This seemingly simple problem is complicated by the presence of incorrect and missing bases (called *base-call errors*) in the output of automatic sequencers. We will first design an algorithm to quantify the degree to which two DNA strings s_1 and s_2 are similar; then we will modify it to find the suffix of s_1 and prefix of s_2 that are most similar. In Chapter 13, we will discuss how to use this information to assemble sequences.

Determining overlap in the presence of errors is one of a large number of *approximate string-matching* problems. As in Chapter 2, dynamic programming provides the basis of our solution.

3.2 Alignments and Similarity

To measure similarity, we will first introduce the concept of *alignment.* Next, we will define a scoring scheme for alignments. Finally, we will define the similarity of two strings to be the score of the highest-scoring alignment of the strings. Although we are introducing similarity in the context of DNA sequences, the concept is also applicable to amino acid sequences. For variety, we will use strings of Roman letters (often English words or phases) in our examples.

Although it is certainly possible to give a formal mathematical definition of "alignment", it is easier to proceed with examples. Here are seven alignments of the strings HOUSE and HOME:

$$\begin{bmatrix} \text{HOUSE} \\ \text{HOME-} \end{bmatrix}, \quad \begin{bmatrix} \text{HOUSE} \\ \text{-HOME} \end{bmatrix}, \quad \begin{bmatrix} \text{----HOUSE} \\ \text{HOME-----} \end{bmatrix},$$

$$\begin{bmatrix} \text{HOUS-E} \\ \text{-HOME-} \end{bmatrix}, \quad \begin{bmatrix} \text{HOUSE} \\ \text{HO-ME} \end{bmatrix}, \quad \begin{bmatrix} \text{HOUSE} \\ \text{HOM-E} \end{bmatrix}, \quad \begin{bmatrix} \text{H-OUSE} \\ \text{HOM-E-} \end{bmatrix}. \tag{3.1}$$

From looking at these examples, we see that an alignment can be viewed as a placement of one string over another with the following properties:

- the original letters of both strings must appear in their original order;
- *gaps,* signified by "−", can be added to either string or to both strings;

- the lengths (counting gaps) of the upper and lower strings in the alignment must be equal.

We will "score" alignments[3] as follows:

- if either character in a column is a gap, that column scores -2;
- if the letters are identical, the column scores $+1$;
- if the letters are different, the column scores -1;
- the score of the alignment is the sum of its columns' scores.

Thus, the scores of the seven alignments in (3.1) are:

$$1 + 1 + (-1) + (-1) + (-2) = -2,$$
$$(-2) + (-1) + (-1) + (-1) + 1 = -4,$$
$$(-2) \times 9 = -18,$$
$$(-2) + (-1) + (-1) + (-1) + (-2) + (-2) = -9,$$
$$1 + 1 + (-2) + (-1) + 1 = 0,$$
$$1 + 1 + (-1) + (-2) + 1 = 0,$$
$$1 + (-2) + (-1) + (-2) + (-1) + (-2) = -7.$$

Now we can define the *similarity of two strings s and t*, written $\text{sim}(s, t)$: it is the highest score attainable by any possible alignment of s and t.

Although we have not listed *all* possible alignments of HOUSE and HOME, the fifth and sixth of the seven listed attain the maximum possible score. Thus, $\text{sim}(\text{HOUSE}, \text{HOME}) = 0$. It should be clear that, for any string s, the quantity $\text{sim}(s, s) = |s|$, the length of s. If Λ represents the empty string, then

$$\text{sim}(s, \Lambda) = -2|s|. \tag{3.2}$$

Also, if s and t are two strings with lengths $|s|$ and $|t|$, where $|s| \leq |t|$, then

$$-|s| - 2(|t| - |s|) \leq \text{sim}(s, t) \leq |s| - 2(|t| - |s|).$$

How many alignments must be constructed to find the similarity of two strings with lengths m and n when $m \leq n$? Suppose that there are k_1 columns with matching characters and k_2 columns with mismatched (nongap) characters. Then the score of the alignment is

$$k_1 - k_2 - 2((m - k_1 - k_2) + (n - k_1 - k_2)) = 5k_1 + 3k_2 - 2m - 2n,$$

since there are $m - k_1 - k_2$ characters paired with gaps in the shorter strings and $n - k_1 - k_2$ characters in the longer. In fact, the score of an alignment is determined

---

[3] These column scores are convenient examples; different numerical values are used in different biological contexts.

by the total number of pairs $k = k_1 + k_2$, the k paired letters of the longer string and the k paired letters of the shorter string. If we consider the very last of the alignments of (3.1), we have $k = 3$ and paired letters HOS from HOUSE and HME from HOME. The left-to-right property of alignments requires that we pair H to H, O to M, and S to E, so $k_1 = 1$ and $k_2 = 2$. By our formula, the score is $(5 \times 1) + (3 \times 2) - (2 \times 5) - (2 \times 4) = -7$, which agrees with our previous direct computation. The total number of ways of choosing the same number of letters from each string is

$$\sum_{k=0}^{m} \binom{m}{k}\binom{n}{k}. \tag{3.3}$$

To get a rough idea of the size of this number, let's assume that $m = n$ and that n is even. Using Sterling's approximation $n! \approx \sqrt{2\pi n}(n/e)^n$, we have

$$\binom{n}{n/2} = \frac{n!}{(n/2)!\,(n/2)!} \approx \frac{\sqrt{2\pi n}}{\pi n}\frac{n^n}{(n/2)^n} \approx \sqrt{\frac{2}{\pi n}}2^n.$$

The summand $\binom{n}{k}\binom{n}{k}$ is largest when $k = \lfloor n/2 \rfloor$, so

$$\frac{2}{\pi n}4^n \approx \binom{n}{n/2}^2 \leq \sum_{k=0}^{n}\binom{n}{k}\binom{n}{k} \leq n\binom{n}{n/2}^2 \approx \frac{2}{\pi}4^n.$$

To be concrete, the number of different alignments of two strings of length 100 is exactly

90548514656103281165404177077484163874504589675413336841320 $\approx 9 \times 10^{57}$.

So, even when n is only moderately large, the number of possible alignments of two strings of length n is very large indeed![4]

Luckily, we do not have to consider *all* possible alignments just to find the *best*. In fact, we will be able to get by with considering only about mn different alignments of the two strings and their substrings. To see how this is possible, let's see how it is possible to build up the best alignment beginning with the right ends of the two strings HOUSE and HOME.

There are really only three possibilities:

$$\left(\begin{matrix} \text{some align-} & \text{HOUS} \\ \text{ment of} & \text{HOM} \end{matrix}\right)\begin{bmatrix} \text{E} \\ \text{E} \end{bmatrix},$$

$$\left(\begin{matrix} \text{some align-} & \text{HOUSE} \\ \text{ment of} & \text{HOM} \end{matrix}\right)\begin{bmatrix} - \\ \text{E} \end{bmatrix}, \qquad \left(\begin{matrix} \text{some align-} & \text{HOUS} \\ \text{ment of} & \text{HOME} \end{matrix}\right)\begin{bmatrix} \text{E} \\ - \end{bmatrix}.$$

Since we are looking for the best (highest-scoring) alignment overall and since the columns of an alignment are scored independently, nothing can be gained by considering

---

[4] Exercise 2 deals with the case $m \neq n$.

	:Λ	:H	:HO	:HOM	:HOME
Λ	$0{:}\begin{bmatrix} \\ \end{bmatrix}$	$-2{:}\begin{bmatrix} H \\ - \end{bmatrix}$	$-4{:}\begin{bmatrix} HO \\ -- \end{bmatrix}$	$-6{:}\begin{bmatrix} HOM \\ --- \end{bmatrix}$	$-8{:}\begin{bmatrix} HOME \\ ---- \end{bmatrix}$
H	$-2{:}\begin{bmatrix} - \\ H \end{bmatrix}$	$+1{:}\begin{bmatrix} H \\ H \end{bmatrix}$	$-1{:}\begin{bmatrix} HO \\ H- \end{bmatrix}$	$-3{:}\begin{bmatrix} HOM \\ H-- \end{bmatrix}$	$-5{:}\begin{bmatrix} HOME \\ H--- \end{bmatrix}$
HO	$-4{:}\begin{bmatrix} -- \\ HO \end{bmatrix}$	$-1{:}\begin{bmatrix} H- \\ HO \end{bmatrix}$	$+2{:}\begin{bmatrix} HO \\ HO \end{bmatrix}$	$0{:}\begin{bmatrix} HOM \\ HO- \end{bmatrix}$	$-2{:}\begin{bmatrix} HOME \\ HO-- \end{bmatrix}$
HOU	$-6{:}\begin{bmatrix} --- \\ HOU \end{bmatrix}$	$-3{:}\begin{bmatrix} H-- \\ HOU \end{bmatrix}$	$0{:}\begin{bmatrix} HO- \\ HOU \end{bmatrix}$	$+1{:}\begin{bmatrix} HOM \\ HOU \end{bmatrix}$	$-1{:}\begin{bmatrix} HOME \\ HO-U \end{bmatrix}$
HOUS	$-8{:}\begin{bmatrix} ---- \\ HOUS \end{bmatrix}$	$-5{:}\begin{bmatrix} H--- \\ HOUS \end{bmatrix}$	$-2{:}\begin{bmatrix} HO-- \\ HOUS \end{bmatrix}$	$-1{:}\begin{bmatrix} HO-M \\ HOUS \end{bmatrix}$	$0{:}\begin{bmatrix} HOME \\ HOUS \end{bmatrix}$
HOUSE	$-10{:}\begin{bmatrix} ----- \\ HOUSE \end{bmatrix}$	$-7{:}\begin{bmatrix} H---- \\ HOUSE \end{bmatrix}$	$-4{:}\begin{bmatrix} HO--- \\ HOUSE \end{bmatrix}$	$-3{:}\begin{bmatrix} HO--M \\ HOUSE \end{bmatrix}$	$0{:}\begin{bmatrix} HO-ME \\ HOUSE \end{bmatrix}$

Figure 3.1: Dynamic programming table for similarity of HOUSE and HOME.

poor alignments of the pairs (HOUS, HOM), (HOUSE, HOM), and (HOUS, HOME). So the best alignment of the whole strings will be one of the following:

$$\left(\begin{matrix} \textit{best} \text{ align-} & \text{HOUS} \\ \text{ment of} & \text{HOM} \end{matrix}\right)\begin{bmatrix} E \\ E \end{bmatrix},$$

$$\left(\begin{matrix} \textit{best} \text{ align-} & \text{HOUSE} \\ \text{ment of} & \text{HOM} \end{matrix}\right)\begin{bmatrix} - \\ E \end{bmatrix}, \qquad \left(\begin{matrix} \textit{best} \text{ align-} & \text{HOUS} \\ \text{ment of} & \text{HOME} \end{matrix}\right)\begin{bmatrix} E \\ - \end{bmatrix}.$$

Likewise, when determining the best alignment of HOUS and HOM, we need only consider the *best,* not every, alignment of the pairs (HOU, HO), (HOUS, HO), and (HOU, HOM), ...

In a different form,

$$\text{sim}(\text{HOUSE, HOME}) = \max \left\{ \begin{matrix} 1 + \text{sim}(\text{HOUS, HOM}), \\ -2 + \text{sim}(\text{HOUSE, HOM}), \\ -2 + \text{sim}(\text{HOUS, HOME}) \end{matrix} \right\};$$

in general,

$$\text{sim}(s_1 s_2 \ldots s_n, t_1 t_2 \ldots t_m)$$

$$= \max \left\{ \begin{matrix} p(s_n, t_m) + \text{sim}(s_1 s_2 \ldots s_{n-1}, t_1 t_2 \ldots t_{m-1}), \\ -2 + \text{sim}(s_1 s_2 \ldots s_n, t_1 t_2 \ldots t_{m-1}), \\ -2 + \text{sim}(s_1 s_2 \ldots s_{n-1}, t_1 t_2 \ldots t_m) \end{matrix} \right\}, \qquad (3.4)$$

where $p(s_n, t_m) = +1$ if $s_n = t_m$ and -1 if $s_n \neq t_m$.

Thus, to find the similarity of two strings, it will be enough for us to find the (or just one, if there are more than one) highest-scoring alignment of every possible prefix of one string with every possible prefix of the other. The prefixes of HOUSE are Λ (the empty string), H, HO, HOU, HOUS, and HOUSE; the prefixes of HOME are Λ, H, HO, HOM, and HOME. The results can be organized in a table like the one displayed as Figure 3.1.

3.3   Alignment and Similarity in Perl

We will need a few new features of Perl to write our program for alignment and similarity in Perl.

Assignment to Lists. We have already seen that Perl allows assignments to list variables like this:

```
@animals = ("DOG","CAT","MOUSE");
```

Unlike most well-known programming languages, Perl also allows assignments to *lists of variables*:

```
($canine,$feline,$murine) = ("DOG","CAT","MOUSE");
```

is equivalent to

```
$canine = "DOG"; $feline = "CAT"; $murine = "MOUSE";
```

We do not need a "temporary variable" to swap the values of two Perl variables; we may simply write

```
($x,$y) = ($y,$x);
```

We can also write statements like

```
($canine,@noncanine) = ("DOG","CAT","MOUSE");
```

which has the same effect as

```
$canine = "DOG";
@noncanine = ("CAT","MOUSE");
```

We are now in a better position to understand Perl's way of passing arguments to functions. Perl simply makes a list of all the arguments' values and names this list @_ in the subroutine. If we wish (and we usually do), we can give the arguments names with an assignment like

my ($x,$y,$z) = @_;

We also have the flexibility to write functions like this one, which accepts any positive number of numerical arguments and returns the largest:

```
sub max {
    my ($m,@l) = @_;
    foreach my $x (@l) { $m = $x if ($x>$m) }
    return $m;
}
```

Substring Extraction. The substring extraction operator **substr** takes three operands. The first is a string. The second, called the *offset*, gives the distance of the starting position from one of the ends of the string: if nonnegative, the left end; if negative, the right end. The third operand may be omitted; if so, the substring extends to the end of the original string. If the third operand is positive, this many characters are taken beginning at the starting position; if negative, this many are dropped from the end of the original string. A substring may be the left-hand side of an assignment. The operation of **substr** is summarized by the following code fragment:

```
my $s = "ABCDEFGHIJK";
print substr($s,3), " * ", substr($s,−5), "\n";
print substr($s,3,3), " * ", substr($s,−5,3), "\n";
print substr($s,3,−3)," * ", substr($s,−5,−3), "\n";
substr($s,−5,3) = lc $s;
print $s, "\n";
```

which prints:

```
    DEFGHIJK * GHIJK
    DEF * GHI
    DEFGH * GH
    ABCDEFabcdefghijkJK
```

(The **lc** operator shifts its string operand to lower case.)

Multidimensional Arrays. In Perl, multidimensional arrays can be created without specifying fixed dimensions.[5] In our program, we will have an $n \times m$ scoring matrix M declared and used like this:

---

[5] Beginning with Perl 5; early versions of Perl provided no direct support for multidimensional arrays.

```
my @M;              ## similarity scoring matrix
$M[$i][$j] = ... ;   ## row i, column j
```

Since Perl cannot know the eventual dimensions of arrays created in this way, some or all of the array may have to be copied as Perl reorganizes storage to accommodate growing arrays. If we know in advance that an array will be two-dimensional with $n rows and $m columns (for example), we can make the program a bit faster by inducing a large allocation "up front" like this:

```
for (my $i=$n−1; $i>=0; $i−−) { $M[$i][$m−1] = 0; }
```

The Needleman–Wunsch Algorithm. Now we are ready to use the observations of the previous section to write a program to perform alignment and compute similarity. First, we need a subroutine p that returns +1 when its arguments are identical and −1 when they are different.

```
sub p {
    my ($aa1,$aa2) = @_;
    return ($aa1 eq $aa2) ? 1 : −1;
}
```

We will now write similarity, which fills the matrix @M and returns the similarity score of its arguments $s and $t. $M[$i][$j] will contain the similarity score of the first $i letters of $s and the first $j letters of $t. If either $i or $j is 0, then one prefix is the empty string and $M[$i][$j] is computed by (3.2). Otherwise, we follow the pattern of (3.4). Filling the matrix row by row is sufficient to ensure that all values are available when needed:

```
sub similarity {
    my($s,$t) = @_;
    my ($i,$j);
    foreach $i (0..length($s)) { $M[$i][0] = $g * $i; }
    foreach $j (0..length($t)) { $M[0][$j] = $g * $j; }
    foreach $i (1..length($s)) {
        foreach $j (1..length($t)) {
            my $p = p(substr($s,$i−1,1),substr($t,$j−1,1));
            $M[$i][$j] = max($M[$i−1][$j]+$g,
                            $M[$i][$j−1]+$g,
                            $M[$i−1][$j−1]+$p);
```

```
          }
      }
      return ($M[length($s)][length($t)]);
}
```

Although we could have done so, we did not fill our array with alignments. Instead, we recorded only the score of the best alignment. Once the array is filled in with the scores, we can retrace our steps to construct the best alignment of the two strings. The advantage to this is that we construct and store alignment strings for no more than $m + n$ pairs of prefixes instead of for all mn pairs.

To recover the actual alignment, we will look at the entry in the lower right corner of the array, the adjacent entries, and the last two letters of each string to decide how the overall score was arrived at: Did it come from pairing the last two letters of each string, from pairing the last letter of the first string with a gap symbol, or from pairing the last letter of the second string with a gap? Once we determine this, we can back up through the array and the string and repeat the process until we reach the edge of the array.

Our recursive subroutine getAlignment takes the two strings of letters as arguments and assumes that the matrix @M has been filled. It returns a list of the same two strings with gap symbols inserted to achieve the best possible alignment.

```
sub getAlignment {
    my($s,$t) = @_;
    my ($i,$j) = (length($s),length($t));
    return ("-" x $j, $t) if ($i==0);        ## Case 0a
    return ($s, "-" x $i) if ($j==0);        ## Case 0b
    my ($sLast,$tLast) = (substr($s,−1),substr($t,−1));

    if ($M[$i][$j]==$M[$i−1][$j−1]+p($sLast,$tLast)) {       ## Case 1
        ## last letters are paired in the best alignment
        my ($sa,$ta) = getAlignment(substr($s,0,−1),substr($t,0,−1));
        return ($sa.$sLast,$ta.$tLast);
    } elsif ($M[$i][$j]==$M[$i−1][$j]+$g) {                  ## Case 2
        ## last letter of the first string is paired with a gap
        my ($sa,$ta) = getAlignment(substr($s,0,−1),$t);
        return ($sa.$sLast,$ta."-");
    } else {    ## Case 3: last letter of the second string is paired with a gap
        my ($sa,$ta) = getAlignment($s,substr($t,0,−1));
        return ($sa."-",$ta.$tLast);
    }
}
```

If we align the strings HOUSE and HOME, then the sequence of calls and returns of getAlignment is as follows:

```
call getAlignment(HOUSE,HOME) -- Case 1
    call getAlignment(HOUS,HOM) -- Case 1
        call getAlignment(HOU,HO) -- Case 2
            call getAlignment(HO,HO) -- Case 1
                call getAlignment(H,H) -- Case 1
                    call getAlignment(,) -- Case 0a
                    getAlignment(,) returns (,)
                getAlignment(H,H) returns (H,H)
            getAlignment(HO,HO) returns (HO,HO)
        getAlignment(HOU,HO) returns (HOU,HO-)
    getAlignment(HOUS,HOM) returns (HOUS,HO-M)
getAlignment(HOUSE,HOME) returns (HOUSE,HO-ME)
```

3.4 Exercises

1. Section 3.2 lists seven alignments of HOUSE and HOME. Use formula (3.3) to determine the total number of distinct alignments of these two strings.

2. *(Mathematical)* Derive an estimate or an exact formula in closed form for the number of different alignments of a string of length m with one of length n, where $m \neq n$.

3. *(Mathematical)* Since formula (3.3) regards alignments as distinct only if their letter pairs are different, it considers

$$\begin{bmatrix} AX-B \\ A-YB \end{bmatrix} \quad \text{and} \quad \begin{bmatrix} AX-B \\ A-YB \end{bmatrix}$$

to be two ways of writing the same alignment. This is reasonable, since they have the same score and there is no biological distinction. Still, we might like to know how many alignments there are if we count such pairs as distinct. Devise a recurrence relation that describes the number of pairs of strings of letters and gap symbols that represent alignments of a string of length m and a string of length n. Put the formula into closed form. (Remember that no column may contain two gap symbols.)

4. Our alignment program prints out only *one* alignment. In some cases (as with HOUSE and HOME), there is more than one best alignment. Modify the program to print out *all* alignments that achieve the highest possible score.

   Hints: Modify the recursive subroutine alignment to return a list of an even number of strings instead of two strings. If @a is the list ("DOG","CAT") and @b is ("COW","GOAT"), then the list ("DOG","CAT","COW","GOAT") will be assigned to @c by the statement @c = (@a,@b);.

5. Perl's motto is "There's always more than one way to do it." Rewrite the alignment program so that it doesn't use the **substr** operator. Try two different approaches:

    (a) use pattern matching – in particular, the ^ and/or $ patterns;
    (b) use **split**, **join**, and/or **splice**.

6. Our program computes the similarity of two DNA sequences. But the original motivation for writing the program (p. 32) was to determine whether two sequences *overlap* and by how much. To measure the overlap of two strings s and t, we consider all $(m + 1)(n + 1)$ possible pairs consisting of a suffix of s and a prefix of t, and we define the *overlap score* overlap(s, t) to be the highest similarity score among these pairs. For example, if $s = $ XXXXXABCDE and $t = $ ABDEYYYYYY, then overlap$(s, t) = $ sim(ABCDE, ABDE) = 2.

    One way to compute overlap$(s, t) = $ sim(ABCDE, ABDE) = 2 is with the following subroutine:

```
sub overlap {
    my($s,$t) = @_;
    my $best = 0;
    foreach my $i (1..length($s)) {
        foreach my $j (1..length($t)) {
            $best = max($best,
                            similarity(substr($s,-$i),substr($t,0,$j)));
        }
    }
    return $best;
}
```

However, this method is highly inefficient. A more efficient method is to make two small modifications to similarity. The first is to replace

```
    return ($M[length($s)][length($t)]);
```

with

```
    my $best = 0;
    foreach my $j (0..length($t)) {
        $best = max($best,$M[length($s)][$j]);
    }
    return $best;
```

What is the other modification? *Hint:* It involves changing only a single line of code.

7. Biologists regard the simple linear gap penalty we employed as unrealistic. Since the events introducing gaps are rare, an alignment that exhibits a single gap of length 2 (i.e., two consecutive gap characters) is, all else being equal, a much more likely expression of the actual course of evolution than an alignment with two separate gaps of length 1. Therefore, biologists often employ *affine* gap penalties, in which the penalty paid for k consecutive gap symbols is $h + gk$ ($h + g$ is sometimes called the *gap-opening penalty* and g the *gap-extension penalty*).

    Modify the similarity and alignment programs of this chapter to find the optimal alignment of two strings when affine gap penalties are applied. To do this, you will need to replace the array @M by three arrays, @Mbb, @M_b, and @Mb_. $Mbb[$i][$j] will hold the score of the best alignment of the first $i bases of $s and first $i bases of $t that has two bases in the rightmost column. $M_b[$i][$j] will hold the score of the best alignment with a gap symbol over a base in the rightmost column, and $Mb_[$i][$j] will hold the score of the best alignment with a base over a gap symbol in the rightmost column. $M_b[$i][$j] will be updated by the statement

```
$M_b[$i][$j] = max($M_b[$i][$j−1]+$g,
                   $Mbb[$i][$j−1]+$h+$g,
                   $Mb_[$i][$j−1]+$h+$g)
```

    Write similar statements to update $Mbb and $Mb_. Also, modify getAlignment to print out the best-scoring alignment.

8. Rewrite getAlignment without recursion. *Hint:*

```
sub getAlignment {
    my ($s,$t) = @_;
    my ($i,$j) = (length($s),length($t));
    my ($sRight,$tRight);
    while ($i>0 && $j>0) {
        ...
    }
    ...
}
```

3.5 Complete Program Listings

The program of this chapter is included in the software distribution as chap3/similarity.pl.

3.6 Bibliographic Notes

Needleman and Wunsch (1970) described the first sequence similarity algorithm. Anderson (1981) was first to describe the "shotgun" sequencing technique. The application of the technique was speeded up by Mullis's (1990) invention of polymerase chain reaction. Laboratory protocols for PCR and sequencing are given in Palumbi (1996) and Hillis et al. (1996a).

The completion of the sequencing of the human genome was announced in the journal *Science* (Pennisi 2001).

## CHAPTER FOUR

Predicting Species: Statistical Models

Suppose we are given a strand of DNA and asked to determine whether it comes from corn (*Zea mays*) or from fruit flies (*Drosophila melanogaster*). One very simple way to attack this problem is to analyze the relative frequencies of the nucleotides in the strand. Even before the double-helix structure of DNA was determined, researchers had observed that, while the numbers of Gs and Cs in a DNA were roughly equal (and likewise for As and Ts), the relative numbers of G + C and A + T differed from species to species. This relationship is usually expressed as *percent* GC, and species are said to be GC-*rich* or GC-*poor*. Corn is slightly GC-poor, with 49% GC. Fruit fly is GC-rich, with 55% GC.

We examine the first ten bases of our DNA and see: `GATGTCGTAT`. Is this DNA from corn or fruit fly?

First of all, it should be clear that we cannot get a definitive answer to the question by observing bases, especially just a few. Corn's protein-coding sequences are distinctly GC-rich, while its noncoding DNA is GC-poor. Such variations in GC content within a single genome are sometimes exploited to find the starting point of genes in the genome (near so-called C*p*G *islands*). In the absence of additional information, the best we can hope for is to learn whether it's "more likely" that we have corn or fly DNA, and how much more likely.

In the language of statistics, we have the outlines of two *statistical models,* one (denoted $\mathcal{M}_Z$) for corn DNA and one (denoted $\mathcal{M}_D$) for fruit-fly DNA. We would like to know the *likelihood* of each model given the particular *observation* of the ten given bases. If this observation assigns a higher likelihood to the corn model than the fly model then it is more likely we have corn DNA, and vice versa.

Our two models are members of a *family* of models. We will discuss the common features of all the models in the family a little later. However, the members of this family are distinguished from each other by the probabilities of the four bases, denoted by $(p(A), p(C), p(G), p(T))$. These numbers are the *parameters* of the family.[1] To summarize our models, we might write

---

[1] Actually, the probability of any single base suffices to define the model. If $p(A) = \alpha$, then $p(T) = \alpha$ and $p(C) = p(G) = 0.5 - \alpha$.

$$\mathcal{M}_Z = \mathcal{M}(p_Z(A), p_Z(C), p_Z(G), p_Z(T)) = \mathcal{M}(0.255, 0.245, 0.245, 0.255)$$

and

$$\mathcal{M}_D = \mathcal{M}(p_D(A), p_D(C), p_D(G), p_D(T)) = \mathcal{M}(0.225, 0.275, 0.275, 0.225).$$

We define the *likelihood of model $\mathcal{M}$ given observation $\mathcal{O}$*, written $\mathcal{L}(\mathcal{M} \mid \mathcal{O})$, to be $\Pr(\mathcal{O} \mid \mathcal{M})$, the probability of the observation $\mathcal{O}$ under the model $\mathcal{M}$. Thus, the likelihoods of our models given our observation GATGTCGTAT are

$$\mathcal{L}(\mathcal{M}_Z \mid \mathcal{O}) = \Pr(\mathcal{O} \mid \mathcal{M}_Z)$$
$$= p_Z(G) \cdot p_Z(A) \cdot p_Z(T) \cdot p_Z(G) \cdot p_Z(T) \cdot p_Z(C) \cdot p_Z(G)$$
$$\cdot p_Z(T) \cdot p_Z(A) \cdot p_Z(T)$$
$$= (0.245)(0.255)(0.255)(0.245)(0.255)(0.245)(0.245)(0.255)$$
$$\times (0.255)(0.255)$$
$$\approx 9.9 \times 10^{-7}$$

and

$$\mathcal{L}(\mathcal{M}_D \mid \mathcal{O}) = \Pr(\mathcal{O} \mid \mathcal{M}_D)$$
$$= (0.275)(0.225)(0.225)(0.275)(0.225)(0.275)(0.275)(0.225)$$
$$\times (0.225)(0.225)$$
$$\approx 7.4 \times 10^{-7}.$$

The observation favors corn by *odds* of about $(9.9 \times 10^{-7})/(7.4 \times 10^{-7}) \approx 1.34$.

To increase our confidence, we might wish to observe more bases. If we observe the 20-base sequence $\mathcal{O}' = $ GATGTCGTATATACGACATC, the likelihoods become $\mathcal{L}(\mathcal{M}_Z \mid \mathcal{O}') \approx 9.8 \times 10^{-13}$ and $\mathcal{L}(\mathcal{M}_D \mid \mathcal{O}') \approx 5.5 \times 10^{-13}$; the odds have increased to about 1.78. For "typical" sequences, likelihoods will decrease roughly exponentially with the length of the observed sequence, while odds will either increase or decrease exponentially.

If we foresee repeating this exercise for many and/or long sequences of bases, it will be worthwhile to simplify the computation of likelihoods as much as possible. We can avoid working directly with very small likelihoods (and possible computer underflow errors[2]) by *adding up the logarithms* of the factors above to compute *log likelihoods*. Since $\log(0.225) \approx -1.492$, $\log(0.245) \approx -1.406$, $\log(0.255) \approx -1.366$, and $\log(0.275) \approx -1.291$, we can compute

---

[2] *Underflow errors* occur when the result of an arithmetic operation is smaller than the smallest number representable in the computer's number system.

$$\log(\mathcal{L}(\mathcal{M}_Z \mid \mathcal{O})) = \log(p_Z(\text{G})) + \log(p_Z(\text{A})) + \log(p_Z(\text{T})) + \log(p_Z(\text{G}))$$
$$+ \log(p_Z(\text{T})) + \log(p_Z(\text{C})) + \log(p_Z(\text{G}))$$
$$+ \log(p_Z(\text{T})) + \log(p_Z(\text{A})) + \log(p_Z(\text{T}))$$
$$= 4\log(0.245) + 6\log(0.255)$$
$$\approx 4(-1.406) + 6(-1.366)$$
$$= -13.820.$$

In a similar fashion, we compute

$$\log(\mathcal{L}(\mathcal{M}_D \mid \mathcal{O})) \approx 4(-1.291) + 6(-1.492) = -14.116.$$

As before, the observation favors corn by odds of approximately

$$\exp((-13.820) - (-14.116)) = e^{0.296} \approx 1.34.$$

The number 0.296 is called a *log likelihood ratio, log of odds,* or *lod score* of the observation. Logarithms have the additional advantages that (a) the sign of the result tells us which model is favored and (b) the magnitude of favorable and unfavorable scores can be compared directly. In this case, a positive lod score of $+s$ favors $\mathcal{M}_D$ by the same odds that a negative score of $-s$ favors $\mathcal{M}_Z$.

A second useful insight is that we can compute lod scores directly rather than by computing log likelihoods and then subtracting. An A or a T contributes $\log(0.255/0.225) = \log(0.255) - \log(0.225) \approx 0.126$ to the lod score; a C or a G contributes $\log(0.245/0.275) = \log(0.245) - \log(0.275) \approx -0.115$. Thus, the lod score of our observation is about $4(-0.115) + 6(0.126) = 0.296$.[3]

In all of these computations we have used *natural* logarithms, that is, logarithms to base $e \approx 2.718$. Scores expressed in natural logarithms are said to be expressed in *nats*. The natural logarithm, often abbreviated ln, is usually the most convenient to use in continuous mathematics; it arises "naturally" in many situations, such as the indefinite integral $\int \frac{dx}{x} = \ln x$. *Binary* or base-2 logarithms, abbreviated lg, are often favored in discrete mathematics and the analysis of algorithms. Scores using binary logarithms are said to be expressed in *bits*. Thus, an A contributes $\lg(0.255/0.225) \approx 0.182$ bit to the lod score. Binary and natural logarithms differ by a constant factor: 1 bit ≈ 0.693 nat and 1 nat ≈ 1.443 bit. *Decimal* or base-10 logarithms, giving scores in *digits,* are also possible. The notation "log" will mean either that the base of logarithms is insignificant or that base e is to be assumed.

In many situations, it is desirable to have integer scores. In a computer, integers take up less memory than floating-point numbers, and integer arithmetic is faster, too. To obtain integer scores, lod scores are usually multiplied by a scale factor before

---

[3] Actually, $\log(0.255/0.225) \approx 0.1252$ and $\log(0.245/0.275) \approx -0.1155$. Our computers will calculate all of these quantities with greater accuracy than is necessary in these examples.

rounding. This allows the relative values of the scores to be maintained at any desired level of accuracy. For example, if a binary ($\log_2$) score is multiplied by 2 and then rounded, it is said to be expressed in "half-bits"; a base-e score multiplied by 5 and rounded is in "fifth-nats".

Entropy and Relative Entropy. Finally, let's consider what log likelihood we obtain for $\mathcal{M}_Z$ if our observation is the "Platonic ideal" of corn DNA, "perfectly typical" in the distribution of its n bases. On average, we should see about $(0.255)n$ As, $(0.245)n$ Cs, $(0.245)n$ Gs, and $(0.255)n$ Ts. The log likelihood of model $\mathcal{M}_Z$ given this hypothetical random observation is therefore

$$0.255n(-1.366) + 0.245n(-1.406) + 0.245n(-1.406) + 0.255n(-1.366)$$

$$= -1.39n.$$

Thus, on average, observing another base will contribute 1.39 nats to the log likelihood. This number, computed by the formula

$$H(\mathcal{M}) = -\log(p(\text{A})^{p(\text{A})} p(\text{C})^{p(\text{C})} p(\text{G})^{p(\text{G})} p(\text{T})^{p(\text{T})})$$

$$= -(p(\text{A})\log p(\text{A}) + p(\text{C})\log p(\text{C}) + p(\text{G})\log p(\text{G}) + p(\text{T})\log p(\text{T})),$$

is called the *entropy* of model $\mathcal{M}$.[4] Since probabilities are always between 0 and 1, the negative sign ensures that entropy is always a nonnegative number.

We could likewise consider the typical contribution to the log-of-odds score $\ln(\mathcal{L}(\mathcal{M}_Z \mid \mathcal{O})/\mathcal{L}(\mathcal{M}_D \mid \mathcal{O}))$ of a sequence of n bases from corn. This is

$$0.255n(0.126) + 0.245n(-0.115) + 0.245n(-0.115) + 0.255n(0.126) = 0.008n.$$

To have hundred-to-one odds that a "perfect" corn sequence is from corn rather than from fruit fly, we would need about n observations, where n satisfies

$$e^{0.008n} = 100.$$

From this, we see that

$$n = \ln(100)/(0.008) \approx 576.$$

The quantity 0.008 is the *relative entropy* of $\mathcal{M}_Z$ with respect to $\mathcal{M}_D$, computed in general by the formula

$$H(\mathcal{M}_Z \mid \mathcal{M}_D)$$

$$= -\log\left(\left(\frac{p_Z(\text{A})}{p_D(\text{A})}\right)^{p_Z(\text{A})} \left(\frac{p_Z(\text{C})}{p_D(\text{C})}\right)^{p_Z(\text{C})} \left(\frac{p_Z(\text{G})}{p_D(\text{G})}\right)^{p_Z(\text{G})} \left(\frac{p_Z(\text{T})}{p_D(\text{T})}\right)^{p_Z(\text{T})}\right)$$

$$= -\sum_{b \in \{\text{A},\text{C},\text{G},\text{T}\}} p_Z(b) \log \frac{p_Z(b)}{p_D(b)}.$$

---

[4] The H here is not "capital h" but rather "capital η"; the word *entropy* comes from German *Entropie*, manufactured from the Greek roots $\eta\nu + \tau\rho\sigma\pi\sigma\varsigma$.

As the asymmetry of this formula suggests, $H(\mathcal{M}_1 \mid \mathcal{M}_2) \neq H(\mathcal{M}_2 \mid \mathcal{M}_1)$ in general.

Entropy measures the *information content* of an observation. Distributions with low entropy exhibit less uncertainty. If yesterday's weather forecast gave a 95% (or 5%) chance of rain for today, there would be little need to look out the window before deciding whether to take an umbrella for a day in the city. Since the entropy of the day's precipitation model is only

$$-(0.95 \lg(0.95) + 0.05 \lg(0.05)) = 0.29 \text{ bit,}$$

relatively little new information is gained from an observation. On the other hand, a 50% forecast has entropy

$$-(0.5 \lg(0.5) + 0.5 \lg(0.5)) = 1 \text{ bit,}$$

so an observation is quite useful under this model.

Maximum Likelihood Models. Now suppose that no model of corn or fruit fly DNA is given in advance. Instead, we are given a *training set* of corn DNA and asked to deduce the best model. Assuming that type of model has been fixed, the best strategy is to choose values of the parameters that make the likelihood of the model as large as possible if the training set is regarded as an observation. This choice gives the *maximum likelihood model* for the training set. Suppose the training set consists of one million bases: 255,000 A, 245,000 C, 245,000 G, and 255,000 T. It seems logical to match the parameters of the model to the observed frequencies of the bases by setting $p(A) = 255,000/1,000,000 = 0.255, \ldots$. It can be proven that, if the observed frequencies are $\hat{p}(A)$, $\hat{p}(C)$, $\hat{p}(G)$, and $\hat{p}(T)$, then the choice of parameters $p(A) = \hat{p}(A), \ldots$ is in fact the one that maximizes the likelihood

$$p(A)^{n\hat{p}(A)} p(C)^{n\hat{p}(C)} p(G)^{n\hat{p}(G)} p(T)^{n\hat{p}(T)}.$$

This is equivalent to minimizing

$$-(\hat{p}(A) \log p(A) + \hat{p}(C) \log p(C) + \hat{p}(G) \log p(G) + \hat{p}(T) \log p(T)).$$

Mathematical Independence: What's Wrong with the Picture. Earlier we pushed aside a discussion of the common assumption of all the models in the family we have considered. There are mainly two. First, our knowledge of the double-stranded nature of DNA led us to assume that $p(A) = p(T)$ and $p(C) = p(G)$ for our models. Second, our *ignorance* led us to assume that each nucleotide in the sequence is an independent realization of the model's probability distribution. In other words, we assumed that the same distribution holds throughout the genome of the species modeled, and that (for example) a C is neither more nor less likely to follow an A than to follow a C, G, or T. An assumption of independence almost always makes the mathematics of a model easier, and it is reasonable in the absence of an accurate understanding of the precise dependencies among random elements being analyzed. The purpose of this chapter is to explain statistical models in a simple context; we will not, in the end, have an especially accurate tool for discerning the species from which a DNA strand came.

4.1   Perl Subroutine Libraries

For species prediction in Perl, we will be defining a Perl subroutine lg for base-2 logarithms. Since we will need lg, Chapter 2's max, and a few similar subroutines repeatedly in the future, we are going to take a brief side-trip into the creation of reusable libraries of Perl subroutines.

In Perl, libraries are called *packages*. Typically, each package is defined in its own source file. Each package has a unique name, and the source file name is normally formed by appending .pm to the package name.

Our package is found in the file Util.pm and begins with the lines

---

```
package Util;
require Exporter;
@ISA = qw(Exporter);
@EXPORT = qw(max min sum lg log10 pow round);
```

The first line gives the package name – Util. Packages and programs that wish to use the subroutines that Util provides must include the statement **use** Util;. For the moment, we will regard the second and third lines as a bit of mandatory magic. The fourth line gives the names of the subroutines that Util will define and make available for other packages to call. Such subroutines are said to be *exported*. Util could (but does not) define other subroutines; such unexported subroutines could be called by code within the source of the Util package, but not by users of the Util package.

The fourth line uses the **qw** operator, which is just a shorthand notation for a list of strings. This line has exactly the same effect as

---

```
@EXPORT = ("max","min","sum","lg","log10","pow","round");
```

---

or

```
@EXPORT = split /\s+/,"max min sum lg log10 pow round";
```

After these lines comes code for the seven exported routines, beginning with max as seen in Chapter 3 and similar code for min (for finding the smallest number in a list) and sum (for adding up all the numbers in a list).

Shifting Arguments. The code for lg and round illustrates a common application of Perl's **shift** operator to handling function arguments. Recall that **shift** takes a single operand, which is a list variable. It removes the first element from the list and returns the removed element as its value. If no operand is explicitly given, then the argument list @_ is assumed. Thus,

```perl
sub lg { log(shift)/log(2.0); }
```

applies **log** to the first argument of lg; it is equivalent to both

```perl
sub lg {
    my ($x) = @_;
    return(log($x)/log(2.0));
}
```

and

```perl
sub lg { return(log($_[0])/log(2.0)); }
```

Our complete utility library – file Util.pm – is listed below.

```perl
package Util;
require Exporter;
@ISA = qw(Exporter);
@EXPORT = qw(max min sum lg log10 pow round);

use strict;

sub max {
    my ($m,@l) = @_;
    foreach (@l) { $m=$_ if $_>$m };
    $m;
}

sub min {
    my ($m,@l) = @_;
    foreach (@l) { $m=$_ if $_<$m };
    $m;
}
```

```perl
sub sum {
    my ($s,@l) = @_;
    foreach (@l) { $s += $_ };
    $s;
}
```

```perl
sub lg { log(shift)/log(2.0); }
```

```perl
sub log10 { log(shift)/log(10.0); }
```

```perl
sub pow {
    my ($x,$y) = @_;
    exp($y * log($x));
}
```

```perl
sub round { int(0.5+shift); }
```

4.2 Species Prediction in Perl

Our program for species prediction is quite brief and requires little comment. Two of the first three lines,

```perl
#!/usr/bin/perl -I. -I../perllib
use strict;
use Util;
```

pertain to our use of a Perl package. The first line includes the -I option. This option informs Perl where to look for user-defined packages. The first -I asks Perl to look in the directory where the program is running, which may or may not be where it resides. The second option asks Perl to look in the run-time directory's sister directory `../perllib` – that is, a directory named `perllib` found in the run-time directory's parent . . . If many -I options are given, Perl will search through the directories from left to right and then continue with other directories that have been defined as part of the installation of the Perl system.

The software distribution accompanying this book places code used in just one chapter in a directory for that chapter and places shared modules in a directory named `perllib`. The specification just given will work so long as the programs in the distribution are run in the directory in which they reside. To run them from other directories, it is necessary to replace `../perllib` with the absolute location of the `perllib` directory.

Not until the statement **use** Util; is executed does the program import the subroutines exported by Util.

Mapping Lists. We apply Perl's **map** operator twice: first to compute a list of individual lod scores for the four bases A, C, G, and T; and later to sum up the lod scores of the bases of each DNA strand in input. The **map** operator's second operand is a list, and **map** returns a new list of the same length. The first operand is an expression involving $_, and the result list is formed by computing the value of this expression with the elements of the operand list substituted one by one for $_. For example,

```perl
my @a = map(-$_ * $_, (1,3,5,7,9));
```

has the same effect as

```perl
my @a = (-1,-9,-25,-49,-81);
```

A second form of the **map** operator allows a *block* to be given in place of the expression. A block is a series of one or more statements surrounded by braces. Thus, the example above could also be written

```perl
my @a = map {-$_ * $_} (1,3,5,7,9);
```

Although **map** can always be used in place of **foreach**, it is more efficient to use **foreach** if the list of results is not required. (The second use of **map** in the species prediction program could easily be rewritten with **foreach**.)

```perl
#!/usr/bin/perl -I /home/dwyer/book/perl
use strict;
use Util;

### entropy: computes entropy in bits of a distribution
### given as list of probabilities
sub entropy {
    my $eta = 0;
    foreach (@_) { $eta -= $_ * lg($_); }
    return $eta;
}

### read name, probabilities of A,C,G,T from DATA
my ($species1,@model1) = split /\s+/,<DATA>;
my ($species2,@model2) = split /\s+/,<DATA>;
```

compute, print entropy for each species
print "Entropy for $species1 is ", entropy(@model1), " bits.\n";
print "Entropy for $species2 is ", entropy(@model2), " bits.\n";

compute lod scores for A, C, G, T
my @lod = **map**(lg($model1[$_]/$model2[$_]),(0..3));

make a species prediction for each line of the input file
while (**my** $dna= <STDIN>) {
    chomp($dna);
    $dna =~ **tr**/acgtACGT/01230123/;
    my $score = sum(**map**($lod[$_], **split** //,$dna));
    print "lod=$score bits favors ";
    print "$species1 over $species2.\n" **if** $score>=0;
    print "$species2 over $species1.\n" **if** $score<0;
}
__END__
corn .255 .245 .245 .255
fruit fly .225 .275 .275 .225

4.3 Exercises

1. Compute the entropy of the model $\mathcal{M}_D$ in both bits and nats.

2. Compute the entropy of the following models:
   (a) $\mathcal{M}(0.25, 0.25, 0.25, 0.25)$;
   (b) $\mathcal{M}(1, 0, 0, 0)$.

   Hint: Since $x \log x \to 0$ as $x \to 0$ from above, treat $0 \log 0$ as 0.

3. Verify that $H(\mathcal{M}_Z \mid \mathcal{M}_D) \neq H(\mathcal{M}_D \mid \mathcal{M}_Z)$ by computing $H(\mathcal{M}_D \mid \mathcal{M}_Z)$.

4. Compute $H(\mathcal{M}_D \mid \mathcal{M}_D)$ and $H(\mathcal{M}_Z \mid \mathcal{M}_Z)$.

5. Use the **map** operator to write a statement that assigns to @L1 a list of n 1s, where n is the length of list @L.

6. Repeat the previous exercise using the **x** operator instead of **map**.

4.4 Complete Program Listings

Files chap4/Util.pm and chap4/whichspecies.pl are included in the software distribution.

4.5　Bibliographic Note

The book by Durbin et al. (1998) deals extensively with statistical modeling for sequences, as well as computational methods for the same, without addressing programming issues directly.

## CHAPTER FIVE

Substitution Matrices for Amino Acids

In Chapter 3 we considered the problem of aligning DNA sequences that had been read from the same source but with errors introduced by laboratory procedures. Rather arbitrarily, we assigned a reward of $+1$ for a match and penalties of -1 and -2 for mismatches and gaps. In this chapter, we will examine how to align protein sequences that differ as a result of evolution itself rather than owing to experimental error. The outcome will be a method for constructing *substitution matrices* for scoring alignments. Potentially, these matrices can assign a different reward or penalty for each of the 210 possible unordered pairs of amino acids that may appear in a column of an alignment.

The function of a protein is determined by its shape and charge distribution, not by the exact sequence of amino acids. During DNA replication, various *mutations* can alter the protein produced by a gene. Some types of mutations are:

- *point mutations,* in which the machinery of replication randomly substitutes an incorrect nucleotide for the correct one;
- *indels,* or insertions and deletions, in which extra bases are randomly inserted or bases are omitted;
- *translocations,* in which longer pieces of DNA – possibly including one or more entire genes – are moved from one part of the chromosome to another part, or to another chromosome; and
- *duplications,* in which long pieces of DNA are copied and integrated into a chromosome.

Although translocations and duplications have played an important role in evolution, we will be concerned only with point mutations in developing a statistical model of substitution matrices.

Some point mutations are silent: if a mutation changes a codon from CCC to CCT, it still codes for the amino acid proline, and the ribosome is likely to be none the wiser.[1] Other mutations change the amino acid sequence but without altering the

---

[1] Actually, we will see in Chapter 14 that "silent" mutations can, in fact, at least slow down or speed up the ribosome.

overall structure of the protein in any essential way. Still other mutations are disad-vantageous or even lethal, such as the three-base deletion that causes cystic fibrosis. And a few mutations are even beneficial – without them, evolution cannot progress.

There are actually two steps in the genesis of a substitution of amino acids. The first step is a mutation of one or more bases in the DNA. Changes involving a single nucleotide are more likely than changes involving two or more. For example, any two codons for leucine (L) and asparagine (A) differ by at least two bases, whereas either tyrosine (Y) codon can be mutated to phenylalanine (F) by a single nucleotide sub-stitution. The second step is *acceptance* of the mutation, which can occur only if the mutation has no life-threatening effect on the organism. Substituting a hydrophobic L for a hydrophilic A is likely to alter protein structure, since hydrophobic residues tend to turn away from the cytoplasm and toward the inside of a protein whereas hydrophilic residues prefer contact with the cytoplasm. Because random structural changes to any working system are more likely to have negative than positive effects, we expect observations of such substitutions to be relatively rare when comparing re-lated proteins of living species. On the other hand, Y and F have similar chemical properties, and substitutions are observed frequently. In both of these examples, the information-theoretic properties of the genetic code and chemical properties of the amino acids favor the same outcome: no mutation for A and L, and mutation for Y and F. For other residue pairs, the code favors mutation while chemistry precludes acceptance, or vice versa.

Proteins that have evolved from a common source are said to be *homologous.* Be-cause of accumulated mutations, homologous proteins do not have identical amino acid sequences. Homology can be recognized by similarity of sequence, similar three-dimensional structure, and similar function. Before the advent of high-throughput DNA sequencing, it was generally less difficult to determine the structures of proteins through crystallographic techniques than to determine their amino acid sequences. At present, this relationship is reversed, and the three-dimensional structures of most protein sequences are not known definitively. Therefore, the first evidence for ho-mology of proteins now typically results from *sequence comparison.*

Early attempts at sequence comparison focused on *percent identity* and used align-ment scoring schemes similar to the one described in Chapter 3. However, it was soon observed in closely studied groups of proteins (like the globins) that proteins with genuine homology might have only a few identical residues. It was also known that amino acids could be grouped in different ways according to chemical properties and that certain residues regularly substituted for others in homologous proteins.[2] Eventually, this led to the creation of substitution matrices for scoring alignments. Rather than simply counting matches and mismatches, these matrices assign a dif-ferent reward or penalty for aligning each of the 210 possible unordered pairs of amino acids. This chapter explains the statistical basis of substitution matrices and includes a program to build a substitution matrix from a set of pairwise protein se-quence alignments.

---

[2] A common grouping scheme is: C, STPAG, NDEQ, HRK, MILV, FYW.

5.1 More on Homology

Biologists recognize two sorts of homology: *orthology* and *paralogy*. The globin family provides illustrations of each.

The globins are involved in the storage and transport of oxygen. *Hemoglobin* transports oxygen through the bloodstream of vertebrates. *Myoglobin* stores oxygen in muscle tissue. *Leghemoglobin* in soybean plants transports oxygen to colonies of symbiotic nitrogen-fixing bacteria in underground nodules of the plant. All globins are able to capture an iron atom,[3] to which oxygen atoms readily bind.

The hemoglobin in humans and the hemoglobin in rats are *orthologous*; they are descendants of a common ancestor gene by *speciation,* the emergence of new species. They perform the same function of oxygen transport in different species. Hemoglobin and myoglobin in humans are *paralogous*; their common ancestor gene was duplicated within a single species, the two copies mutated independently, and the genes became specialized to perform related but distinct functions. Orthology and paralogy are not mutually exclusive; turkey hemoglobin and human myoglobin are both orthologous and paralogous.[4]

Words of Caution. It often said – but never correctly – that two proteins with 35% sequence identity are "35% homologous", a concept akin to "35% pregnant". Proteins are correctly said to be homologous if and only if their genes descend from a common ancestor gene. Similarity of two sequences provides only *evidence,* not proof, that their structures are similar. Some homologous proteins show no statistically significant sequence similarity. Conversely, some sequences with statistically significant sequence similarity show no structural similarity and are not homologous. Homology is a biological property that cannot be definitively established without due consideration of protein structure and function.

5.2 Deriving Substitution Matrices from Alignments

In the remainder of this chapter, we will learn how to create simple substitution matrices from alignment data. Then we will learn about the PAM matrices – a standard series of matrices scaled for sequences at different evolutionary distances – and their construction.

In the previous chapter, we considered statistical models of DNA sequences and explored the concepts of likelihood and entropy. In this section, we develop statistical models of *alignments*. This time, we are not given the parameters of the models a priori. Instead, our task is to deduce the parameters from a given set of *training data*. Our goal is to find parameters that assign a high probability to data that is "like"

---

[3] More correctly, a *heme group* including an iron atom.

[4] Actually, several distinct paralogous human hemoglobin proteins exist, each identified by a Greek letter. Adult hemoglobin consists of a complex of two α- and two β-hemoglobins; δ-hemoglobin, a fetal form, is present only in primates and resulted from a duplication of the α-hemoglobin gene about 50 million years ago.

the training data. To accomplish this, we find the *maximum likelihood model* for the training data; that is, the set of model parameters that assigns the highest possible probability to the training set itself.

Since we are developing a model of alignments, our training set also consists of alignments that we believe to be both correct and representative. By "correct" we mean that the aligned sequence pairs are genuinely homologous; by "representative" we mean that the sequences range over the whole of the set of sequences we seek to model and are not concentrated in a single group. Generally speaking, it is easier to assure ourselves of the alignments' correctness than of their representativeness. This is because we usually develop a statistical model precisely because we *don't know* the whole set of sequences. Thus, developing statistical models is sometimes an iterative process in which each step yields a better training set for the next.

We actually develop two models of alignment. One describes strings that are randomly aligned without considering whether or not they are homologous. The other describes alignments based on homology. We regard our training set as simply a collection of unordered pairs of residues. Because of some difficult technicalities, we ignore pairs involving the gap character (indels) and develop a model of substitutions only. Therefore, the parameters of the models are the 210 probabilities

$$
\begin{aligned}
\big(p(\mathrm{A}, \mathrm{A}), \quad & p(\mathrm{A}, \mathrm{C}), \quad \ldots \quad p(\mathrm{A}, \mathrm{Y}), \\
& p(\mathrm{C}, \mathrm{C}), \quad \ldots \quad p(\mathrm{C}, \mathrm{Y}), \\
& \qquad \ddots \qquad \vdots \\
& \qquad \qquad p(\mathrm{Y}, \mathrm{Y})\big)
\end{aligned}
$$

corresponding to unordered pairs of amino acid residues.

We assume that the sequences being aligned are real protein sequences. In any set of proteins sequences, each of the amino acid residues occurs with its own distinct *background frequency*. We denote these frequencies by $(q_\mathrm{A}, q_\mathrm{C}, \ldots, q_\mathrm{Y})$. We can estimate these frequencies by counting the number of occurrences of each residue in our training set and then dividing by the total number of occurrences of all residues.

First, let $\mathcal{M}_R$ be a model for alignments of two random strings. (By "random strings" we mean strings in which each position is filled independently according to the background frequencies.) How often should we expect to see the pair G-K, for example? There are two ways to form a G-K pair: either the G or the K can be in the first string. Since the two strings are assumed to be unrelated and therefore statistically independent, the answer must be

$$p_R(\mathrm{G}, \mathrm{K}) = q_\mathrm{G} \cdot q_\mathrm{K} + q_\mathrm{K} \cdot q_\mathrm{G} = 2 q_\mathrm{G} q_\mathrm{K}.$$

There is only one way to form the pair if its two elements are identical; thus,

$$p_R(\mathrm{G}, \mathrm{G}) = q_\mathrm{G}^2.$$

Following these two examples gives us all 210 parameters of $\mathcal{M}_R$.

For the model $\mathcal{M}_H$ of alignments of homologous strings, we rely on estimates derived from the columns of the alignments of training data, ignoring columns with gaps. To determine $p_H(G, K)$, we simply count the number of times this pair occurs in all columns of the training set and then divide by the total number of (nongap) columns. Since we are counting actual pairs, we don't need to handle pairs of identical residues any differently.

If we model only ungapped alignments, the log-of-odds score of strings $s = s_1 s_2 \ldots s_n$ and $t = t_1 t_2 \ldots t_n$ is

$$\log\left(\frac{\mathcal{L}(\mathcal{M}_H \mid s, t)}{\mathcal{L}(\mathcal{M}_R \mid s, t)}\right) = \log\left(\frac{\Pr(s, t \mid \mathcal{M}_H)}{\Pr(s, t \mid \mathcal{M}_R)}\right)$$

$$= \log\left(\frac{\prod_i p_H(s_i, t_i)}{\prod_i p_R(s_i, t_i)}\right)$$

$$= \log\left(\prod_i \left(\frac{p_H(s_i, t_i)}{p_R(s_i, t_i)}\right)\right)$$

$$= \sum_i \log\left(\frac{p_H(s_i, t_i)}{p_R(s_i, t_i)}\right)$$

$$= \sum_i \mathrm{lod}(s_i, t_i),$$

where, for $R_1, R_2 \in \mathcal{R} = \{\mathtt{A}, \mathtt{C}, \ldots, \mathtt{Y}\}$,

$$\mathrm{lod}(R_1, R_2) = \log\left(\frac{p_H(R_1, R_2)}{2q_{R_1}q_{R_2}}\right) \quad \text{for } R_1 \neq R_2,$$

$$\mathrm{lod}(R_1, R_1) = \log\left(\frac{p_H(R_1, R_1)}{(q_{R_1})^2}\right). \tag{5.1}$$

If we tabulate these log-of-odds scores for each of the 210 possible unordered pairs of residues, then we can quickly score any gapless alignment. (Normally, the scores are scaled and then rounded to integers to make them even easier and faster to work with.) Scoring schemes for alignments with gaps generally combine these lod scores for pairs of amino acids with gap scores developed empirically.

Pseudocounts. The method just described has a serious weakness when the training set is small. Suppose that, for example, no C-P pairs occur in the training set. This leads to the estimates $p_H(\mathtt{C}, \mathtt{P}) = 0$ and

$$\mathcal{L}\left(\mathcal{M}_H \mid \begin{bmatrix} \mathtt{THESETWOPROTEINSARENEARLYIDENTICAL} \\ \mathtt{THESETWOCROTEINSARENEARLYIDENTIPAL} \end{bmatrix}\right) = 0.$$

Normally, we would not put so much confidence in the representativeness of the training set to allow the absence of a single pair to completely negate the effect of many high-scoring pairs.

To resolve this difficulty, it is common to "seed" the frequency counts with so-called *pseudocounts*. Each table entry gets "one for free" – or even more than one if the training set is quite small or otherwise particularly untrustworthy. This ensures that the model assigns a positive probability and a finite lod score to every pair.

5.3   Substitution Matrices in Perl

Our program for computing protein substitution matrices in Perl will take as input a file containing a number of alignments presumed to reflect genuine homologies. Each alignment will be in the format produced by Chapter 3's implementation of the Needleman–Wunsch algorithm. To facilitate later developments, we will create another Perl subroutine library ("package") called PAMBuilder, which will reside in the file PAMBuilder.pm. Below is a brief main program using the subroutines of this library to produce a substitution matrix.

```
#!/usr/bin/perl -I. -I/home/dwyer/book/perl
use strict;
use PAMBuilder;

my $A = collectCounts( );
printMatrix($A,"Raw Counts");
my $A = addPseudocounts($A,1);
printMatrix($A,"Raw Counts plus Pseudocounts");
my $P = freqsFromCounts($A);
printMatrix($P,"Training Set Frequencies");
my $L = lodsFromFreqs($P);
printMatrix($L,"LOD scores in half-bits");
```

The function of collectCounts is to assemble raw residue and residue-pair counts from the alignments in STDIN. addPseudocounts increases each of these counts by a specified number – in this case, 1. freqsFromCounts converts a matrix of integer counts into a matrix of floating-point frequencies between 0 and 1. lodsFromFreqs($P) uses frequency data to create a matrix of log-of-odds scores.

We will represent the various sorts of matrices in use by Perl hashes. The hashes will have three kinds of keys.

- The key type will be associated with a description of the type of information in the matrix: raw counts, frequencies, lod scores, or mutation rates.
- Single-character keys like a, c,… will be associated with background counts or frequencies of the 20 individual amino acids in the training set.
- Two-character keys like aa, mi,… will be associated with counts of frequencies of the 400 possible amino acid pairs observable in the training set.

References. Creating a *reference* to a data item such as a hash, scalar, or list[5] allows us to give a subroutine access to the data object without making a new copy.

---

[5] Or even a subroutine or a filehandle.

To avoid expensive copying and wasted space, the routines in this package will operate on *references to hashes* rather than directly on hashes. That is, they will take references to hashes as arguments and return references to hashes as values. The reference itself is a *scalar* and can be stored in a scalar variable. Here is a fragment of code that creates a reference to the scalar variable $y, stores the reference in $x and $z, and manipulates the underlying scalar through $x, $y, and $z:

```
my $y = 3;
my $x = \$y;
my $$x = 7;
print "$y,$$x\n";      ## prints 7,7
$y = 2;
print "$y,$$x\n";      ## prints 2,2
my $z = $x;
$$z = 9;
print "$y,$$x\n";      ## prints 9,9
```

It may be helpful to think of a reference as a "nickname" for a data object. A person has only one full, legal name but may have many nicknames.[6]

Recall that hashes function as collections of (key, value) pairs. In Chapter 1, we used a hash to translate codon keys into amino acid values. We can also create references to hashes with \% and *dereference* a hash reference with %$, as in this example:

```
my %y = (dog=>3,cat=>7);
my $x = \%y;
$$x{cat}++;
print "$y{cat},$$x{cat}\n";          ## prints 8,8
$$x{mouse} = 5;
print "$y{mouse},$$x{mouse}\n";      ## prints 5,5
delete $$x{cat};
print (sort keys %y), "\n";          ## prints 'dogmouse'
print (sort keys %$x), "\n";         ## prints 'dogmouse'
```

A hash reference can be used in any context where the hash's original name fits. Like any other scalar, it can be passed to and returned by subroutines and stored in lists and hashes. (Perl's **keys** operator, which appears in the last two lines of the example, has a hash as its operand and returns a list of the keys presently in the hash in no particular order; the **sort** operator copies and sorts its list operand into lexicographic

---

[6] The analogy is not perfect, since Perl permits the creation of "anonymous" data objects accessible only by their reference nickname.

order.[7]) The first subroutine in our package is collectCounts. Its function is to assemble raw residue and residue-pair counts from the alignments in the default input source STDIN. It creates a new hash of the appropriate type for raw counts, then reads the input file two lines at a time, walking down the two strings of each alignment simultaneously to count residues and residue pairs. Each pair is counted twice, in both orders. This is not strictly necessary, but it turns out to be convenient when we study the PAM matrices in the next section.

```perl
sub collectCounts {
    my %nu = (type=>'counts');
    while (chomp(my $seq1 = <STDIN>)) {
        chomp(my $seq2 = <STDIN>);
        foreach my $i (0..((length $seq1)−1)) {
            my ($r1,$r2) = (substr($seq1,$i,1),substr($seq2,$i,1));
            next if ($r1.$r2) =~ "-";      ## skip if either "residue" is gap
            $nu{$r1}++; $nu{$r2}++;
            $nu{$r1.$r2}++; $nu{$r2.$r1}++;
        }
    }
    return \%nu;
}
```

Counts are stored in the new hash, and a reference to the new hash is returned to the calling routine. Because of the returned reference, the hash lives on and is accessible after the return of the subroutine even though it is no longer accessible through its original "full legal" name %nu! The result of the subroutine call can be assigned to a scalar variable, passed to another subroutine, or used in any other scalar context.

More Loop Control. This package also illustrates the use of one of Perl's loop control operators. **last** immediately terminates the *innermost* loop in which it occurs. When we iterate over all pairs of residues with two nested loops, we normally process each unordered pair twice, once in each order. To avoid this, we could use **last** as follows:

```perl
foreach my $r1 (@residues) {
    foreach my $r2 (@residues) {
        ## process pair here
        last if ($r1 eq $r2);
    }
}
```

---

[7] See Chapter 7 for more on **sort**.

The **next** operator terminates the current iteration of the loop, but does not terminate the loop entirely. Instead, iteration proceeds immediately to the next item. We
could process only nonidentical pairs like this:

```
foreach my $r1 (@residues) {
    foreach my $r2 (@residues) {
        next if ($r1 eq $r2);
        ## process nonidentical pair here
    }
}
```

Both **last** and **next** can be used to control **foreach**-, **while**-, **do while**-, and **for**-
loops. They can also control a loop other than the innermost if the loop is labeled.
However, this is rarely necessary and often confusing.

Optional Arguments. The next subroutine, addPseudocounts, takes a hash reference as an argument and returns a reference to a new hash. The entries of the new
hash are equal to the sum of the old entries and a pseudocount. The pseudocount
is an optional argument; its default value is 1. If only one argument is passed to
addPseudocounts, then that argument (hopefully a hash reference) is assigned to
$M by Line 1 and $pseudo remains undefined. Then Line 2, which is equivalent to
$pseudo = $pseudo || 1, will assign 1 to $pseudo.

```
sub addPseudocounts {
    my ($M,$pseudo) = @_;    ## $pseudo optional;              ## 1
    $pseudo ||= 1;    ## default is 1                          ## 2
    checkType($M,'counts');                                     ## 3
    my %nu = %$M;                                               ## 4
    foreach my $r1 (@residues) {                                ## 5
        foreach my $r2 (@residues) {                           ## 6
            $nu{$r1} += $pseudo; $nu{$r2} += $pseudo;          ## 7
            $nu{$r1.$r2} += $pseudo; $nu{$r2.$r1} += $pseudo;  ## 8
            last if ($r1 eq $r2);                              ## 9
        }                                                      ## 10
    }                                                          ## 11
    return \%nu;                                               ## 12
}
```

Line 3 of addPseudocounts calls a brief subroutine that checks that the hash
referred to by $M is of the correct type by examining $$M{type}.

Line 4 copies the contents of the argument hash into the new hash. %nu is an
identical (for the moment) twin of %$M, not a "nickname".

Lines 5 and 6 iterate over the list of 20 residues defined for use only within this package by the statement

---

my @residues = **split** //,"acdefghiklmnpqrstvwy";

---

near the beginning of the package definition. Lines 7 and 8 alter only the copy, not the original hash.

The next subroutine, freqsFromCounts, expects a reference to a matrix of counts as input and returns a reference to a matrix of frequencies. When the subroutine returns control to the main program, we have

$$\$nu\{a\} = q_A, \ \$nu\{c\} = q_C, \ \ldots, \ \$nu\{y\} = q_Y$$

and

$$\$nu\{aa\} = p_H(A, A), \ \$nu\{ac\} = p_H(A, C), \ \ldots, \ \$nu\{yy\} = p_H(Y, Y).$$

This was accomplished by finding the total number of residues and residue pairs and dividing the raw counts by these totals.

```
sub freqsFromCounts {
    my ($M) = @_;
    checkType($M,'counts');
    my %nu = (type=>'frequencies');
    my $total = 0;
    foreach my $r (@residues) { $total += $$M{$r}; }
    foreach my $r (@residues) { $nu{$r} = $$M{$r} / $total; }
    $total = 0;
    foreach my $r1 (@residues) {
        foreach my $r2 (@residues) {
            $total += $$M{$r1.$r2};
        }
    }
    foreach my $r1 (@residues) {
        foreach my $r2 (@residues) {
            $nu{$r1.$r2} = $$M{$r1.$r2}/$total;
        }
    }
    return \%nu;
}
```

Subroutine lodsFromFreqs takes a reference to a matrix as its argument and returns a reference to a new matrix of log-of-odds scores expressed in half-bits. Subroutine printMatrix takes a reference to a matrix as its argument and prints the matrix in a tabular form. Both subroutines display a nested loop structure similar to freqs-FromCounts. They are contained in the complete module listing in the software distribution.

5.4 The PAM Matrices

Although it may be worthwhile in some cases, it is uncommon to create substitution matrices from scratch for particular species or applications. We usually rely on standard matrices from the PAM and BLOSUM families. The BLOSUM matrices are a significant improvement upon, but not a total departure from, the older PAM matrices. The basic modeling procedure elaborated in the preceding section was used to produce both families; the main differences lie in the construction of the training sets. Since the BLOSUM construction depends on multiple alignments,[8] we will limit our attention to the PAM family in this chapter.

PAM stands for *percent accepted mutation*. The PAM1 matrix, upon which the entire family was based, models sequences that differ on average by one mutation for each 100 residues. The PAM approach is to build a training set of alignment sequence pairs separated by very short evolutionary time periods. Models of longer evolution are constructed by simulating repeated short evolutions.

Unlike the training sets described in the previous section, the PAM training set included no alignments of two modern, observed protein sequences. Instead, observed sequences of a particular group (e.g., myoglobins) were placed at the leaves of an evolutionary tree, or *phylogeny*.[9] From the modern sequences and the chronology of the divergence of species described by the evolutionary trees, sequences were inferred for all ancestral species so as to minimize the total number of mutations in each phylogeny.[10] Finally, the training set was formed by aligning each sequence (whether observed or inferred) with its immediate ancestor in the phylogeny.

To provide even greater confidence that only sequences at small evolutionary distance were compared, any protein group in which two observed sequences were less than 85% identical was split into smaller groups. (This also gave greater confidence in the correctness of the alignments, which had to be constructed using pencil and paper.) The derivation of the published PAM matrices was based on analysis of a total of 1572 substitutions observed in 71 groups of closely related proteins.

The PAM approach focuses on mutation rates rather than pair frequencies. The mutation rate $m(R_1, R_2)$ is the probability that residue R_1 changes to R_2 in the evolutionary period of interest – more precisely, the probability that a residue position containing R_1 at the beginning of the period contains R_2 at its end. The probability that

---

[8] See Chapters 9 and 10.

[9] See Chapter 11. Published descriptions do not make clear whether the phylogenies used were constructed by computer or human analysis of molecular data or whether they simply represented contemporary scientific consensus based on older taxonomic methods.

[10] Such a criterion for inferring ancestral sequences is said to be *parsimonious*. In general, parsimony favors the simplest possible explanation of data. As medical students are advised: "When you hear hoofbeats, think horses, not zebras."

R_1 is unchanged is $m(R_1, R_1)$, so that, for any fixed R_1, the sum $\sum_{R_2 \in \mathcal{R}} m(R_1, R_2) = 1$, where $\mathcal{R}$ is the set of all 20 residues. Given the probabilities $p_H(R_1, R_2)$ and $q(R_1)$ defined in Section 5.2, we can compute mutation rates by

$$m(R_1, R_2) = \frac{p_H(R_1, R_2)}{\sum_{R_2 \in \mathcal{R}} p_H(R_1, R_2)} = \frac{p_H(R_1, R_2)}{q(R_1)}.$$

We will also be interested in the average mutation rate of a random sequence. This is a weighted average of the mutation rates of individual residues defined by

$$\rho = \sum_{R_1 \in \mathcal{R}} \Pr\{R_1 \text{ occurs}\} \cdot \Pr\{\text{some mutation occurs} \mid R_1 \text{ occurs}\}$$

$$= \sum_{R_1 \in \mathcal{R}} q(R_1)(1 - m(R_1, R_1)) = 1 - \sum_{R_1 \in \mathcal{R}} p_H(R_1, R_1). \qquad (5.2)$$

In the PAM approach, the emphasis is upon obtaining a training set with small ρ. The criterion of 85% identity guarantees that $\rho < 0.15$, but in practice it will be much smaller. To obtain the mutation rate of exactly 0.01 needed for the PAM1 matrix, all mutation rates are scaled by a factor of $0.01\rho^{-1}$, giving

$$m_1(R_1, R_2) = (0.01\rho^{-1})m(R_1, R_2) \quad \text{for } R_1 \neq R_2,$$
$$m_1(R_1, R_1) = 1 - (0.01\rho^{-1})(1 - m(R_1, R_1)).$$

Log-of-odds scores can be derived from these mutation rates by first transforming back to probabilities by the formula

$$p_1(R_1, R_2) = m_1(R_1, R_2) \cdot q(R_1)$$

before applying (5.1). These lod scores form the matrix known as PAM1.

Now suppose that a protein evolves for a period twice as long as the one modeled by the PAM1 matrix. What are the corresponding mutation rates? To compute $m_2(R_1, R_2)$, we note that the position occupied by R_1 at the beginning of the period could be occupied by *any* of the 20 residues at the halfway point. If R_3 represents the residue at the halfway point, then

$$m_2(R_1, R_2) = \sum_{R_3 \in \mathcal{R}} m_1(R_1, R_3) \cdot m_1(R_3, R_2).$$

This equation is, in essence, the same as the one defining matrix multiplication: If $S = A \times B$, then $S_{ij} = \sum_k A_{ik} B_{kj}$. In fact, if M_1 represents the mutation rate matrix corresponding to PAM1, then

$$M_k = (M_1)^k = M_1 \times M_1 \times \cdots \times M_1 \quad (k \text{ factors})$$

defines the mutation rate matrix corresponding to a period k times as long.[11] This matrix can be transformed to the PAMk matrix by converting first to pair frequencies and then to lod scores.

---

[11] The mathematically informed will recognize a model of evolution by a *Markov chain*.

	A	R	N	D	C	Q	E	G	H	I	L	K	M	F	P	S	T	W	Y	V
A	2	-2	0	0	-2	0	0	1	-1	-1	-2	-1	-1	-3	1	1	1	-6	-3	0
R	-2	6	0	-1	-4	1	-1	-3	2	-2	-3	3	0	-4	0	0	-1	2	-4	-2
N	0	0	2	2	-4	1	1	0	2	-2	-3	1	-2	-3	0	1	0	-4	-2	-2
D	0	-1	2	4	-5	2	3	1	1	-2	-4	0	-3	-6	-1	0	0	-7	-4	-2
C	-2	-4	-4	-5	12	-5	-5	-3	-3	-2	-6	-5	-5	-4	-3	0	-2	-8	0	-2
Q	0	1	1	2	-5	4	2	-1	3	-2	-2	1	-1	-5	0	-1	-1	-5	-4	-2
E	0	-1	1	3	-5	2	4	0	1	-2	-3	0	-2	-5	-1	0	0	-7	-4	-2
G	1	-3	0	1	-3	-1	0	5	-2	-3	-4	-2	-3	-5	0	1	0	-7	-5	-1
H	-1	2	2	1	-3	3	1	-2	6	-2	-2	0	-2	-2	0	-1	-1	-3	0	-2
I	-1	-2	-2	-2	-2	-2	-2	-3	-2	5	2	-2	2	1	-2	-1	0	-5	-1	4
L	-2	-3	-3	-4	-6	-2	-3	-4	-2	2	6	-3	4	2	-3	-3	-2	-2	-1	2
K	-1	3	1	0	-5	1	0	-2	0	-2	-3	5	0	-5	-1	0	0	-3	-4	-2
M	-1	0	-2	-3	-5	-1	-2	-3	-2	2	4	0	6	0	-2	-2	-1	-4	-2	2
F	-3	-4	-3	-6	-4	-5	-5	-5	-2	1	2	-5	0	9	-5	-3	-3	0	7	-1
P	1	0	0	-1	-3	0	-1	0	0	-2	-3	-1	-2	-5	6	1	0	-6	-5	-1
S	1	0	1	0	0	-1	0	1	-1	-1	-3	0	-2	-3	1	2	1	-2	-3	-1
T	1	-1	0	0	-2	-1	0	0	-1	0	-2	0	-1	-3	0	1	3	-5	-3	0
W	-6	2	-4	-7	-8	-5	-7	-7	-3	-5	-2	-3	-4	0	-6	-2	-5	17	0	-6
Y	-3	-4	-2	-4	0	-4	-4	-5	0	-1	-1	-4	-2	7	-5	-3	-3	0	10	-2
V	0	-2	-2	-2	-2	-2	-2	-1	-2	4	2	-2	2	-1	-1	-1	0	-6	-2	4

Figure 5.1: The PAM250 substitution matrix.

PAM30, PAM70, PAM150, and PAM250 matrices are most commonly used. The various PAM matrices have sometimes been compared to a set of camera lenses, each of which focuses best on evolutionary relationships at a certain distance. Just as inexpensive "focus-free" cameras have a fixed "infinite" focal length, the PAM250 matrix, focused roughly on the most distant detectable relationships, has been the most popular evolutionary lens. It is easy to detect the presence of a photographer's thumb in a photo with any lens, even if it is not in focus. Likewise, PAM250 (shown in Figure 5.1) is usually successful at bringing out less distant evolutionary relationships, even if not in precise detail.

It may be surprising that it is possible to detect similarities between sequences with 250% accepted mutation, since this seems to imply that every element of the sequence has mutated more than twice. However, if we assume that the positions for mutation are selected independently and uniformly, then about $n \ln n$ mutations are required to select each position at least once[12] – 570 PAMs for a sequence of length 300. But this is only half the story, since selective pressures cause the distribution of *accepted* mutations to be far from uniform. In practice, PAM250 corresponds to identity near 20%.

---

[12] Probabilists call this the *coupon collector's problem*.

5.5   PAM Matrices in Perl

To implement the derivation of PAM matrices in Perl, we will add a few more subroutines to our PAMBuilder package. Using subroutines from Section 5.3, the following program reads alignments from STDIN and computes a frequency matrix. Next, it uses ratesFromFreqs to compute a mutation rate matrix and rescaleRates to compute another matrix of mutation rates with an average mutation rate of 1%. This 1% mutation rate matrix is repeatedly multiplied by itself using multiplyRates, and the results are converted to lod scores by lodsFromFreqs; averageMutationRate is then called to verify that mutation rates are as expected.

```perl
#!/use/bin/perl -I/home/dwyer/book/perl -I.

use strict;
use PAMBuilder;

my $A = collectCounts();
printMatrix($A,"Raw Counts");
my $A = addPseudocounts($A);
printMatrix($A,"Raw Counts plus Pseudocounts");
my $P = freqsFromCounts($A);
printMatrix($P,"Frequencies");
my $M = ratesFromFreqs($P);
my $PAMM = rescaleRates($M,0.01);
for (my $i=1; $i<=256; $i+=$i) {
    my $rho = averageMutationRate($PAMM);
    print "Mutation rate of PAM$i is $rho.\n";
    printMatrix(lodsFromFreqs(freqsFromRates($PAMM)),
            "PAM$i (half-bits)");
    $PAMM = multiplyRates($PAMM,$PAMM);
}
```

Three of these subroutines – ratesFromFreqs, freqsFromRates, and rescaleRates – are so similar to ratesFromCounts that no further comment is warranted. We will give some attention to multiplyRates and averageMutationRate.

Subroutine multiplyRates multiplies two matrices of mutation rates. Its arguments are references to two hashes, and its value is a reference to a new hash. It distinguishes itself from the preceding routines mainly by its third nested loop, which computes the sum of products required to find one entry of the product of two matrices.

```perl
sub multiplyRates {
    my ($M1,$M2) = @_;
```

```
    checkType($M1,'mutation rates');
    checkType($M2,'mutation rates');
    my %nu = (type=>'mutation rates');
    foreach my $r1 (@residues) {
        foreach my $r2 (@residues) {
            my $prod;
            foreach my $r (@residues) {
                $prod += $$M1{$r1.$r} * $$M2{$r.$r2};
            }
            $nu{$r1.$r2} = $prod;
        }
        $nu{$r1} = $$M1{$r1};
    }
    return \%nu;
}
```

Subroutine averageMutationRate is capable of computing the average mutation rate of both frequency matrices and mutation rate matrices. It makes use of the full features of checkType to do so. The type description $type passed to checkType is not just a simple string but rather a Perl pattern consisting of one or more allowed matrix types separated by |. The Perl pattern matcher is used to match this pattern against the type of the matrix $M, saving the matched substring (if any) in $1. If $1 is defined then it is returned on the last line; otherwise, the Perl interpreter proceeds to evaluate the **die**-statement, which prints an error message and immediately aborts the program.

```
sub averageMutationRate {
    my ($M) = @_;
    my $type = checkType($M,'mutation rates|frequencies');
    my $rate = 1;
    if ($type eq "frequencies") {
        foreach my $r (@residues) { $rate -= $$M{$r.$r}; }
    } else {
        foreach my $r (@residues) { $rate -= $$M{$r} * $$M{$r.$r}};
    }
    return $rate;
}

sub checkType {
    my ($M,$type) = @_;
    $$M{type} =~ /($type)/;
    $1 or die "Matrix is of type $$M{type}; $type is required.";
}
```

5.6 Exercises

1. A UFO recently crashed near Lake Raleigh. Upon inspection, it was found to contain well-preserved specimens of several life forms from the planet Koolman. Analysis of their proteins showed that they consist of only four amino acids: K, L, M, and N. Further analysis established a number of homologies and produced a number of alignments with a high percentage of identical residues. Specifically, the following residue pairs were observed in the alignments:

K	L	M	N	
1	12	2	4	K
	20	20	8	L
		11	16	M
			6	N

   Use these counts to estimate background frequencies and to derive a log-of-odds matrix for scoring alignments of proteins from the UFO's home planet.

2. Modify our program to print the rows and columns by group – that is, in the order `cstpagndeqhrkmiyfyw`.

3. Show that the two expressions in (5.2) are equal.

4. We have seen how any set of (background and) substitution probabilities can be transformed into a log-of-odds scoring matrix. This process can naturally be inverted, implying that every scoring matrix corresponds to some underlying model of substitution probabilities regardless of whether that model was articulated explicitly during the development of the scores.

   (a) What substitution probabilities P_H underlie our ± 1 model for scoring DNA alignments? Assume that background probabilities are uniform (i.e., that each is 0.25). Don't forget that the unit of the scores (half-bits, quarter-nats, etc.) is unknown; this means that the base of logarithms in (5.1) is one of the variables to be solved for.

   (b) Repeat part (a) but use the default scoring scheme of the PHRAP sequence assembly program: $+1$ for a match, -2 for a mismatch. (PHRAP is discussed in Chapter 13.)

   (c) Refer to technical specifications of sequencing machinery to determine whether either one of these sets of probabilities conveys a realistic estimate of the probabilities that two sequencing reads of the same sequence will call the same base in a fixed position. If not, try to develop a better scoring matrix for aligning DNA sequences that differ because of sequencing errors.

5. Rewrite the inner loop of multiplyRates with **map** and sum (from **package** Util).

5.7 Complete Program Listings

The software distribution includes the module `chap5/PAMBuilder.pm` and its driver `chap5/buildpam.pl`.

5.8 Bibliographic Notes

Altschul (1991) provides a good explanation of the mathematical theory of substitution matrices in general. Dayhoff, Schwartz, and Orcutt (1978) outline the derivation of the PAM matrices. Henikoff and Henikoff (1992) describe the construction of the BLOSUM series.

## CHAPTER SIX

Sequence Databases

Once DNA fragments have been sequenced and assembled, the results must be properly identified and labeled for storage so that their origins will not be the subject of confusion later. As the sequences are studied further, annotations of various sorts will be added. Eventually, it will be appropriate to make the sequences available to a larger audience. Generally speaking, a sequence will begin in a database available only to workers in a particular lab, then move into a database used primarily by workers on a particular organism, then finally arrive in a large publicly accessible database. By far, the most important public database of biological sequences is one maintained jointly by three organizations:

- the National Center for Biotechnology Information (NCBI), a constituent of the U.S. National Institutes of Health;
- the European Molecular Biology Laboratory (EMBL); and
- the DNA Databank of Japan (DDBJ).

At present, the three organizations distribute the same information; many other organizations maintain copies at internal sites, either for faster search or for avoidance of legal "disclosure" of potentially patent-worthy sequence data by submission for search on publicly accessible servers. Although their overall formats are not identical, these organizations have collaborated since 1987 on a common annotation format. We will focus on NCBI's GenBank database as a representative of the group.

A new release of the entire GenBank database is prepared every two months, and updates are issued daily. Release 127 (December 2001) contained more than 1.7×10^7 sequence entries comprising 1.6×10^{10} nucleotides and nearly 5.8×10^{10} bytes (58 Gigabytes) in ASCII format; this represented a 50% increase over Release 121 a year earlier. The database has 17 "divisions". Some divisions are taxonomic, such as BCT (bacterial), ROD (rodent), and PLN (plant); others are technological, such as EST (expressed sequence tags) and HTG (high-throughput genomic).

Many scientific journals now require that sequences dealt with in their pages be submitted to the public databases and referred to by *accession number* rather than included in full in the journal. Bulk submissions come regularly from a number of

```
>gi|532319|pir|TVFV2E|TVFV2E envelope protein
ELRLRYCAPAGFALLKCNDADYDGFKTNCSNVSVVHCTNLMNTTVTTGLLLNGSYSENRT
QIWQKHRTSNDSALILLNKHYNLTVTCKRPGNKTVLPVTIMAGLVFHSQKYNLRLRQAWC
HFPSNWKGAWKEVKEEIVNLPKERYRGTNDPKRIFFQRQWGDPETANLWFNCHGEFFYCK
MDWFLNYLNNLTVDADHNECKNTSGTKSGNKRAPGPCVQRTYVACHIRSVIIWLETISKK
TYAPPREGHLECTSTVTGMTVELNYIPKNRTNVTLSPQIESIWAAELDRYKLVEITPIGF
APTEVRRYTGGHERQKRVPFVXXXXXXXXXXXXXXXXXXXXXXXXXVQSQHLLAGILQQQKNL
LAAVEAQQQMLKLTIWGVK
```

Figure 6.1: A protein sequence in FASTA format.

high-throughput sequencing centers around the world. It is clear that the size of Gen-Bank will continue to increase rapidly.

This chapter begins with a brief excursion into the FASTA file format. Next, we will consider the GenBank format in some detail. We will study the annotation used to describe the locations of subsequences and exploit the power of Perl's pattern-matching facilities to extract subsequences automatically. Then, we will develop a sequence reader package that is able to examine an input file and adapt itself to read-ing sequences in whichever format it finds in the file – be it FASTA, GenBank, or a simple one-sequence-per-line format. Along the way, we will learn about Perl's features to support object-oriented programming.

6.1 FASTA Format

FASTA format was designed for the FASTA database searching program, an alterna-tive to the BLAST method that we will describe in Chapter 7. Annotation in FASTA files is minimal; each sequence is preceded by a one-line "description" marked by the presence of the greater-than symbol > in the first column. Descriptions are often divided into fields by the vertical bar |, but there is no universally accepted standard for the structure of the description and in some circumstances the description line contains only the initial >.

The sequence itself may occupy many lines. When writing FASTA files, it is ad-visable to limit the length of individual lines to 72 or 80 characters. Characters in sequences may be either upper or lower case. A typical entry for a single protein se-quence is shown in Figure 6.1.

6.2 GenBank Format

A GenBank file consists of one or more *entries* or *accessions,* each containing three parts – *headers, features,* and *sequence* – in that order.

The sequence section is easiest to describe. A typical beginning is

```
BASE COUNT 19783 a 15709 c 15910 g 19157 t
ORIGIN
        1 tttactctgc ttcgctgaat tttcgggttg atgtaaaacg tctgcaactg cgtgaattgc
       61 atcaacagag ccgggggagca gccggcagca gaacactgag tctgctgatg cgtcagtcgg
```

The first line reports the total numbers of the four nucleotides in the sequence that follows. The second line reports the beginning of the sequence per se. Subsequent lines present the sequence itself. Each line presents 60 nucleotides in groups of ten separated by a single space. The nucleotides are preceded by the index of the first in the line. The total length of the line is 75 characters.

The header and feature sections consist of 80-character lines. (See Figure 6.2, pp. 76–7.) The header section consists of header keywords and subkeywords in columns 1–10 and supporting information in columns 13–80 and gives information about the sequence as a whole. Header lines give accession and version numbers of the sequence; the source organism and its complete taxonomic classification; and bibliographic entries for journal articles related to the sequence.

The feature section, often referred to as the *Feature Table* or *FT,* identifies regions of biological significance in the sequence. Each feature annotation includes a *feature key,* a *location* (exact or approximate), and zero or more *qualifiers.* The feature key appears in columns 6–20 and the location appears in columns 22–80 of the first line describing the feature. Subsequent lines may contain qualifiers, beginning in column 22.

The FT format permits more than sixty different feature keys and provides for future growth in the set.[1] However, according to one analysis, only eight keys account for more than 80% of features:

- CDS, for "coding sequence", identifies subsequences that encode (or are believed to encode) the amino acid sequence of a protein.
- source identifies the biological source of a subsequence. Every entry must include at least one source feature key; a sequence assembled from several sources may be documented by several source features, although this occurs infrequently.
- gene identifies a gene. Genes may span many smaller features, such as introns, exons, 3′ and 5′ untranslated regions, and promoters.
- repeat_region identifies regions where a single short sequence is repeated many times. We discuss these further in Chapter 15.
- misc_feature is intended for the annotation of miscellaneous rare features such as "5′ recombination hotspot". Its frequent appearance (as well as a proscriptive comment in the official document defining the format) suggests that it is more often used merely to attach a label or comment to a location that lacks independent biological significance (in this situation, the − "pseudokey" should be used).
- exon and intron identify exons and introns. Exons encompass coding sequences (CDS) as well as 5′ and 3′ untranslated regions (5′UTR and 3′UTR).
- rRNA identifies DNA coding for ribosomal RNAs.

Each feature may be followed by one or more *qualifiers.* Qualifiers are proceeded by "/" and (usually) followed by "=" and a value. Many qualifiers exist, but – just as for feature keys – a small number predominate. For a particular feature key, some

---

[1] Local installations may define additional key names beginning with an asterisk; these names are reserved permanently for local use.

qualifiers may be required and others optional. Nearly 85% of qualifiers of CDS features come from the following list:

- db_xref provides a cross-reference to related information in another database. The value of this qualifier has two parts: a database identifier such as SWISS-PROT or taxon, and an accession or entry number within the database (e.g., /db_xref="SWISS-PROT:P28763").
- codon_start indicates the relationship of the reading frame to the beginning of the coding sequence. In eukaryotic genes, a single codon may be split across two exons. This qualifier indicates whether the first whole codon begins in the first, second, or third position of the coding sequence.
- product gives the name of the protein coded for, for example, /product= "galactokinase".
- gene gives a short symbol representing the gene; for example, /gene="galK".
- transl_table specifies a nonstandard table for translating codons to amino acids. Twenty-three values (one standard and 22 others) are recognized; for example, /transl_table=14 specifies the Flatworm Mitochondrial Code.[2]
- note gives a miscellaneous note; for example, /note="This one kept us busy.".

The most common qualifiers of source are:

- organism, the only required qualifier, is given in binomial form (for example, /organism="Homo sapiens").
- db_xref, as for CDS.
- tissue_type, as in /tissue_type="liver".
- strain, clone, and clone_library identify the specific lineage of the cells from which the DNA was obtained.

Strictly speaking, GenBank is a *nucleotide* database; protein sequences are stored in a database named GenPept in the same format. However, this distinction is often lost in both speech and writing. An actual GenBank entry, edited to retain only a few typical feature table annotations, is shown in Figure 6.2.

6.3   GenBank's Feature Locations

As previously mentioned, GenBank feature descriptions include indicators of their precise (or approximate) location within the sequence entry. The language of location specifications is fairly rich; for example, the specification

    join(<13.>25,84..92,complement(138.141))

refers to a subsequence formed by concatenating three separate items: a single base from a range whose exact boundaries are unknown but that covers at least positions

---

[2] We have until now ignored the fact that there are some very slight variations in the interpretation of the 64 codons among species and among organelles within species.

```
LOCUS       NC_001972   70559 bp    DNA    circular BCT       28-DEC-2000
DEFINITION  Yersinia pestis plasmid pCD1, complete sequence.
ACCESSION   NC_001972
VERSION     NC_001972.1  GI:10955757
KEYWORDS    .
SOURCE      Yersinia pestis.
  ORGANISM  Plasmid Yersinia pestis
            Bacteria; Proteobacteria; gamma subdivision; Enterobacteriaceae;
            Yersinia.
REFERENCE   1  (bases 43318 to 44840)
  AUTHORS   Leung,K.Y. and Straley,S.C.
  TITLE     The yopM gene of Yersinia pestis encodes a released protein having
            homology with the human platelet surface protein GPIb alpha
  JOURNAL   J. Bacteriol. 171 (9), 4623-4632 (1989)
  MEDLINE   89359090
   PUBMED   2670888
REFERENCE   2  (bases 1 to 70559)
  AUTHORS   Perry,R.D., Straley,S.C., Fetherston,J.D., Rose,D.J., Gregor,J. and
            Blattner,F.R.
  TITLE     DNA sequencing and analysis of the low-Ca2+-response plasmid pCD1
            of Yersinia pestis KIM5
  JOURNAL   Infect. Immun. 66 (10), 4611-4623 (1998)
  MEDLINE   98427122
   PUBMED   9746557
REFERENCE   3  (bases 43318 to 44840)
  AUTHORS   Straley,S.C.
  TITLE     Direct Submission
  JOURNAL   Submitted (26-APR-1993) Microbiology and Immunology, University of
            Kentucky, MS415 Medical Center, Lexington, KY 40536-0084, USA
REFERENCE   4  (bases 1 to 70559)
  AUTHORS   Perry,R.D., Straley,S.C., Fetherston,J.D., Rose,D.J., Gregor,J. and
            Blattner,F.R.
  TITLE     Direct Submission
  JOURNAL   Submitted (25-JUN-1998) Microbiology and Immunology, University of
            Kentucky, MS415 Medical Center, Lexington, KY 40536-0084, USA
COMMENT     PROVISIONAL REFSEQ: This record has not yet been subject to final
            NCBI review. The reference sequence was derived from AF074612.
FEATURES             Location/Qualifiers
     source          1..70559
                     /organism="Yersinia pestis"
                     /plasmid="pCD1"
                     /strain="KIM5"
                     /db_xref="taxon:632"
     gene            57..368
                     /gene="Y0001"
     CDS             57..368
                     /gene="Y0001"
                     /note="o103; 43 pct identical (0 gaps) to 100 residues of
                     an approx. 200 aa protein GENPEPT: gi|537126, orf_o198
                     Escherichia coli"
                     /codon_start=1
                     /transl_table=11
                     /product="unknown"
                     /protein_id="NP_047561.1"
                     /db_xref="GI:10955758"
                     /translation="MHQQSRGAAGSRTLSLLMRQSGYNVVRWLARRLMRECGLASRQP
                     GKPRYRGEREVSLASPDLLKRQFKPSEPNRVWSGYISYIKVNGGWCYLALVIDLYSFH
                     W"
     gene            665..1033
                     /gene="nuc"
                     /note="Y0002"
```

```
      CDS               665..1033
                        /gene="nuc"
                        /codon_start=1
                        /transl_table=11
                        /product="endonuclease"
                        /protein_id="NP_047562.1"
                        /db_xref="GI:10955759"
                        /translation="MDTKLLQHTPIGTMVDYRPVNTKSGGKRLRRCPDFVIHYRMDLL
                        VNAGIPVRTVNSFKALHDKVIIVDGKNTQMGSFNFSQAAVQSNSENVLIIWGDFTVVQ
                        AYLQYWQSRWNKGTDWRSSY"
...
    misc_RNA            complement(1560..1649)
                        /note="antisense RNA"
                        /product="copA"
...
    repeat_region       11939..15801
                        /rpt_family="IS285/IS100"
...
BASE COUNT     19783 a   15709 c   15910 g   19157 t
ORIGIN
        1 tttactctgc ttcgctgaat tttcgggttg atgtaaaacg tctgcaactg cgtgaattgc
       61 atcaacagag ccggggagca gccggcagca gaacactgag tctgctgatg cgtcagtcgg
      121 gttataacgt ggtgcgctgg ctggcccgca ggctgatgcg ggaatgtggt ctggcgagtc
...
    70441 cgttaacacg gctgaaaaca aaatggccag tggtcgaatt gtgccgcctg ctcaaaataa
    70501 cgcgcagtgt ttactctgct tcgctgaatt ttcgggttga tgtaaaacgt ctgcaactg
//
```

Figure 6.2: Abridged GenBank entry.

13 through 25; the nine bases in positions 84 through 92; and the complement of a single base from one of the positions 138 through 141.

Fortunately, much of the complexity of the notation is associated with *inexact* specifications, and many (if not most) programming tasks are concerned with extracting and processing substrings that are *exactly* specified. Processing an exact location specification to produce the subsequence it identifies is a good exercise in the sort of string processing that Perl makes easy. Exact locations have no single dots or less-than or greater-than symbols; they contain only `join`, `complement`, `..`, and integers.

Formal language theory (a course required of – but not savored by – most undergraduate computer science majors) tells us that Perl's patterns are good for matching *regular languages,* whereas its nested parentheses make the language of GenBank locations a member of the more complex family of *context-free languages.* This indicates that we will not be able to decompose GenBank locations into their natural constituent pieces for processing (i.e., to *parse* them) without first understanding the theory of context-free languages and a method for parsing them.

Fortunately, we will be able to finesse this point by combining Perl's pattern-matching capabilities with another feature, the **eval** operator. The **eval** operator allows us to build up the text of a small Perl program in a string and then to execute it and obtain a result. Our strategy will be to transform the string containing the Gen-Bank location descriptor into an executable Perl program by making a few simple textual substitutions. The GenBank location itself will never be parsed; the altered version will be processed by the full power of the Perl interpreter itself rather than by Perl's pattern-matcher alone.

The subroutine we write, extract_feature, transforms the location in six steps.

1. $location = **lc** $location; shifts all letters to lower case, because the GenBank format is not case-sensitive but the Perl interpreter is.

2. $location =˜ **s**/,/./g; changes commas to dots. g indicates a *global* substitution, meaning that *all* commas (not just the first) must be changed. In the GenBank format, commas separate items that are to be concatenated by join; the dot denotes concatenation in Perl.

3. $location =˜ **s**/join/ident/g; changes every occurrence of the word join to ident. Since the strings to be joined will be concatenated by the dots inserted in the previous step, we will write a Perl function ident that simply returns its one and only argument.

4. $location =˜ **s**/(\d+)/N$1/g; tags each number in the location string with a capital N. This will help us distinguish unprocessed numbers in the original string from numbers inserted as part of the translation process. Since the entire string was shifted to lower case, we can be sure that any upper-case N does indeed tag a number. The pattern \d matches a decimal digit, and the plus sign indicates "one or more". The parentheses cause the substring matched by \d+ to be saved in the special variable $1. The replacement string is the string matched with N added to the front.

5. $location =˜ **s**/N(\d+)\.\.N(\d+)/**substr**(\$\$s,$1−1,$2+1−$1)/g; replaces pairs of numbers separated by two dots with a **substr** construct that extracts the corresponding locations from the string $$s. The backslashes preceding the dots and the dollar signs are to cancel the special significance these characters normally have in patterns. In the replacement string, we obtain the offset by subtracting 1 from the first number; this accounts for the fact that GenBank indexes strings from 1 but Perl indexes from 0.

6. $location =˜ **s**/N(\d+)/**substr**(\$\$s,$1−1,1)/g; transforms any remaining numbers from the original location into **substr** constructs with length 1. By matching the N-tags, we guarantee that the numbers transformed are not part of some replacement string from the previous step.

If we trace the effects of these substitutions step by step on an example location specification, we observe:

```
join(18..24,COMPLEMENT(join(5..10,6..11)),27,40..45)
join(18..24.complement(join(5..10.6..11)).27.40..45)
ident(18..24.complement(ident(5..10.6..11)).27.40..45)
join(N18..N24.complement(join(N5..N10.N6..N11)).N27.N40..N45)
ident(substr($$s,18-1,24+1-18).complement(ident(substr($$s,5-1,10+1-5).
      substr($$s,6-1,11+1-6))).N27.substr($$s,40-1,45+1-40))
ident(
    substr($$s,18-1,24+1-18).complement(ident(substr($$s,5-1,10+1-5).
    substr($$s,6-1,11+1-6))).substr($$s,27-1,1).substr($$s,40-1,45+1-40))
```

For the final string to be successfully evaluated, we must define the subroutines complement and ident, but these are easy.

```
sub complement {
    my ($s) = @_;
    $s =~ tr/acgtACGT/tgcaTGCA/;
    return reverse $s;
}
sub ident {
    my ($s) = @_;
    return $s;
}
```

Subroutine extract_feature, which accepts either a string or a reference to a string as its second argument, appears below.

```
sub extract_feature {
    my ($location,$s) = @_;
    print @_, "\n";
    ## allow $s to be either a string or a reference to a string
    unless (ref $s) { my $t=$s; $s=\$t; }
    $location = lc $location;
    $location =~ s/,/\./g;
    $location =~ s/join/ident/g;
    $location =~ s/(\d+)/N$1/g;
    $location =~ s/N(\d+)\.\.N(\d+)/substr(\$\$s,$1-1,$2+1-$1)/g;
    $location =~ s/N(\d+)/substr(\$\$s,$1-1,1)/g;
    return eval($location);
}
```

6.4   Reading Sequence Files in Perl

In this section, we will develop a set of Perl packages able to read sequences from several different sources with different formats. To do so, we will exploit Perl's support for *object-oriented programming*.

To begin, we will create a sequence reader *class* called SimpleReader that reads from files with a simple format of one sequence per line. It will be possible for many *objects* or *instantiations* of this class to exist at the same time, each reading from a different source.

Next, we will create a *superclass* SeqReader for SimpleReader and define two new *subclasses* of SeqReader – FastaReader and GenBankReader – that *inherit* some methods from SeqReader but redefine others. When a program attempts to instantiate it, the new SeqReader package will determine the format of the input file from its first line and will then create the appropriate type of reader object. Thus, an application program will be able to open and read sequences from a sequence file with no specific knowledge of its format.

6.4.1   Object-Oriented Programming in Perl

Although object-oriented programming languages entered the mainstream only with the advent of C++ in the late 1980s, nearly all of its main features were elaborated in the mid-1960s in a language called SIMULA. The SIMULA language was designed for discrete-event simulations (to model, e.g., the flow of passengers, freight, and aircraft through an airport), and the general applicability of its features was not immediately recognized. Since the mid-1990s, object-oriented languages like C++ and Java have *become* the mainstream, and introductory programming courses in most universities are now taught in one of these languages. A brief outline of the object-oriented approach follows for readers who may have learned to program in an older language like C or Pascal.

In earlier approaches, definitions of data structures and definitions of the procedures that manipulate them were syntactically distinct. Functions and procedures *operated on* data entities received as arguments. If an application had to deal with triangles, squares, and circles, and if it had to be able to draw, compute area, and compute perimeter of each, then there were likely to be three procedures (draw, area, and perimeter) with three cases (triangle, square, and circle) in each. If a new shape like "ellipse" had to be added, modifications in all three procedures were required.

In object-oriented languages, definitions of procedures to manipulate data structures are integrated into the definition of the data structure itself. For our geometric problem, three *classes* – Triangle, Square, and Circle – would be separately defined, each with its own *methods* named draw, area, and perimeter. Individual triangles are created as distinct *objects* of the class Triangle. A statement to a draw an object $t is likely to take the form $t−>draw(5), meaning "Shape stored in $t, please use *your* draw method to draw yourself with five-pixel-thick lines!", rather than the form draw($t,5), meaning "Subroutine draw, please figure out what kind of figure $t is and draw it with five-pixel-thick lines". Objected-oriented practitioners say that the program's knowledge about triangles is *encapsulated* by the class Triangle. If ellipses are added later, it is by defining a new class Ellipse encapsulating all knowledge of ellipses; no modification of existing code is required. (Of course, if we add a new operation – say, meanWidth – then we will have to add code to all the classes, but we still avoid modifying any existing methods.)

Another important concept in object-oriented languages is *inheritance*. Classes may be defined as *subclasses* of other classes, meaning that objects of the subclass have all the data items and methods of the superclass as well as other data and methods unique to the subclass. A class Person with data items like name, address, and date of birth might have subclasses Employee, with data like title, salary, and hire date; Customer, with data related to purchases; and Supplier, with data related to sales.

Perl's support of object-oriented programming is, in some respects, merely rudimentary. Perl does not allow a class definition to declare variables that belong to each object of the class. Instead, each object possesses a single hash in which to store the data items it encapsulates. More significantly, C++ and Java allow access to the data items encapsulated in an object to be restricted to the methods of the same object. This allows the programmer of the class to enforce invariants on the data in each object. For example, it makes no sense for an Employee object to have a negative salary. By requiring "outsiders" to manipulate this number through methods she supplies, the Java or C++ programmer of an Employee class can reserve for herself the opportunity to intercept attempts to make the value negative. Perl supplies no such capability, relying only on the good intentions of the outsiders to prevent such direct manipulation of an object's data. These points notwithstanding, Perl's object-oriented facilities are quite adequate for many purposes.

6.4.2 The SimpleReader Class

The "simple" file format read by SimpleReader objects has one sequence per line. Each sequence may be preceded by an optional identifier and colon; there is no other annotation. The SimpleReader class has five methods: new, to open a file and create a new object to read it; fileName, to report the file from which the reader reads; readSeq, to read and return the next sequence from the file; seqId, to return the identifier of the most recent sequence; and close, to close the file and release space. The following short program illustrates how we will create and use a SimpleReader object.

```perl
use SimpleReader;
my $r = SimpleReader->new("file.txt");
while (my $seq = $r->readSeq()) {
    print "\n\n", $r->fileName();
    print " ", $r->seqId(), "\n";
    print "$seq\n";
}
$r->close();
```

Perl does not include a syntactic construct for delimiting the boundaries of a class definition. Instead, each class must be defined in its own file, with suffix .pm. The first line of SimpleReader.pm defines the name of the class or package.

```perl
package SimpleReader;
```

Likewise, Perl defines no "default constructors", nor does it give special meaning to the word new. Instead, it is merely *conventional* that each class defines a subroutine named new. (Some programmers prefer to give constructors the same name as the package.)

Now, suppose we want to create a new reader for some text file called foo.txt. We can do this with the statement:

```
my $r1 = SimpleReader−>new("foo.txt");
```

We can create a second reader for file bar.txt by either of these statements:

```
my $r2 = SimpleReader−>new("bar.txt");
my $r2 = $r1−>new("bar.txt");
```

These statements cause the following to occur.

1. First, Perl determines the relevant package by looking at what appears to the left of the *dereference operator* −> . If it is a variable (like $r1), then the class of the object determines the package.
2. Next, Perl finds the subroutine named to the right of the arrow in the file for the package, in this case, new.
3. Finally, Perl shifts the reference or package name on the left of the −> into the left end of the argument list @_ and then invokes the subroutine with the extended list. Thus, in this specific case, although the invocation of new appears to involve only one argument, the subroutine definition in SimpleReader.pm extracts *two* arguments from @_. Many programmers choose to name the first argument $this or $self, but (unlike C++ and Java) Perl itself gives these names no special meaning.

The complete method new appears below.

```
use strict;
use FileHandle;

sub new
{
    my ($this,          ## literal "SimpleReader" or ref to existing SimpleReader
        $fileName)      ## string giving operating system's name for file
        = @_;
    my $fh = *STDIN;
    if ($fileName ne "STDIN") {
```

```
    $fh = FileHandle->new("<$fileName")
        or die "can't open $fileName ";
  }
  my $buff = <$fh>;
  chomp $buff;
  my $hash = {fh=>$fh,              ## save filehandle
                buff=>$buff,          ## save line read for next read
                fileName=>$fileName};  ## save filename
  bless $hash;
  return $hash;
}
```

The first few lines of new are concerned with opening the specified input file and establishing access through a *filehandle*. FileHandle is a standard Perl package. For the moment, we will ignore the details and simply note that, once these lines are executed, $fh contains a reference to an object that reads a line from a file when <$fh> is evaluated. In fact, the line **my** $buff = <$fh>; reads the first line of the file.

The line beginning **my** $hash = ... creates a hash to hold the data describing the open file. Objects are typically represented by hashes in Perl; the methods associated with the objects access the individual data items of an object by key.[3]

The **bless** operator on the fourteenth line tags a reference with a string. Although a second argument can be supplied, by default, the tag is the name of the package in which the **bless** operator is found. Thus, the "object reference" returned by new is just a *reference to a hash, tagged with the name of the package in which its methods are to be found.* In the hash can be found a filehandle under key fh, a "buffer" under buff, and the file's name under fileName. The dereference operator −> relies on the tag on its left operand to determine where to look for the method on its right. If we add the statement

```
  print "$hash\n";
```

both before and after the blessing, we get output similar to

```
    HASH(0x80cb9c8)
    SimpleReader=HASH(0x80cb9c8)
```

The hexadecimal number 0x80cb9c8 is the memory address where the hash is stored.

Programs that use the SimpleReader package must (or at least should) access the hash through four different methods. The fileName method provides the simplest

---

[3] It is also possible to use arrays, but this complicates inheritance.

possible example. Although it appears to be called with no arguments (as $r −> file-Name( )) in our program, the object reference is shifted into the argument list. Thus, $this is a reference to a hash, and both $this−>{fileName} and $$this{fileName} will return the file name string stored in the hash.

```perl
sub fileName {
    my ($this) = @_;
    $this−>{fileName};
}
```

Method seqId works the same way.

```perl
sub seqId {
    my ($this) = @_;
    $this−>{seqId};
}
```

The close method is only slightly more complicated. It invokes the close method of the filehandle object, then frees up some memory by removing all entries from the hash. (The statement **undef** $this; would not free up any memory, since the code calling this method presumably retains a reference to the hash.)

```perl
sub close {
    my ($this) = @_;
    $this−>{fh}−>close( );
    delete @{$this}{keys %$this};
}
```

The most interesting method is readSeq. SimpleReaders always read ahead one line and store the line under buff in the hash. (This may seem pointless at the moment, but it will be useful soon.) So, to pass the next sequence to the application, we need to process the line already saved in the buffer, then read a new line for the buffer. To process the line, we attempt to split it on the first colon to separate the identifier from the sequence. If this is unsuccessful, we assume that the identifier was omitted and assign the whole line to $seq. We stash the identifier in the hash, and return the sequence.

```perl
sub readSeq {
    my ($this) = @_;
    return "" unless $this−>{buff};
```

```perl
    my ($id,$seq) = ($this->{buff} =~ m/(.*):(.*)$/);
    $this->{seqId}=$id;
    $seq ||= $this->{buff};
    $seq =~ s/[^a-z]//g;
    my $fh = $this->{fh};
    $this->{buff} = <$fh>;
    chomp $this->{buff};
    return $seq;
}
```

Finally, the last line in any .pm file should be

```perl
1;
```

Otherwise, the program using the package may assume that it failed to load correctly.

6.4.3   Hiding File Formats with Method Inheritance

In the last section, we developed a sequence reader for files with one sequence per line. We will begin this section by developing a readSeq method for FASTA files and another for GenBank files, both of which involve sequences spanning many lines. Then we will explain how Perl's support for inheritance allows us to avoid redefining methods common to readers for our three formats.

A sequence entry in a FASTA file consists of one line of identification information beginning with ">" followed by any number of lines of sequence. Strictly speaking, each line of sequence should contain 80 characters, but there is no particular advantage to having our reader be inflexible on this point, so we allow lines of any length. In general, we cannot know that we have reached the end of an entry until we read the identification line of the next entry. This is where our one-line buffer shows its usefulness. Our reading method is shown below:

```perl
sub readSeq {
    my ($this) = @_;
    return () unless $this->{buff};
    my $id = $this->{buff};
    $id =~ s/^\>\s*//;
    $this->{seqId} = $id;
    my ($seq,$tbuff);
    my $fh = $this->{fh};
    while (($tbuff=<$fh>) && ($tbuff !~ /^\>/)) {
        chomp $tbuff;
        $tbuff = lc $tbuff;
```

```
        $tbuff =~ s/[^a-z]//g;
        $seq .= $tbuff;
    }
    chomp $tbuff;
    $this->{buff} = $tbuff;
    return $seq;
}
```

The fourth line removes the opening right arrow and leading whitespace from the line of identifying information. The **while**-loop repeatedly reads lines and appends them to the sequence until either the end of the file or a new identification line is found. The identification line of the next sequence is stored away in the hash under buff to be processed with the next sequence. The sequence is filtered and returned as the value of the method.

The remainder of the methods of a FastaReader (close, seqId, etc.) are virtually identical to a SimpleReader's. We can take advantage of the similarities by putting the common method in a *superclass* of these two readers called SeqReader and having SimpleReader, FastaReader, and (later) GenBankReader *inherit* them. To make FastaReader a subclass of SeqReader, we begin file `FastaReader.pm` with the following lines:

```
package FastaReader;
require SeqReader;
@ISA = qw(SeqReader);
```

The third line gives a list of packages from which FastaReader inherits methods.[4] It says that every FastaReader "is a" SeqReader, and has all the methods defined for SeqReaders in the file `SeqReader.pm`. Thus, the workings of the dereference operator are a little more complicated than was stated in the last section.

1. First, Perl extracts the package tag on the reference on the left.
2. Next, Perl looks for a subroutine matching the name on the right in the file for the package.
3. If no matching subroutine is found in the package, then Perl checks to see whether the list @ISA is defined in the current package. If not, the method is not defined for this object and the program halts with an error.
4. If @ISA is defined, then Perl goes to the file for the package named there to find the subroutine.
5. If this fails, Perl looks for a definition of @ISA in *that* package, and so on, until it either finds the subroutine or @ISA is undefined.

---

[4] So-called multiple inheritance can be specified by listing more than one package; however, we will be able to avoid the subtle issues involved with multiple inheritance in this book.

We can now define our superclass SeqReader in `SeqReader.pm` as follows:

```perl
package SeqReader;
use strict;
use FileHandle;
use FastaReader;
use GenBankReader;
use SimpleReader;

sub new
{
    my ($this,           ## literal "SeqReader" or ref to an existing SeqReader
        $fileName)       ## string giving operating system's name for file
        = @_;
    my $fh = *STDIN;
    if ($fileName ne "STDIN") {
        $fh = new FileHandle "<$fileName"
            or die "can't open $fileName ";
    }
    my $buff = <$fh>;
    chomp $buff;
    my $hash = {fh=>$fh,                 ## save filehandle
                buff=>$buff,             ## save line read for next read
                fileName=>$fileName};    ## save filename
    my $reader = $this->verifyFormat($hash)
        or die "can't open $fileName with a " . (ref($this) || $this);
    return $reader;
}
sub fileType {
    my ($this) = @_;
    my ($t) = (ref($this) =~ /^(.*)Reader/);
    $t;
}

## The following as in the previous section:
sub fileName { ... }
sub seqId { ... }
sub close { ... }
1;
```

Little has changed. The new method fileType returns the type of input file the reader is reading. This will allow a program to open a file without knowing its type and then ask for its type. It uses Perl's **ref** operator, which extracts from a reference the tag attached by **bless**. In this method, the tag is then subjected to pattern matching and Simple, Fasta, or GenBank is returned. The construction (**ref**($this) || $this) accounts for the fact that new may be invoked with either a class name (a string) or a reference to an existing object of the class to be instantiated. In the first case, **ref** will return an empty string.

The subclasses SimpleReader, FastaReader, and GenBankReader define only two methods: readSeq, which we have already seen, and verifyFormat. Method verifyFormat will examine the first line of the file, which is stored under buff in the hash, to determine whether the file is of the type read by each subclass. If so, verifyFormat will bless the reference into the subclass; if not, the empty string will be returned. Although we want to allow users of our readers to be able to open a file without knowing its type, we also want them to be able to require a particular type if appropriate for a certain programming task. The user will accomplish the first by invoking SeqReader—>new and the second by invoking FastaReader—>new, GenBankReader—>new, or SimpleReader—>new.

Method FastaReader—>verifyFormat is defined in `FastaReader.pm` by

```perl
sub verifyFormat {
    my ($this,$hash) = @_;
    return (bless $hash) if $$hash{buff} =~ m/^\>/;
    return "";
}
```

Methods GenBankReader—>verifyFormat and SimpleReader—>verifyFormat are similar.

Method SeqReader—>verifyFormat in file `SeqReader.pm` is defined by

```perl
sub verifyFormat {
    my ($this,$hash) = @_;
    return(FastaReader—>verifyFormat($hash)
        or GenBankReader—>verifyFormat($hash)
        or SimpleReader—>verifyFormat($hash));
}
```

This method tries each of the three file formats in turn; it then returns the first blessed reference it receives from any of the verifyFormat methods. No reference is ever blessed directly into the SeqReader class; this is an example of what object-oriented programmers call an *abstract class,* which exists not to be instantiated but only to provide methods for other classes to inherit.

It is noteworthy that the three types of readers are able to share a single constructor SeqReader—>new. This works because the method call FastaReader—>new sets $this equal to FastaReader, whereas calling SeqReader−>new sets $this to SeqReader. This causes the appropriate verifyFormat method to be called later on in new.

6.5 Exercises

1. GCG (sometimes called "the Wisconsin package") is an extensive suite of software tools for computer sequence analysis. It stores sequences in its own format; the sequences per se are stored as in GenBank, in blocks of ten letters with several blocks per line and guide numbers in the leftmost column. The variety of annotations stored is similar to GenBank's, but the format is somewhat different. An example of a GCG single-sequence file is shown in Figure 6.3.

   Investigate the details of GCG's file format, and create a module named GCGReader.pm to read GCG files.

2. Often it is useful to be able to treat two or more sequence files as a single file for purposes of reading the sequences into a program. If the files are of the same type, they can simply be concatenated into a single large file (e.g., with the Unix command cat) and the large file can be read with a SeqReader. However, this won't work if some files are in FASTA format but others are in GenBank format. Even when the files have the same type, it may be desirable to avoid the time and space required to create the single large file. Extend the SeqReader class in one or more of the following ways to allow the user to create a single SeqReader that reads sucessively from multiple files.

   (a) Allow the user to list any number of file names when creating a SeqReader. For example, the SeqReader object returned by

   ```
   my $inSeqs = new SeqReader("turkey.fasta","cat.gb","mouse.fasta");
   ```

   should have a readSeq method that, on successive calls, returns first each sequence entry in turkey.fasta, then each entry in cat.gb, then those in mouse.fasta – all without its user's being aware of the file boundaries.

   (b) Allow the user to pass a directory name to a SeqReader, rather than just a file name. For example, the SeqReader object returned by

   ```
   my $inSeqs = new SeqReader("turkey");
   ```

   should have a readSeq method that, on successive calls, returns every sequence in every file in the directory turkey, again without its user's being aware of the file boundaries.

```
!!AA_SEQUENCE 1.0
P1;S12513 - delta-conotoxin TxVIA precursor - cone shell (Conus
  textile)
N;Alternate names: conotoxin IA; King-Kong peptide (KK-0)
C;Species: Conus textile (cloth-of-gold cone)
C;Date: 19-Mar-1997 #sequence_revision 11-Apr-1997 #text_change
  10-Sep-1997
C;Accession: S12513; A30103; S19553
R;Woodward, S.R.; Cruz, L.J.; Olivera, B.M.; Hillyard, D.R.
  EMBO J. 9, 1015-1020, 1990
A;Title: Constant and hypervariable regions in conotoxin propeptides.
A;Reference number: S12513; MUID:90214607
A;Accession: S12513
A;Molecule type: mRNA
A;Residues: 1-78 <WOO>
A;Cross-references: EMBL:X53283; NID:g10887; PID:g10888
R;Hillyard, D.R.; Olivera, B.M.; Woodward, S.; Corpuz, G.P.; Gray,
  W.R.; Ramilo, C.A.; Cruz, L.J.
Biochemistry 28, 358-361, 1989
A;Title: A molluscivorous Conus toxin: conserved frameworks in
  conotoxins.
A;Reference number: A30103; MUID:89207553
A;Accession: A30103
A;Molecule type: protein
A;Residues: 52-78 <HIL>
R;Fainzilber, M.; Gordon, D.; Hasson, A.; Spira, M.E.; Zlotkin, E.
Eur. J. Biochem. 202, 589-595, 1991
A;Title: Mollusc-specific toxins from the venom of Conus textile
  neovicarius.
A;Reference number: S19553; MUID:92104183
A;Accession: S19553
A;Molecule type: protein
A;Residues: 52-78 <FAI>
C;Superfamily: omega-conotoxin
C;Keywords: neurotoxin; sodium channel inhibitor; venom
F;1-22/Domain: signal sequence #status predicted <SIG>
F;23-51/Domain: propeptide #status predicted <PRO>
F;52-78/Product: delta-conotoxin TxVIA #status experimental <MAT>
F;53-68,60-72,67-77/Disulfide bonds: #status predicted
   A30103  Length: 78  September 1, 1999 12:47  Type: P  Check: 2364  ..
        1   MKLTCMMIVA VLFLTAWTFA TADDPRNGLG NLFSNAHHEM KNPEASKLNK
       51   RWCKQSGEMC NLLDQNCCDG YCIVLVCT
```

Figure 6.3: Sample GCG single-sequence file.

(c) Allow the user to pass a pattern for file names rather than a fixed file name. For example, the SeqReader object returned by

```perl
my $inSeqs = new SeqReader("duck\d\d\d.fasta");
```

should have a readSeq method that, on successive calls, returns every sequence in every file with a name like duck037.fasta (i.e., file names consisting of duck plus three digits plus the extension .fasta).

To complete this exercise, you should investigate the Perl operators **opendir**, **readdir**, -d, -t, and **glob**.

3. Add an annotation parser to GenBankReader.pm. The parser should convert the annotation of the most recently read sequence to a list form. For example, the annotation of the *Yersinia* sequence in Figure 6.2 should be returned in the form

```
[
    ["source","1..70559",
        ["organism","Yersinia pestis"],
        ["plasmid","pCD1"],
        ...,
        ["db_xref","taxon:632"]],
    ["gene","57..368",
        ["gene","Y0001"]],
    ["CDS","57..368",
        ["gene","Y0001"],
        ["note","o103; 43 pct..."],
        ...,
        ["translation","MHQQS...YSFHW"]]
    ["gene","665..1033",
        ["gene","nuc"],
        ["note","Y0002"]],
    ["CDS","665..1033",
        ["gene","nuc"],
        ["codon_start",1],
        ...,
        ["translation","MDTKL...WRSSY"]]
    ["misc_RNA","complement(1560..1649)",
        ["note","antisense RNA"],
        ["product","copA"]],
    ["repeat_region","11939..15801",
        ["rpt_family","IS285/IS100"]]
]
```

6.6   Complete Program Listings

Directory chap6/ of the software includes package files SeqReader.pm, Simple-Reader.pm, FastaReader.pm, and GenBankReader.pm, as well as two short driver programs, simple.pl and read.pl.

6.7   Bibliographic Notes

Books by Bishop (1998) and Bishop and Rawlings (1988) deal extensively with database. Benson et al. (1996) describe GenBank. Bairoch and Apweiler (1996) describe the SWISS-PROT protein database.

Statistics on the occurrence of GenBank features and qualifiers come from Allen Morgan's MS thesis (1999).

## CHAPTER SEVEN

Local Alignment and the BLAST Heuristic

In Chapters 3 and 5 we saw how to develop an alignment scoring matrix and, given such a matrix, how to find the alignment of two strings with the highest score. In Chapter 6, we learned about some of the large genomic databases available for reference. Perhaps the most commonly performed bioinformatic task is to search a large protein sequence database for entries whose similarity scores may indicate homology with some query sequence, often a newly sequenced protein or putative protein.

The Needleman–Wunsch algorithm described in Chapter 3 constructs *global* alignments – alignments of the *entireties* of its two input sequences. In practice, homologous proteins are not similar over their entire lengths. This is because of differences in the importance of different segments of the sequence for the function of the protein. A typical protein has one or more *active sites* that play crucial roles in the chemical reactions it catalyzes. Acceptance of mutations in the midst of an active site is infrequent, since such mutations are likely to disrupt the protein's function. The segments intervening between the active sites help give the protein its peculiar shape but do not form strong bonds with other molecules as the protein performs its function. Mutations in these regions are more easily tolerated and thus are more common. Hemoglobin provides a good example; it easily tolerates mutations on its outer surface, but mutations affecting the active sites in its interior can destroy its ability to hold the iron-binding heme group essential to its role as oxygen carrier.

To account for these differences in importance of the proteins' different segments, database searching methods nearly always employ *local alignment* algorithms. A local alignment of two sequences is an alignment of any substring of one sequence with any substring of the other sequence. The *local similarity* of two sequences is the highest score attainable by any local alignment of the sequences. By appropriately combining scores of several local alignments, we can also recognize the frequently occurring situation of two related proteins that share a common set of active sites but in a different order in the primary structure.

7.1　The Smith–Waterman Algorithm

From our definition of the local similarity score of two strings as the score of the best alignment of any substring of one string with any substring of the other, it is clear that we could compute local similarity using the global similarity subroutine similarity( ) of Chapter 3 as follows:

```
sub reallySlowLocalSimilarity {
    my($s,$t) = @_;
    my ($is,$ls,$it,$lt);
    my $best = 0;
    foreach $ls (1..length($s)) {              ## every possible length in $s
        foreach $is (0..length($s)−$ls) {      ## every possible start in $s
            foreach $lt (1..length($t)) {      ## every possible length in $t
                foreach $it (0..length($t)−$lt) {   ## every possible start in $t
                    $best=max($best,similarity(substr($s,$is,$ls),
                                               substr($t,$it,$lt)));
}}}}}
```

However, it is not too difficult to see that the method just described will require time proportional to the third power of the product of the lengths of the two sequences, that is, the *sixth* power of the length of the input![1] Fortunately, it is possible to modify the Needleman–Wunsch algorithm to compute local alignments efficiently. The result is commonly known as the Smith–Waterman algorithm.

---

[1]　The running time of this subroutine is proportional to

$$\sum_{l_s=1}^{|s|} \sum_{i_s=0}^{|s|-l_s} \sum_{l_t=1}^{|t|} \sum_{i_t=0}^{|t|-l_t} l_s l_t = \left(\sum_{l_s=1}^{|s|} \sum_{i_s=0}^{|s|-l_s} l_s \right) \left(\sum_{l_t=1}^{|t|} \sum_{i_t=0}^{|t|-l_t} l_t \right).$$

For the first factor, we have

$$\sum_{l_s=1}^{|s|} \sum_{i_s=0}^{|s|-l_s} l_s = \sum_{l_s=1}^{|s|} (|s| - l_s + 1) l_s$$

$$= (|s| + 1) \sum_{l_s=1}^{|s|} l_s - \sum_{l_s=1}^{|s|} l_s^2$$

$$= \frac{(|s| + 1)(|s|)(|s| - 1)}{2} - \frac{2|s|^3 + 3|s|^2 + |s|}{6}$$

$$\approx \frac{|s|^3}{6}.$$

Handling the second factor similarly, we have a running time of $(|s|^3 |t|^3)/36$. If both strings have length $n/2$, then the total input length is n and the running time is proportional to $(n/2)^6/36 = n^6/2304$. If the input size is increased by a factor of α, the running time increases by a factor of α^6. Computer scientists call this an $O(n^6)$ ("order n to the sixth") algorithm.

In Chapter 3's Needleman–Wunsch program, $M[$i][$j] was filled with the score of the best (global) alignment of the first $i letters of $s and the first $j letters of $t. In our Smith–Waterman program, $M[$i][$j] will contain the score of the best alignment of any *suffix* (possibly empty) of the first $i letters of $s and any *suffix* of the first $j letters of $t. Since every substring is a suffix of some prefix, this scheme accounts for all possible alignments of substrings.

The first key to our modification is the observation that the best alignment for $M[$i][$j] could be formed from alignments of shorter strings in any of the three ways possible for global alignments, but that a fourth possibility also exists: we can align the two *empty suffixes* for a score of 0. This observation leads to two changes:

1. we add a fourth argument, 0, to the call to max( ) in the inner loop of the program;
2. we delete the two lines

```
foreach $i (0..length($s)) { $M[$i][0]=$g * $i; }
foreach $j (0..length($t)) { $M[0][$j]=$g * $j; }
```

which assign negative values to the first row and column of @M, and instead rely on Perl's policy of treating uninitialized values as 0.

The second key to the modification is that the best local alignment could end anywhere in the two strings. So we add code to find the largest entry in @M and return this as the local similarity score. The resulting code follows.

```
# The Smith–Waterman Algorithm
sub localSimilarity {
    my($s,$t) = @_;
    my ($i,$j,$best);
    foreach $i (1..length($s)) {
        foreach $j (1..length($t)) {
            my $p = p(substr($s,$i−1,1),substr($t,$j−1,1));
            $M[$i][$j] = max(0,
                            $M[$i−1][$j]+$g,
                            $M[$i][$j−1]+$g,
                            $M[$i−1][$j−1]+$p);
            $best = max($M[$i][$j],$best);
        }
    }
    return ($best);
}
```

It should be clear that our modifications have not changed the running time of this program; it is still proportional to the (first power of) the product of the lengths of the two sequences.[2]

7.2 The BLAST Heuristic

The running time of the Smith–Waterman algorithm makes it impractical for use in routine searches of protein databases, which often contain hundreds of millions of residues. Ideal in these applications would be a method with running time proportional not to the product but rather to the *sum* of the lengths of the query and database sequences. Unfortunately, the only methods approaching this speed are *heuristic* in nature. This means that, though they often find and score the best possible alignment or at least a very good alignment, they are not guaranteed to do so. Often, heuristic methods are used as a "screen": the small proportion of database sequences to which the heuristic assigns high scores are then rescored with the slower Smith–Waterman algorithm. This assures that sequences passing the screen are given the correct local similarity score; on the other hand, there is no absolute guarantee that all sequences with high Smith–Waterman scores pass through the screen.

The two most commonly used heuristic similarity-searching programs are FASTA and BLAST.[3] Each works by carrying out a significant amount of *preprocessing* on the query sequence before considering any database sequences. The information tabulated in the preprocessing phase makes it possible to eliminate most letters (perhaps 99%) in the database in constant time. Since the remaining letters may require processing time proportional to the length of the query, the overall running time is still proportional to the product of the lengths of the query and the database. In the average case, however, the constant of proportionality is a small fraction that can be estimated from the scoring scheme. We will study the BLAST heuristic in some detail.

The BLAST heuristic relies on two fundamental assumptions. The first is that most high-scoring local alignments contain one or more high-scoring pairs of three-letter substrings called *words*. By quickly identifying locations in the two strings that contain high-scoring word pairs, BLAST collects "seeds" from which high-scoring alignments can be grown.

The second assumption is that homologous proteins contain significant *gap-free* alignments. This assumption greatly simplifies the process of growing alignments from the three-letter seeds.[4] The Smith–Waterman algorithm can easily be modified

---

[2] If we increase the length of each string by a factor of α then the running time increases by a factor of α^2, so this is an $O(n^2)$ ("order n-squared" or "quadratic") algorithm.

[3] Actually, each is a whole family of programs including members for matching proteins, nucleotides, and combinations.

[4] The very earliest versions of BLAST did not consider gaps in any form. The most recent version identifies promising database sequences according to the scores of gap-free alignments, then constructs alignments with gaps on the promising sequences using an optimized Smith–Waterman algorithm.

```
>gi|4836688|gb|AAD30521.1|AF132040_1 (AF132040) red opsin [Felis catus]

>gi|12644063|sp|O35478|OPSG_SCICA GREEN-SENSITIVE OPSIN (GREEN CONE PHOTORECEPTOR)
 gi|4836696|gb|AAD30525.1|AF132044_1 (AF132044) green opsin [Sciurus carolinensis]
          Length = 364

 Score =  211 bits (539), Expect = 2e-54
 Identities = 105/120 (87%), Positives = 111/120 (92%)

Query: 1    MTQRWGPQRLAGGQPHAGLEDSTRASIFTYTNSNATRGPFEGPNYHIAPRWVYHVTSAWM 60
            M QRW PQRLAGGQP   EDST++SIFTYTNSNATRGPFEGPNYHIAPRWVYH+TS WM
Sbjct: 1    MAQRWDPQRLAGGQPQDSHEDSTQSSIFTYTNSNATRGPFEGPNYHIAPRWVYHITSTWM 60

Query: 61   IFVVIASVFTNGLVLAATMKFKKLRHPLNWILVNLAVADLAETIIASTISVVNQIYGYFV 120
            I VVIASVFTNGLVL ATMKFKKLRHPLNWILVNLA+ADLAET+IASTISVVNQ+YGYFV
Sbjct: 61   IIVVIASVFTNGLVLVATMKFKKLRHPLNWILVNLAIADLAETVIASTISVVNQLYGYFV 120


...


>gi|5359720|gb|AAD42779.1|AF133907_1 (AF133907) opsin; SWS2 [Anolis carolinensis]
          Length = 363

 Score = 48.3 bits (114), Expect = 4e-05
 Identities = 29/84 (34%), Positives = 46/84 (54%)

Query: 36   TRGPFEGPNYHIAPRWVYHVTSAWMIFVVIASVFTNGLVLAATMKFKKLRHPLNWILVNL 95
            T PF P H+   ++  +A+M +++ V N L +  T K+KKLR LN+ILVNL
Sbjct: 29   TLSPFLVPQTHLGNPSLFMGMAAFMFILIVLGVPINVLTIFCTFKYKKLRSHLNYILVNL 88

Query: 96   AVADLAETIIASTISVVNQIYGYF 119
            +V++L  + ST +   YF
Sbjct: 89   SVSNLLVVCVGSTTAFYSFSNMYF 112
```

Figure 7.1: Abridged BLAST output. The complete output includes alignments with and without gaps, but only gap-free alignments are shown.

to find the best gap-free alignment by deleting two terms from the maximum in its inner loop:

$M[$i][$j] = max(0, $M[$i−1][$j−1]+$p);$

However, the running time of the resulting program is still proportional to the *product* of the length of the query and the database. BLAST avoids computing scores for most values of $i and $h altogether, because no value will be computed unless a high-scoring pair of words is nearby.

Figure 7.1, a highly abridged BLAST output, shows two of the dozens of gap-free local alignments found when BLAST was run on the first 120 amino acids of a house-cat's red opsin (a protein involved in color vision) and the `nr` nonredundant database of 832,540 protein sequences.[5] These two alignments (and most others returned) involve homologous (both orthologous and paralogous) proteins from the gray squirrel

---

[5] April 5, 2001. The sequences contained 3.0×10^9 residues.

	A	R	N	D	C	Q	E	G	H	I	L	K	M	F	P	S	T	W	Y	V
A	4	-1	-2	-2	0	-1	-1	0	-2	-1	-1	-1	-1	-2	-1	1	0	-3	-2	0
R	-1	5	0	-2	-3	1	0	-2	0	-3	-2	2	-1	-3	-2	-1	-1	-3	-2	-3
N	-2	0	6	1	-3	0	0	0	1	-3	-3	0	-2	-3	-2	1	0	-4	-2	-3
D	-2	-2	1	6	-3	0	2	-1	-1	-3	-4	-1	-3	-3	-1	0	-1	-4	-3	-3
C	0	-3	-3	-3	9	-3	-4	-3	-3	-1	-1	-3	-1	-2	-3	-1	-1	-2	-2	-1
Q	-1	1	0	0	-3	5	2	-2	0	-3	-2	1	0	-3	-1	0	-1	-2	-1	-2
E	-1	0	0	2	-4	2	5	-2	0	-3	-3	1	-2	-3	-1	0	-1	-3	-2	-2
G	0	-2	0	-1	-3	-2	-2	6	-2	-4	-4	-2	-3	-3	-2	0	-2	-2	-3	-3
H	-2	0	1	-1	-3	0	0	-2	8	-3	-3	-1	-2	-1	-2	-1	-2	-2	2	-3
I	-1	-3	-3	-3	-1	-3	-3	-4	-3	4	2	-3	1	0	-3	-2	-1	-3	-1	3
L	-1	-2	-3	-4	-1	-2	-3	-4	-3	2	4	-2	2	0	-3	-2	-1	-2	-1	1
K	-1	2	0	-1	-3	1	1	-2	-1	-3	-2	5	-1	-3	-1	0	-1	-3	-2	-2
M	-1	-1	-2	-3	-1	0	-2	-3	-2	1	2	-1	5	0	-2	-1	-1	-1	-1	1
F	-2	-3	-3	-3	-2	-3	-3	-3	-1	0	0	-3	0	6	-4	-2	-2	1	3	-1
P	-1	-2	-2	-1	-3	-1	-1	-2	-2	-3	-3	-1	-2	-4	7	-1	-1	-4	-3	-2
S	1	-1	1	0	-1	0	0	0	-1	-2	-2	0	-1	-2	-1	4	1	-3	-2	-2
T	0	-1	0	-1	-1	-1	-1	-2	-2	-1	-1	-1	-1	-2	-1	1	5	-2	-2	0
W	-3	-3	-4	-4	-2	-2	-3	-2	-2	-3	-2	-3	-1	1	-4	-3	-2	11	2	-3
Y	-2	-2	-2	-3	-2	-1	-2	-3	2	-1	-1	-2	-1	3	-3	-2	-2	2	7	-1
V	0	-3	-3	-3	-1	-2	-2	-3	-3	3	1	-2	1	-1	-2	-2	0	-3	-1	4

Figure 7.2: The BLOSUM62 scoring matrix.

and the green anole (a lizard). In the first, the entire query string is aligned to a prefix of the database string; in the second, a substring is aligned to a substring.

7.2.1 Preprocessing the Query String

The goal of the preprocessing stage is to build a hash table, which we will call the *query index*. The keys of the hash table are the $20 \times 20 \times 20 = 8000$ possible three-letter words. The value associated with each word is a list of positions in the query string of all words that give a high score when aligned with the key word. With the default BLOSUM62 scoring matrix (Figure 7.2), the threshold for a "high score" is usually 11. No matter how high the threshold is set, an exact three-letter match is always treated as high-scoring.

For example, if the query is CINCINNATI, the table entry for CNN will be (1, 4, 5), since

$$s_{CC} + s_{NI} + s_{NN} = 9 + (-3) + 6 = 12 \quad \text{(positions 1 and 4)},$$
$$s_{CI} + s_{NN} + s_{NN} = (-1) + 6 + 6 = 11 \quad \text{(position 5)}.$$

With the BLOSUM62 matrix, 54 words score 11 or better when aligned with CIN:

```
CAN CCN CDN CEN CFN CGN CHN CIA CID CIE CIG CIH CIK CIM CIN CIP
CIQ CIR CIS CIT CIY CKN CLD CLE CLG CLH CLK CLN CLQ CLR CLS CLT
CMD CMH CMN CMS CNN CPN CQN CRN CSN CTN CVD CVE CVG CVH CVK CVN
CVQ CVR CVS CVT CWN CYN
```

whereas only three form high-scoring pairs with ATI:

```
    ATI ATL ATV
```

In all, preprocessing of CINCINNATI assigns nonempty entries to 204 of the 8000 possible keys, including:

```
    CCN: 1,4      NCS: 3
    CIN: 1,4      NCT: 3,7
    CNC: 2        NNC: 2,6
    CNN: 1,4,5    NNE: 6
```

7.2.2   Scanning the Target String

The next stage is to scan the target string to identify *hits*: locations of high-scoring word pairs.

For example, to scan the target string PRECINCTS, we successively look up the words PRE, REC, ECI, CIN, INC, NCT, and CTS in the query index. Looking up NCT from position 6 of the target string generates two hits, $(3, 6)$ and $(7, 6)$, since similar words occurred at positions 3 and 7 of the query string.

Each hit is immediately extended to the right and to the left to increase the alignment's score. The alignment is extended to the right to whatever position maximizes the overall alignment score. It is not hard to see that extensions beyond the first negative score are pointless. The alignment is extended to the left in a similar fashion.

In the CINCINNATI–PRECINCTS example, the hit at query position 3 and target position 6 corresponds to the alignment

$$\begin{bmatrix} \text{---ciNCInnati} \\ \text{preciNCTs----} \end{bmatrix}.$$

Since s_{NS} is positive, we can extend the alignment to the right:

$$\begin{bmatrix} \text{---ciNCINnati} \\ \text{preciNCTS----} \end{bmatrix},$$

and, given the exact match to the left, we can also extend to the left,

$$\begin{bmatrix} \text{---CINCINnati} \\ \text{preCINCTS----} \end{bmatrix},$$

to achieve the following local alignment:

$$\begin{bmatrix} \texttt{CINCIN} \\ \texttt{CINCTS} \end{bmatrix}$$

with a score of

$$s_{CC} + s_{II} + s_{NN} + s_{CC} + s_{IT} + s_{NS} = 9 + 4 + 6 + 9 + (-1) + 1 = 28.$$

We have another hit at query position 7 and target position 6:

$$\begin{bmatrix} \texttt{cincinNATi} \\ \texttt{-preciNCTs} \end{bmatrix};$$

however, the score of this alignment cannot be increased by extending it on either left or right. In all, there are six hits to consider (find them!), but the best gapless local alignment is the one with score 28.

7.3   Implementing BLAST in Perl

The main program of our Perl implementation of BLAST calls three major subroutines. fillLod reads in the scoring matrix from a file. The program's outer loop calls preprocessQuery to build the query index for each query with the help of findSimilarWords. After each query index is built, the inner loop begins and scanTarget is called for every database sequence in search of words in the query index. Whenever a hit is found, extendHit is called to extend the gap-free alignment as far as possible. At the end of the outer loop, the best alignments for the current query are identified and printed.

Command Line Access from a Perl Program.   The interface to our Perl version of BLAST will be simplified if we allow the user to put the query and the database in two separate input files and allow these files to be specified on the command line. We will also allow the user to specify a scoring matrix on the command line if desired; we will use BLOSUM62 by default. For example, if searching for the canine homologues of human globins with the PAM250 matrix, the user might have a database of dog proteins in the file dogDB.fasta and a few human globins in file humanGlobins.fasta. She could invoke our BLAST program with the command line

```
% perl blast.pl humanGlobins.fasta dogDB.fasta pam250
```

or, if the default BLOSUM62 scoring scheme is acceptable, with

```
% perl blast.pl humanGlobins.fasta dogDB.fasta
```

In order to implement this interface, we will need to be able to access the command line from the Perl program it invokes. The built-in list @ARGV provides this access. When a Perl program is invoked, **split** is used to break the command line into "words" separated by white space, and the resulting list (minus the program name!) is assigned to the array @ARGV. So, the first command line is equivalent in effect to the assignments

```
$ARGV[0] = "humanGlobins.fasta";
$ARGV[1] = "dogDB.fasta";
$ARGV[2] = "pam250";
```

being made before the first line of the program is executed.

Perl's Built-in Sorting. The simplest form of Perl's built-in **sort** function made a brief appearance in Chapter 5; it sorts list elements lexicographically:

```
@sortedL = sort @L;
```

A more powerful version allows the programmer to specify precisely how two list elements will be compared for sorting. The statement

```
@sortedL = sort { $a cmp $b } @L;
```

is equivalent to the previous version. The comparison method is written as a block, surrounded by braces.[6] The block *always* refers to the two items compared as $a and $b, and it should have a negative value if $a should precede $b, a positive value if $a should follow $b, and a zero value if either ordering is satisfactory. **cmp** is a built-in operator of this type for lexicographic comparison of strings; $<=>$ compares numbers similarly. We can build up arbitrarily complicated blocks, but most comparisons eventually use either **cmp** or $<=>$. Here are some more examples:

```
my @L = qw(Five Four Three Two One five four three two one);
my %h = (Five=>5, Four=>4, Three=>3, Two=>2, One=>1,
         five=>5, four=>4, three=>3, two=>2, one=>1);

print sort { $a cmp $b } @L;
print "\n";      ## sorts alphabetically
                 ## prints: FiveFourOneThreeTwofivefouronethreetwo

print sort { lc($a) cmp lc($b) } @L;
```

---

[6] A third form of **sort** allows a named comparison subroutine: **sort** compareItems @L.

```
print "\n";      ## sorts alphabetically, ignoring case
                 ## prints: FivefiveFourfourOneoneThreethreetwoTwo

print sort { length($b) <=> length($a) } @L;
print "\n";      ## sorts longest string to shortest string
                 ## prints: ThreethreeFiveFourfivefourOneTwotwoone

print sort { substr($a,−2) cmp substr($b,−2)} @L;
print "\n";      ## sorts alphabetically by last 2 letters
                 ## prints: ThreethreeOneoneFourfourfiveFivetwoTwo

print sort { $h{$a} <=> $h{$b} || $b cmp $a } @L;
print "\n";      ## sorts first by value in hash, then lexicographically
                 ## prints: oneOnetwoTwothreeThreefourFourfiveFive
```

Our main program begins by calling subroutine fillLod to read in the scoring matrix from the file named on the command line, if present, or the file blosum62. The rest of the main program opens the query file and reads the query sequences one by one. To handle the first query sequence, it is preprocessed and then the database file is opened, each database sequence is aligned to the query sequence using the BLAST heuristic, and the database file is closed again. The process is repeated for the rest of the sequences in the query file, with the database file being reopened and closed each time. The sequences are read using the SeqReader module described in Chapter 6. Preprocessing and alignment are performed by the subroutines preprocessQuery and scanTarget, respectively.

```
fillLod($ARGV[2] || "blosum62");
my $queryFile = new SeqReader $ARGV[0];
while (my $query = $queryFile−>readSeq()) {
    my $qId = $queryFile−>seqId();
    my $queryIndex = preprocessQuery($query);
    my $targetFile = new SeqReader $ARGV[1];
    my @bestAlignments;
    while (my $target = $targetFile−>readSeq()) {
        my $tId = $targetFile−>seqId();
        print "\nQuery: $qId\nTarget: $tId\n";
        my @newAlignments
            = scanTarget($target,$tId,$query,$qId,$queryIndex);
        ### add new alignments to sorted list
        push @bestAlignments,@newAlignments;
        @bestAlignments = sort { $$b[0] <=> $$a[0] } @bestAlignments;
        ### cut off list at 25 alignments
```

```perl
            $#bestAlignments = min($#bestAlignments,24);
        }
        $targetFile−>close( );
        print "********** 25 Best Alignments for $qId:\n\n";
        foreach (@bestAlignments) {
            print(join("\n", @$_), "\n\n");
        }
    }
}
$queryFile−>close( );
```

The subroutine fillLod assumes that the first line of the file contains 20 column head-
ings, one letter for each of the 20 amino acids. On the subsequent 20 lines, it expects
to find a one-letter row heading followed by the 20 log-of-odds scores in that row.

```perl
my %lod;
sub fillLod {
    my ($matrixFile) = @_;      ## argument is the file name
    open LOD, $matrixFile;
    ### read column headings
    my ($trash,@residues) = split /\s+/, <LOD>;
    while (my $line=<LOD>) {     ## read rest of lines one by one
        ## each line is heading plus 20 scores
        my ($r,@scores) = split /\s+/, $line;
        foreach (0..19) {
            ## save score in hash under residue–pair key
            $lod{$r.$residues[$_]} = $scores[$_];
        }
    }
}
```

The preprocessing subroutine preprocessQuery takes the query string as its argument. It creates the query index table as a hash named %similarPositions and
returns a *hash reference.* By returning a reference to %similarPositions rather than
the hash itself, the subroutine avoids the inefficiency of making a copy of the entire
hash upon return. Later on, the same strategy will be used to avoid copying when
passing the query index table to scanTarget.

To facilitate the construction of the query index table, preprocessQuery maintains a second hash called %similarWords. The keys of this hash are three-letter
words, and the corresponding values are lists of words that form high-scoring pairs
with the key word. More precisely, the corresponding values are *references* to lists
of similar words, since the value stored in a hash table must always be a scalar. (All
references are scalars, even references to nonscalar data objects.)

One possible strategy might be to precompute all the words similar to each of the 8000 possible three-letter words as soon as the scoring matrix is read. Much of this computation is likely to be wasted, since a typical query has only several hundred residues and therefore at most this many words. The strategy we adopt is to delay computing the words similar to INN, for example, until the first time the word INN appears in the query. Once computed, the result is assigned to $similarWords{"INN"} so that it does not need to be recomputed if INN appears a second time in the query string. This strategy is implemented by the statement

```
$similarWords{$word} ||= findSimilarWords($word);
```

The loop

```
foreach (@{$similarWords{$word}}) {
    push @{$similarPositions{$_}}, $i+1;
}
```

iterates over all words similar to the one extracted from the query sequence and then uses the **push** operator to add the current position in the query sequence to the list of positions similar to the word. In general, the **push** operator adds one or more items to the end of a list. For a simpler example, the following code assigns the list (2,4,6,8) to @evens:

```
my @evens;
foreach (1..4) {
    push @evens, (2*$_);
}
```

In our case, the hash %similarPositions contains not lists but instead *references* to lists. Since **push** requires a list, we must *dereference* the values that are stored in %similarPositions with the @{} operator. Likewise, we must dereference the reference $similarWords{$word} to gain access to the words in the list.

Subroutine findSimilarWords is not at all sophisticated. It simply computes the score achieved by pairing each of the 8000 possible words with its argument and then returns a reference to a list of those whose scores are at least 11.

```
my @residues = split //,"acdefghiklmnpqrstvwy";     ## list of 20 amino acids

sub findSimilarWords {
    my ($word) = @_;
```

```
    my ($w1,$w2,$w3) = split //,$word;
    return [] if $lod{$w1.$w1}+$lod{$w2.$w2}+$lod{$w3.$w3}<11;
    my @similar;
    foreach my $r1 (@residues) {
        foreach my $r2 (@residues) {
            my $t = 11−$lod{$w1.$r1}−$lod{$w2.$r2};
            foreach my $r3 (@residues) {
                push @similar, "$r1$r2$r3" if $lod{$w3.$r3}>=$t;
            }
        }
    }
    return \@similar;
}
```

Once the query sequence is preprocessed and a reference to the results stored in $queryIndex, the subroutine scanTarget is called to compare each database sequence to the query sequence. This subroutine considers the target sequence word by word, and it uses the query index table to detect hits.

```
sub scanTarget {
    my ($target,$tId,$query,$qId,$queryIndex) = @_;
    my @alignments;
    for (my $i=0; $i<(length $target)−2; $i++) {
        my $word = substr($target,$i,3);
        my $tPos = $i+1;
        foreach my $qPos (@{$$queryIndex{$word}}) {
            my $alignment = extendHit($query,$qPos,$qId,$target,$tPos,$tId);
            push @alignments, $alignment;
        }
    }
    @alignments = sort { $$b[0] <=> $$a[0] } @alignments;
    $#alignments = min($#alignments,24);
    return @alignments;
}
```

The construction of the iterator of the inner loop of scanTarget may seem complicated, so let's look at it from the inside out, keeping in mind that we want a list of numbers identifying positions in the query string. Since $queryIndex is a reference to a hash, %$queryIndex is a hash and $$queryIndex{$word} is the value corresponding to the key $word. What sort of values are stored in this hash? They are references to lists (they can't be lists themselves, since only scalars can be stored as values in hashes). Hence, to obtain the list value needed in this context, we must apply the list dereferencing operator @{ }.

For each hit detected in the target string, scanTarget calls subroutine extendHit to extend the alignment to the right and to the left as far as is profitable. extendHit returns a reference to a list with three elements. The first is the score achieved by the best alignment grown from the hit. The second and third are strings, one for the query and one for the target. These two strings give the subsequences aligned, their positions within the larger sequences, and the names of the query and target sequences. scanTarget collects these list references in the list variable @alignments. Eventually, all alignments produced for a single target are sorted by score, and the best 25 are retained and returned as the value of the subroutine.

Once returned to the main program, these alignments are added to the list of best alignments from all targets processed so far, the larger list is sorted, and again only the best 25 are retained. This process is repeated for each target until, finally, the best 25 alignments from any target are displayed.

The process of extending hits is straightforward. It can be terminated whenever the score drops below 0.

```
sub extendHit {
    my ($query,$qPos,$qId,$target,$tPos,$tId) = @_;
    my @target = ("-", split //,$target);
    my @query = ("-", split //,$query);
    my ($lo,$hi) = (0,2);
    my $maxscore = my $score
                  = $lod{$target[$tPos].$query[$qPos]}
                   +$lod{$target[$tPos+1].$query[$qPos+1]}
                   +$lod{$target[$tPos+2].$query[$qPos+2]};

    ### try to grow gap-free alignment to right
    my $ilim = min(@target-$tPos, @query-$qPos);
    for (my $i=3; $i<$ilim; $i++) {
        $score += $lod{$target[$tPos+$i].$query[$qPos+$i]};
        last if $score < 0;
        ($maxscore,$hi) = ($score,$i) if $score > $maxscore;
    }

    ### try to grow gap-free alignment to left ... (similar)
    ### return best alignment as triple ...
}
```

7.4 Exercises

1. Did you complete Exercise 6 in Chapter 3, relating to computing the overlap of two sequences? Look at it again.

2. Modify getAlignment of Chapter 3 to return the best local alignment, based on the matrix filled by localSimilarity.

3. Often, the 25 alignments printed by the BLAST program in this chapter are not distinct (i.e., one alignment may appear repeatedly). Explain why, and correct this deficiency.

4. Early versions of BLAST set a higher similarity threshold (14) for words and attempted to extend each and every single hit forwards and backwards. Later versions have lowered the threshold to 11, creating more hits, but extend hits only when two nearby hits have the same *offset*. If i_Q is the location of the hit in the query string and i_T is its location in the target string, then the hit's offset is $d = i_Q - i_T$. ("Nearby" is defined to be within 40 or fewer positions.) This approach was found to give equally good alignments in significantly less time.

    Modify the program in this chapter to implement this strategy. *Hint:* Record the hits in a *hit table*; its keys are offsets, and its values are lists of hit locations in the target string. Since offsets can be negative, it will be easiest to store the hit table in a hash. When a new hit is recorded, check to see if other hits with the same offset have been detected, and, if so, form an alignment running between the hits and attempt to extend its ends.

    Note that the set of all pairs with the same offset value forms a diagonal in the Smith–Waterman matrix for the two strings.

5. Subroutine extendHit has an inefficiency in the lines

    ```
    my @target = ("-", split //,$target);
    my @query = ("-", split //,$query);
    ```

    The problem is that these statements require time proportional to the sum of the lengths of the entire strings, while the rest of extendHit requires time proportional only to (roughly) the length of the alignment found. Also, the work is repeated on each call, whereas the actual values of $target and $query are often repeated from call to call. Can you eliminate this inefficiency?

6. Although it does not dominate the overall running time of our BLAST program, the time spent to compute entries of %similarWords can be reduced by the following observation. Since CMS is in the list of words similar to CIN, also MSC will be in the list for INC, MCS will be in the list for ICN, and so on. Therefore, if the list for CIN must be computed but the list for ICN is already present, then CIN's list could be constructed by permuting each of the elements of ICN's list rather than comparing all 8000 words to CIN. Implement this strategy.

7. What inputs (query plus database) can BLAST handle very quickly – that is, for which inputs does scanTarget require time proportional only to the sum of the

lengths of the query and the target? Give a method for generating one such in-put for any m and n, where m and n are the lengths of the query and the database (respectively).

8. What inputs (query plus database) does BLAST handle very slowly – that is, for which inputs does scanTarget require time proportional to the length of the target times the length of the query? Give a method for generating one such input for any m and n, where m and n are the lengths of the query and the database (respectively).

7.5   Complete Program Listings

The software distribution contains this chapter's mini-BLAST program in file `chap7/blast.pl`.

7.6   Bibliographic Notes

In Smith and Waterman (1981), the authors describe their local alignment algorithm.

The first version of BLAST is described by Altschul et al. (1990), whereas Altschul et al. (1997) describes a more recent version that allows the database to be searched for sequences that are (simultaneously) similar to a group of related sequences.

The FASTA family of database searching programs (see Pearson and Lipman 1988, Pearson 1989) offers an alternative to BLAST that is usually described as "faster but less sensitive". Pearson, FASTA's co-author, has compared the methods (see Pearson 1995).

General issues of GenBank searching are given in Altschul et al. (1994) and Smith (1996). Stryer (1981, p. 107) describes the effects of point mutations on hemoglobin.

## CHAPTER EIGHT

Statistics of BLAST Database Searches

In Chapter 7, we justified the use of fast alignment heuristics like BLAST by our need to quickly align large numbers of sequences in a database with a given *query sequence* to determine which were most similar to the query. In this chapter, we will consider statistical aspects of the set of alignment scores we might encounter when performing such a database search. The distribution of scores obviously depends on the substitution matrix employed (PAM30, BLOSUM62, PAM250, etc.), and a proof of the general result requires rather extensive use of sophisticated mathematical notation. We will avoid this by concentrating on a specific, simple scoring matrix for DNA before outlining the general result.

Like all but the most recent versions of BLAST, we will focus on gapless alignments. The theory of statistical properties of scores of alignments with gaps has been elucidated only approximately and only for special cases. The theory supports the empirical observation that their behavior is similar to the behavior of statistics without gaps.

8.1 BLAST Scores for Random DNA

Suppose that Q is a query sequence of DNA and that D is a sequence from a database. As usual for DNA, we will score $+1$ for matched bases and -1 for mismatched bases in alignments of D and Q. Suppose a BLAST search discovers a local gap-free alignment with a score of 13. Does this suggest that D and Q are in some way related, or could this be better explained by chance? To answer this question, we must define precisely what "by chance" means to us, and then compute – or at least estimate – the probability that a score of 13 or higher occurs under that definition.

The simplest definition of "by chance" is that D and Q are "random" sequences in which each position is equally likely to be A, C, G, or T; and each position is independent of every other. We will be interested in the best score best(i, j) of any local alignment ending at the ith position of D and the jth position of Q. In particular, we want to know the probabilities

$$P_s(i, j) = \Pr\{\text{best}(i, j) \geq s\} \quad \text{for scores} \quad s = 0, 1, 2, \ldots.$$

Since we can always align the two empty strings ending at positions i and j, we know for sure that

$$P_0(i, j) = 1 \quad \text{for all } i \text{ and } j;$$

in fact, for *any* scoring matrix we have

$$P_s(i, j) = 1 \quad \text{for all } i \text{ and } j \text{ and all } s \leq 0,$$

because we can always align the empty substrings ending at the two positions. For our particular simple scoring scheme, we can safely assert that

$$P_s(i, j) = 0 \quad \text{for } i < s \text{ or } j < s,$$

since the score cannot exceed the length of the shorter substring using ± 1 scores.

We know that if D_i and Q_j match then $\text{best}(i, j) \geq 1$. So

$$\Pr\{\text{best}(i, j) \geq 1 \mid D_i = Q_j\} = 1 \quad \text{for all } i \geq 1 \text{ and } j \geq 1.$$

We also know that

$$\Pr\{D_i = Q_j\} = 1/4.$$

On the other hand, it could also be that D_i and Q_j mismatch but that $\text{best}(i - 1, j - 1) \geq 2$. In this case, the best local alignment ending at D_{i-1} and Q_{j-1} could be extended to obtain an alignment with a score greater than or equal to 1. So

$$\Pr\{\text{best}(i, j) \geq 1 \mid D_i \neq Q_j\} = P_2(i - 1, j - 1) \quad \text{for all } i \geq 1 \text{ and } j \geq 1$$

and

$$\Pr\{D_i \neq Q_j\} = 3/4.$$

In fact, because the events "match" and "mismatch" are both mutually exclusive and jointly exhaustive, while the nucleotides themselves are assumed to be independent, we can write

$$\begin{aligned}
P_1(i, j) &= \Pr\{D_i = Q_j\} \cdot \Pr\{\text{best}(i, j) \geq 1 \mid D_i = Q_j\} \\
&\quad + \Pr\{D_i \neq Q_j\} \cdot \Pr\{\text{best}(i, j) \geq 1 \mid D_i \neq Q_j\} \\
&= (1/4) P_0(i - 1, j - 1) + (3/4) P_2(i - 1, j - 1) \quad \text{for } i \geq 1 \text{ and } j \geq 1.
\end{aligned}$$

The same sort of reasoning applies to other positive values of the score, and the following equation holds for all $s > 0$, $i \geq 1$, and $j \geq 1$:

$$P_s(i, j) = (3/4) P_{s+1}(i - 1, j - 1) + (1/4) P_{s-1}(i - 1, j - 1). \tag{8.1}$$

For example, we can compute the value of $P_4(5, 7)$ as follows:

$$
\begin{aligned}
P_4(5, 7) &= (3/4) P_5(4, 6) + (1/4) P_3(4, 6) \\
&= (1/4) P_3(4, 6) \\
&= (1/4)((3/4) P_4(3, 5) + (1/4) P_2(3, 5)) \\
&= (1/16) P_2(3, 5) \\
&= (1/16)((3/4) P_3(2, 4) + (1/4) P_1(2, 4)) \\
&= (1/64) P_1(2, 4) \\
&= (1/64)((3/4) P_2(1, 3) + (1/4) P_0(1, 3)) \\
&= (1/256) P_0(1, 3) \\
&= 1/256.
\end{aligned}
$$

Repeating this computation for $P_4(5, 9)$ or $P_4(13, 5)$ will give the same answer in the same number of steps. The only difference will be that terms of the form $P_s(k, k + 2)$ in the expansion will be replaced by terms of the form $P_s(k, k + 4)$ or $P_s(k + 8, k)$. This should not be too surprising, since a gapless alignment ending at position 5 of one string and position greater than 5 of the other cannot involve more than the last five bases of the longer string.

Surprising or not, the upshot of this observation is that $P_s(i, j)$ really only depends on $m = \min(i, j)$, so that

$$
P_s(i, j) = P_s(j, i) = P_s(m, m).
$$

Then we can restrict our attention to

$$
P_{s,m} = P_s(m, m).
$$

In this notation, equation (8.1) becomes

$$
P_{s,m} = (3/4) P_{s+1, m-1} + (1/4) P_{s-1, m-1} \quad \text{for all } m \geq s.
$$

It turns out that, as m gets larger, $P_{s,m}$ also gets larger, because longer sequences give more possible starting points for alignments. On the other hand, $P_{s,m}$ approaches a limiting value $P_s = \lim_{m \to \infty} P_{s,m}$. Since

$$
\begin{aligned}
\lim P_{s,m} &= (3/4) \lim P_{s+1, m-1} + (1/4) \lim P_{s-1, m-1} \\
&= (3/4) \lim P_{s+1, m} + (1/4) \lim P_{s-1, m},
\end{aligned}
$$

we can study P_s by studying the recurrence relation

$$
P_s = (3/4) P_{s+1} + (1/4) P_{s-1}, \tag{8.2}
$$

or

$$3P_{s+1} - 4P_s + P_{s-1} = 0.$$

This recurrence relation has as its *characteristic equation* the quadratic equation

$$3x^2 - 4x + 1 = 0;$$

its roots are $r_0 = 1$ and $r_1 = 1/3$. When a characteristic equation has two roots r_0 and r_1, solutions to the recurrence relation have the form $c_0 r_0^s + c_1 r_1^s$ for some constants c_0 and c_1. In our case, we have

$$P_s = c_0(1^s) + c_1(1/3)^s.$$

We know that $P_0 = 1$. To determine the correct values of c_0 and c_1, we can compute the value of P_1 by considering the sum

$$\sum_{s\geq 1} P_s.$$

From (8.2) we have

$$\sum_{s\geq 1} P_s = \sum_{s\geq 1}(3/4)P_{s+1} + \sum_{s\geq 1}(1/4)P_{s-1}$$
$$= (3/4)\sum_{s\geq 2} P_s + (1/4)\sum_{s\geq 0} P_s.$$

If we substract $\sum_{s\geq 2} P_s$ from both sides of this equation, we are left with

$$P_1 = (1/4)(P_0 + P_1)$$
$$= 1/4 + P_1/4,$$

or $P_1 = 1/3$. Then we can use the known values of P_0 and P_1 to find the correct values of c_0 and c_1 by solving the linear equations

$$P_0 = c_0(1^0) + c_1(1/3)^0 \quad \text{and} \quad P_1 = c_0(1^1) + c_1(1/3)^1.$$

The final solution is $P_s = (1/3)^s$ for $s \geq 0$. So the probability that a score greater than or equal to s appears at a single fixed position pair (i, j) is no more than $(1/3)^s$, and this estimate is reasonably accurate unless i or j is quite small.

We are more interested in knowing the probability that a score of s or better occurs at *any* of the position pairs in (D, Q). If m and n are the lengths of D and Q (respectively), then there are mn such pairs. Markov's inequality,

$$\Pr\{Z \geq \alpha E[Z]\} \leq 1/\alpha \quad \text{for } Z \geq 0,$$

tells us that we can bound this probability if we know the expected (or average) number of position pairs with scores exceeding s. So let I_{sij} be a random variable whose value is 1 if $best(i, j) \geq s$ and 0 otherwise, and let $Z_{smn} = \sum_{i=1}^{m} \sum_{j=1}^{n} I_{sij}$ be the total number of high-scoring positions. Then

$$E[Z_{smn}] = E\left[\sum_{i=1}^{m} \sum_{j=1}^{n} I_{sij}\right]$$

$$= \sum_{i=1}^{m} \sum_{j=1}^{n} E[I_{sij}]$$

$$= \sum_{i=1}^{m} \sum_{j=1}^{n} P_{s,\,\min(i,j)}$$

$$\leq \sum_{i=1}^{m} \sum_{j=1}^{n} P_s$$

$$= m \cdot n \cdot P_s,$$

and

$$\Pr\{Z_{smn} \geq 1\} = \Pr\{Z_{smn} \geq (m \cdot n \cdot P_s)^{-1} E[Z_{smn}]\} \leq (m \cdot n \cdot P_s).$$

This estimate is just an upper bound on the probability, but it is quite accurate when m and n are only moderately large. The quantity $E[Z_{smn}]$ is conventionally referred to as the "E-value" of score s for query length m and database length n; $\Pr\{Z_{smn} \geq 1\}$ is called the "P-value".

If the scores at the mn different positions were independent, then P-values would obey a thoroughly studied probability distribution known as the *extreme value distribution* and the equation

$$P = 1 - e^{-E}$$

would hold. In reality, the dependence between these mn random variables is weak, and this equation is a very good approximation. Expanding e^{-E} in a power series yields

$$P = 1 - e^{-E}$$

$$= 1 - (1 - E + E^2/2 - E^3/6 + E^4/24 \ldots)$$

$$= E - E^2/2 + E^3/6 - E^4/24 \ldots,$$

and we can see that P and E are about the same when small – within 5% when either is less than 0.1.

Now let us return to our original question: D and Q had a local alignment with a score of 13; should we be impressed, or could it have happened by chance? The answer depends on the lengths of D and Q. Since $3^{-13} \approx 10^{-6}$, a score of 13 is not surprising at all when $m \approx n \approx 1000$. On the other hand, such a score should occur by chance no more than roughly 1 time in 100 when $m \approx n \approx 100$, and in this case we have reasonably strong evidence that D and Q come from the same source.

8.2   BLAST Scores for Random Residues

Let's extend our study of the question: If Q and D are a query sequence and a database sequence, and if BLAST search discovers a local gap-free alignment with a score of 13 (or 37, or 409), is this significant or just chance? In this section, we will consider amino acid sequences with nonuniform background probabilities and scoring matrices.

Fortunately for us, life has just been discovered on another, simpler planet. On the planet Threamino, proteins consist of chains of only three different amino acids, normally abbreviated A, B, and C. Conducting the sort of analysis outlined in Chapter 5, we discover background probabilities of $p_A = 1/3$, $p_B = 1/6$, and $p_C = 1/2$, and we derive the scoring matrix s:

	A	B	C
A	2	−3	−3
B	−3	1	0
C	−3	0	1

Proceeding as before (but at greater speed), we try to determine

$$P_{s,m} = \Pr\{\text{best}(m,m) \geq s\} \quad \text{for } s = 0, 1, 2, \ldots;$$

as before, $P_{0m} = 1$ for all m. Considering $P_{s,m}$, we find nine mutually exclusive cases and proceed by conditioning on these cases:

$$P_{s,m} = \Pr\{Q_m = A \text{ and } D_m = A\} \cdot \Pr\{\text{best}(m,m) \geq s \mid Q_m = A \text{ and } D_m = A\}$$

$$+ \Pr\{Q_m = A \text{ and } D_m = B\} \cdot \Pr\{\text{best}(m,m) \geq s \mid Q_m = A \text{ and } D_m = B\}$$

$$+ \Pr\{Q_m = A \text{ and } D_m = C\} \cdot \Pr\{\text{best}(m,m) \geq s \mid Q_m = A \text{ and } D_m = C\}$$

$$+ \Pr\{Q_m = B \text{ and } D_m = A\} \cdot \Pr\{\text{best}(m,m) \geq s \mid Q_m = B \text{ and } D_m = A\}$$

$$+ \Pr\{Q_m = B \text{ and } D_m = B\} \cdot \Pr\{\text{best}(m,m) \geq s \mid Q_m = B \text{ and } D_m = B\}$$

$$+ \Pr\{Q_m = B \text{ and } D_m = C\} \cdot \Pr\{\text{best}(m,m) \geq s \mid Q_m = B \text{ and } D_m = C\}$$

$$+ \Pr\{Q_m = C \text{ and } D_m = A\} \cdot \Pr\{\text{best}(m,m) \geq s \mid Q_m = C \text{ and } D_m = A\}$$

$$+ \Pr\{Q_m = C \text{ and } D_m = B\} \cdot \Pr\{\text{best}(m,m) \geq s \mid Q_m = C \text{ and } D_m = B\}$$

$$+ \Pr\{Q_m = C \text{ and } D_m = C\} \cdot \Pr\{\text{best}(m,m) \geq s \mid Q_m = C \text{ and } D_m = C\}.$$

Two typical terms can be simplified as follows:

$$\Pr\{Q_m = \text{C and } D_m = \text{B}\} \cdot \Pr\{\text{best}(m, m) \geq s \mid Q_m = \text{C and } D_m = \text{B}\}$$
$$= p_C \cdot p_B \cdot P_{s-s_{CB}, m-1}$$
$$= (1/2)(1/6) P_{s, m-1};$$

$$\Pr\{Q_m = \text{A and } D_m = \text{A}\} \cdot \Pr\{\text{best}(m, m) \geq s \mid Q_m = \text{A and } D_m = \text{A}\}$$
$$= p_A \cdot p_A \cdot P_{s-s_{AA}, m-1}$$
$$= (1/3)(1/3) P_{s-2, m-1}.$$

If we follow this process for every term and then add, the result is

$$P_{s,m} = (1/3)(1/3) P_{s-2, m-1} + (1/3)(1/6) P_{s+3, m-1} + (1/3)(1/2) P_{s+3, m-1}$$
$$+ (1/6)(1/3) P_{s+3, m-1} + (1/6)(1/6) P_{s-1, m-1} + (1/6)(1/2) P_{s, m-1}$$
$$+ (1/2)(1/3) P_{s+3, m-1} + (1/2)(1/6) P_{s, m-1} + (1/2)(1/2) P_{s-1, m-1}$$
$$= (4/9) P_{s+3, m-1} + (1/6) P_{s, m-1} + (5/18) P_{s-1, m-1} + (1/9) P_{s-2, m-1}.$$

Passing to the limit (as we did for (8.2)), we have

$$P_s = (4/9) P_{s+3} + (1/6) P_s + (5/18) P_{s-1} + (1/9) P_{s-2}$$

or

$$8 P_{s+3} - 15 P_s + 5 P_{s-1} + 2 P_{s-2} = 0,$$

as well as the characteristic polynomial equation

$$8x^5 - 15x^2 + 5x + 2 = 0.$$

Unlike quintic equations in general, this one has roots expressible by radicals, but the expressions are complicated.[1] Numerically, the solutions are

$$r_0 = 1,$$
$$r_1 \approx 0.6287441546,$$
$$r_2 \approx -0.2342509346,$$
$$r_3 \approx -0.6972466100 + 1.100567668i,$$
$$r_4 \approx -0.6972466100 - 1.100567668i,$$

---

[1] For example,

$$r_1 = -\frac{1}{4} + \frac{1}{24}\sqrt{6\beta} + \frac{1}{24}\sqrt{-120 - 6\alpha^{1/3} - 240\alpha^{-1/3} + 720\sqrt{6/\beta}},$$

where $\alpha = 10{,}700 + 60\sqrt{31{,}785}$ and $\beta = \alpha^{1/3} - 10 + 40\alpha^{-1/3}$.

where $i^2 = -1$. The theory of recurrence relations tells us that the correct solution has the form

$$P_s = c_0 r_0^s + c_1 r_1^s + c_2 r_2^s + c_3 r_3^s + c_4 r_4^s$$

for some c_i. We can quickly say that $c_0 = c_3 = c_4 = 0$ because $\lim_{s \to \infty} P_s = 0$, but $\lim_{s \to \infty} r_0^s = 1$ and $|r_3^s|, |r_4^s|$ approach infinity. Since $P_0 = 1$, we have $c_1 + c_2 = 1$. We can derive a second constraint on c_1 and c_2 by studying $\sum_{s \geq 1} P_s$ as in the previous section. We see that

$$\sum_{s \geq 1} P_s = (4/9) \sum_{s \geq 4} P_s + (1/6) \sum_{s \geq 1} P_s + (5/18) \sum_{s \geq 0} P_s + (1/9) \sum_{s \geq -1} P_s,$$

giving

$$P_1 + P_2 + P_3 = (1/6)(P_1 + P_2 + P_3)$$
$$+ (5/18)(P_0 + P_1 + P_2 + P_3)$$
$$+ (1/9)(P_{-1} + P_0 + P_1 + P_2 + P_3);$$

since $P_{-1} = P_0 = 1$,

$$P_1 + P_2 + P_3 = 9/8.$$

But this means that

$$c_1(r_1 + r_1^2 + r_1^3) + c_2(r_2 + r_2^2 + r_2^3) = 9/8.$$

Combining this with $c_1 + c_2 = 1$, we conclude that

$$P_s = (0.899)(0.6287)^s + (0.101)(-0.2342)^s.$$

The number crunching performed by the program described in the next section allows us to confirm that, when $s \geq 5$, P_s is approximated to within 1% by

$$P_s \approx (0.9)(0.6287)^s.$$

8.3   BLAST Statistics in Perl

In this section, we develop a Perl program that reads in background probabilities and scores and performs the statistical calculations laid out in Sections 8.1 and 8.2. It finds the recurrence relation implied by the scoring scheme and uses the recurrence to build and display tables of $P_{s,m}$. Examining these tables will allow us to see how quickly $P_{s,m}$ approaches the limit value P_s. The program will find the positive real root r_1 of the characteristic polynomial equation and print a table that illustrates the relationship $P_s \approx c \cdot r_1^s$. In order, its main steps are as follows.

1. Read the scoring matrix and background probabilities.
2. Collect the terms of the recurrence relation implied by the input, and create a Recurrence object.
3. Print the recurrence relation, its characteristic polynomial, and its primary root.
4. Print a table of values of the probabilities $P_{s,m}$ and P_s.

For Step 1, we assume that the first line of the input file contains the background frequencies and that each subsequent line contains one row of the matrix. An input file for the planet Threamino looks like this:

```
0.333333333333333 0.166666666666667 0.5
2 -3 -3
-3 1  0
-3 0  1
```

Subroutine readScoreScheme reads these items into arrays @p and @s and returns references to these arrays.

readScoreScheme has been written especially to demonstrate some of the cryptic shorthand available to Perl programmers using the implicit operand $_. For example, **split** and **print** each appear twice with no explicit operands; in both cases, $_ is used. (**split** also needs a pattern; the default is /\s+/, matching whitespace.)

The diamond operator < > appears twice with STDIN. The first time, the result is assigned explicitly to $_. The second time, the result is assigned implicitly to $_ by **while**(<STDIN>). This illustrates an idiosyncratic feature of Perl's input operator < >. To quote Perl's on-line documentation (the `perlop` man-page),

> Ordinarily you must assign the returned value to a variable, but there is one situation where an automatic assignment happens. If and only if the input symbol is the only thing inside the conditional of a **while** statement (even if disguised as a **for**(;;) loop), the value is automatically assigned to the global variable $_, destroying whatever was there previously. (This may seem like an odd thing to you, but you'll use the construct in almost every Perl script you write.)

Whether or not to exploit this shorthand is a matter of personal preference; its use is certainly cryptic to programmers of other languages.

```perl
sub readScoreScheme {
    ## read background frequencies into @p
    print "Background Frequencies (input)\n";
    $_ = <STDIN>;
    my @p = split;
    print;
```

```perl
## read scoring matrix into @s; remember largest entry
print "\nScoring Matrix (input)\n";
my @s;

while (<STDIN>) {
    print;
    push @s, [split];
}
return (\@p,\@s);
}
```

One item of interest, the average score of a single pair of residues, is defined by the formula

$$\sum_{i=1}^{R}\sum_{j=1}^{R} p_i p_j s_{ij}$$

and computed by this subroutine, which receives references to arrays for background probabilities and scores as arguments:

```perl
sub averagePairScore {     ## compute average score per pair
    my ($p,$s) = @_;
    my $avg;

    foreach my $i (0..$#$p) {
        foreach my $j (0..$#$p) {
            $avg += $$p[$i]*$$p[$j]*$$s[$i][$j];
        }
    }
    return $avg;
}
```

The rest of our task concerns recurrence relations. For this we create a class of Recurrence objects, defined in `Recurrence.pm`. Method Recurrence−>new takes as its argument a list of pairs of index offsets and coefficients defining the recurrence relation. The recurrence defined by a scoring scheme is given by

$$P_s = \sum_{i=1}^{R}\sum_{j=1}^{R} p_i p_j P_{(s-s_{ij})},$$

so our main program forms and passes the needed list with the statements

```perl
my @terms;
foreach my $i (0..$#$p) {
    foreach my $j (0..$#$p) {
        push @terms, $$s[$i][$j], $$p[$i]*$$p[$j];
    }
}
my $Psm=Recurrence->new(@terms);
```

Subsequently, our main program prints out the recurrence and characteristic polynomial in human-readable form by the statements

```perl
print "\nRecurrence Relation is\n";
$Psm->printRecurrence();
print "\nCharacteristic Polynomial is\n";
$Psm->printPolynomial();
my $root=$Psm->polyRoot();
print "\nRelevant root is $root\n";
```

Next, the program prints out some tables of the values of the recurrence:

```perl
print "\nTable of Expected-s[m] and P[s,m]\n";
printTable($Psm, [1..20], [1..22]);
print "\nTable of Expected-s[m] and P[s,m]\n";
printTable($Psm, [1..20], [map(10*$_,(1..22))]);
```

Subroutine printTable is straightforward and can be seen in the complete listing.

Finally, the program prints out a table comparing estimates of P_s computed by finding $P_{s,220}$ to powers of the root of the characteristic polynomial. The ratios in this table give an estimate of the constant c in our previous derivation $P_s \approx c \cdot r^s$. (For the Threamino substitution matrix, we see the ratios $P_{s,220}/(0.6297)^s$ approaching 0.899 as s increases from 1 to 20.)

```perl
print "\nTable of P[s,220] / ($root)^s\n";
foreach my $s (1..20) { printf " %6d", $s; }
print "\n";
my $power=1;
foreach my $s (1..20) {
    $power *= $root;
    printf " %6.4f", $Psm->value($s,220)/$power;
}
print "\n";
```

This code is routine except for the use of the **printf** operator for formatted printing. **printf** has its origins in the C programming language. Its first operand is a *format* containing one or more *directives* for printing the second and subsequent operands. The directives allow rather precise control of features like field width, leading zeros, number of decimal digits, and justification. In this book, only three basic **printf** directives are used:

$%w$d means to print an integer in a field of no fewer than w characters, padded with blanks on the left if necessary;

$%w$s means to print a string in a field of no fewer than w characters, padded with blanks on the right if necessary;

$%w.d$f means to print a number as a floating-point number with d digits to the right of the decimal point in a field of no fewer than w characters, padded with blanks on the left if necessary (so the statement **printf** "pi=%8.4f", 3.14159; is equivalent to **print** "pi=3.1416";).

If the specified field width is too small to allow the integer part of a number to be printed, then the field width is ignored and the number is printed in a larger field.[2]

It is now time to develop the methods of our Recurrence package. Method Recurrence−>new must gather and sum similar terms passed to it in its argument list. It uses a hash %coefficients in which keys are index offsets and values are the cumulating coefficients. A hash is preferable to a list because both positive and negative integers can be used as hash keys. Once the hash is filled, the offsets and coefficients are collected in a list of pairs named @coefficients. A reference to this list is eventually stored in the hash that is blessed into the class Recurrence.

```
sub new
{
    my ($this,@terms) = @_;
    my %coefficients;
    while (@terms) {
        (my $d, my $c, @terms) = @terms;
        $coefficients{0−$d} += $c;
    }
    my @coefficients
        = map([$_,$coefficients{$_}], sort {$a<=>$b} keys %coefficients);
    ...
```

Next, we shall extract from the recurrence relation its characteristic polynomial. The smallest index offset in the recurrence relation (e.g., -2 in $P_s = (4/9)P_{s+3} +$

---

[2] The operator **sprintf** is useful for formatting without printing; the statement $foo = **sprintf** "pi=%8.4f", 3.14159; is equivalent to $foo = "pi=3.1416";.

$(1/6)P_s + (5/18)P_{s-1} + (1/9)P_{s-2})$ corresponds to x^0 in the characteristic polynomial. The polynomial is represented by an array of coefficients, so that $polynomial[3] is the coefficient of x^3, and so forth. (The characteristic polynomial $(4/9)x^5 - (5/6)x^2 + (5/18)x + (1/9)$ is represented by the Perl list (1/9, 5/18, −5/6, 0, 0, 4/9).) We must not forget to move the P_s term on the left-hand side of the recurrence relation to the right-hand side before computing the polynomial's coefficients; this is effected by the statement $polynomial[−$offset] = −1.0;.

```
...
my @polynomial;
my $offset = $coefficients[0][0];
$polynomial[−$offset] = −1.0;
foreach (@coefficients) {
    my ($d,$c) = @$_;
    $polynomial[$d−$offset] += $c;
}
...
```

Finally, we store the coefficients of both the recurrence relation and the polynomial in a hash and then return a reference to the hash. The reference is tagged with Recurrence so that the correct methods can be found when invoked with the object deference operator −>. The empty hash stored under key memory will be explained shortly.

```
...
my $hash = {coefficients => \@coefficients,
            polynomial => \@polynomial,
            memory => { }};
bless $hash, (ref($this) || $this);
}   ## end of new
```

Both Recurrence−>printRecurrence and Recurrence−>printPolynomial are straightforward methods, but Recurrence−>polyRoot merits some comment.

Our characteristic polynomials have special properties arising from the fact the coefficients of the recurrence relation lie between 0 and 1 and add up to 1. In particular: 1 is a root; there is only one root r in the open interval $(0, 1)$; and the polynomial is positive in the interval $(0, r)$ and negative in $(r, 1)$. Method Recurrence−>polyRoot exploits these properties to find the root by *binary search* in the interval $(0, 1)$. In this context, binary search repeatedly halves the range in which the root can lie by examining the sign of the polynomial at the midpoint of the current range. After just a few steps, the range is so small that its endpoints can't be distinguished in the computer's floating-point number representation. The inner loop computes the value of the polynomial for the current value of $x.

```perl
sub polyRoot {     ## not a general routine for solving polynomials!
    my ($this) = @_;
    my $poly = $this->{polynomial};
    my ($neg,$pos) = (1,0);
    while ($neg-$pos > 1E-15) {
        my $x = ($neg+$pos)/2.0;
        my $value = 0;
        my $power = 1;
        foreach my $coeff (@$poly) {
            $value += $coeff * $power;
            $power *= $x;
        }
        if ($value>0) { $pos=$x; } else { $neg=$x; }
    }
    return ($neg+$pos)/2.0;
}
```

It is a clear deficiency of our recurrence package that this method may fail for recurrences not based on probabilities, but a more general method for finding the roots of a polynomial is well beyond the scope of this book.

Recurrence->value returns the value of the recursively defined probability for a specific pair of indices. To avoid repeated recomputation of probabilities, nontrivial values are saved in the hash $this->{memory}. Recurrence->value checks in this hash to see if a value has already been computed before doing anything else. When a new value must be computed, it is saved in this hash before being returned to the caller. Such methods are sometimes called "memo-izing" methods. By computing values as needed instead of filling a large table in advance, we may be able to avoid filling in some entries altogether. On the other hand, we never compute the same entry more than once, as we might if we threw away the table and recomputed each value on demand.

```perl
sub value {
    my ($this,$s,$m) = @_;
    my $mem = $this->{memory};
    return $$mem{$s,$m} if defined($$mem{$s,$m});
    return 1 if $s<=0 && $m>=0;
    return 0 if $m<=0;
```

```
    my $val;
    foreach (@{$this->{coefficients}}) {
        my ($d,$c) = @$_;
        $val += $c * $this->value($s+$d,$m−1)
    }
    return $$mem{$s,$m}=$val;
}
```

8.4   Interpreting BLAST Output

The authors of BLAST describe the E-value of a search for a random string in a random database as satisfying

$$E \approx Kmne^{-\lambda s},$$

where λ is the unique positive real root of the equation

$$\sum_{i,j} p_i p_j e^{\lambda s_{ij}} = 1;$$

we have described the E-value by

$$E \approx mnP_s \approx mnc_1 r_1^s,$$

where r_1 is the unique root in $(0, 1)$ of the equation (in x)

$$\sum_{i,j} p_i p_j x^{-s_{ij}} = 1.$$

Thus, we can equate $K = c_1$ and $r_1 = e^{-\lambda}$ or $\lambda = -\ln r_1$.

BLAST output also rescales each raw alignment score s into a *bit score* s' satisfying

$$E = mn2^{-s'}.$$

By equating these three definitions of E, we can see that

$$\begin{aligned}
s' &= \frac{\lambda s - \ln K}{\ln 2} \\
&= -s \log_2 r_1 - \log_2 c_1 \\
&= \log_2(c_1 r_1^s).
\end{aligned}$$

Bit scores are expressed in a convenient standard unit of information; like E-values, bit scores can be compared directly even when produced by different scoring schemes (e.g., BLOSUM62 vs. PAM250).

```
Sequences producing significant alignments:                    (bits)  Value

gi|4836688|gb|AAD30521.1|AF132040_1 (AF132040) red opsin [Felis ...   235  1e-61
gi|4836694|gb|AAD30524.1|AF132043_1 (AF132043) green opsin [Equu...   222  1e-57
gi|12644063|sp|O35478|OPSG_SCICA GREEN-SENSITIVE OPSIN (GREEN CO...   211  2e-54
...
gi|11417311|ref|XP_004030.1| adrenergic, beta-2-, receptor, surf...    34  0.51
gi|2494947|sp|Q28509|B2AR_MACMU BETA-2 ADRENERGIC RECEPTOR >gi|2...    34  0.51
gi|29373|emb|CAA28511.1| (X04827) beta-adrenergic receptor (AA 1...    34  0.51
gi|2119474|pir||S40691 opsin rh1 - fruit fly (Drosophila pseudoo...    34  0.51
gi|2655177|gb|AAB87898.1| (AF025813) rhodopsin 1 [Drosophila sub...    34  0.51
```

Figure 8.1: Sample BLAST output, excerpt 1.

The first section of a typical BLAST output contains a listing of the database sequences with high-scoring alignments with the query sequence, along with the bit scores and E-values of the alignments, as shown in Figure 8.1.

The second section of BLAST output contains details of some or all of these high-scoring alignments, along with additional statistics. The annotated alignment (see Figure 8.2) includes: the alignment score, first in bits (211) and then in raw form (539 of whatever units were used to express the substitution matrix); the E-value (2×10^{-54}); the number of identical residue pairs (105, or 87%); and the number of residue pairs with positive scores (111, or 92%). The alignment itself is displayed with a third line interposed between the two aligned sequences. On this line, identical aligned pairs are marked with the matching residue, others pairs with positive score are marked with a plus sign, and remaining pairs (zero or negative scores) show a blank space.

The third section (Figure 8.3) includes a number of statistics that relate to the query and database as wholes. Among these are the following.

- The values of λ and K for the substitution matrix used for the alignment. In this case, BLOSUM62 was used; from the values given, we can conclude that the value we called P_s is approximated by

```
>gi|12644063|sp|O35478|OPSG_SCICA GREEN-SENSITIVE OPSIN
(GREEN CONE PHOTORECEPTOR PIGMENT)
gi|4836696|gb|AAD30525.1|AF132044_1 (AF132044) green opsin
[Sciurus carolinensis]
        Length = 364

 Score =  211 bits (539), Expect = 2e-54
 Identities = 105/120 (87%), Positives = 111/120 (92%)

Query: 1    MTQRWGPQRLAGGQPHAGLEDSTRASIFTYTNSNATRGPFEGPNYHIAPRWVYHVTSAWM 60
            M QRW PQRLAGGQP    EDST++SIFTYTNSNATRGPFEGPNYHIAPRWVYH+TS WM
Sbjct: 1    MAQRWDPQRLAGGQPQDSHEDSTQSSIFTYTNSNATRGPFEGPNYHIAPRWVYHITSTWM 60

Query: 61   IFVVIASVFTNGLVLAATMKFKKLRHPLNWILVNLAVADLAETIIASTISVVNQIYGYFV 120
            I VVIASVFTNGLVL ATMKFKKLRHPLNWILVNLA+ADLAET+IASTISVVNQ+YGYFV
Sbjct: 61   IIVVIASVFTNGLVLVATMKFKKLRHPLNWILVNLAIADLAETVIASTISVVNQLYGYFV 120
```

Figure 8.2: Sample BLAST output, excerpt 2.

```
Database: nr
  Posted date:  Apr 5, 2001 10:53 AM
Number of letters in database: 3,042,457,330
Number of sequences in database:  832,540

Lambda     K       H
  0.323    0.135    0.426

Gapped
Lambda     K       H
  0.267   0.0410    0.140

Matrix: BLOSUM62
Gap Penalties: Existence: 11, Extension: 1

length of query: 120
length of database: 209,778,411
effective HSP length: 49
effective length of query: 71
effective length of database: 177,144,362
effective search space: 12577249702
effective search space used: 12577249702
```

Figure 8.3: Sample BLAST output, excerpt 3.

$$P_s \approx (0.135) \exp(-0.323s) \approx (0.135)(0.724)^s.$$

The raw score $s = 539$ of the alignment in Figure 8.2 corresponds to the bit score

$$s' = \frac{\lambda s - \ln K}{\ln 2} \approx \frac{(0.267)(539) + 2.002}{0.693} \approx 211.$$

- The entropy of the scoring scheme (H, as in Chapter 4).
- Empirical estimates of λ, K, and H for scoring of alignments with gaps.[3]
- *Effective lengths* for the query and the database. These statistics seek to take into account the errors caused by estimating P_{sm} by P_s; this estimate is quite accurate when m is moderately large, but P_{sm} can be significantly less when m is small. To compensate, the values of m and n can be adjusted downward in the computation of E-values. BLAST's authors state that the effect of this correction is usually negligible for sequences longer than 200 residues. The *effective search space* is the product of the effective lengths of the query and the database.

8.5   Exercise

1. Work out a formula for P_s for a three-residue protein chemistry with background probabilities $p_A = 1/3$, $p_B = 1/6$, and $p_C = 1/2$ and with scoring matrix

---

[3] Recall that current versions of BLAST apply the Smith–Waterman algorithm to database sequences that have high-scoring gap-free alignments.

	A	B	C
A	1	−3	−3
B	−3	1	0
C	−3	0	1

8.6 Complete Program Listings

Directory chap8/ of the software distribution includes the recurrence package file Recurrence.pm and the program file blaststats.pl.

8.7 Bibliographic Notes

Karlin and Altschul (1990, 1993) outlined the general theory of ungapped alignments behind this chapter. It has been developed further with the help of other authors (see Dembo, Karlin, and Zeitouni 1994; Altschul and Gish 1996). Most recently, Mott and Tribe (1999) have developed an approximate theory for gapped alignments.

## CHAPTER NINE

Multiple Sequence Alignment I

Once a family of homologous proteins has been identified, it is often useful to arrange their sequences in a *multiple alignment* such as the one in Figure 9.1.

A multiple alignment is useful for constructing a so-called *consensus sequence,* which – while probably differing from every individual sequence in the family – is nonetheless a better representative of the family than any of its actual members. Multiple alignments can also form the basis of more abstract statistical models of the protein family called *profiles.*

By examining which elements of the consensus are present in most or all family members and which exhibit a greater degree of variability, we can also find clues to the protein's function. *Highly conserved regions* are likely to have been conserved because they form active sites crucial to function, while more variable regions are more likely to have merely structural roles.

We have already seen in Chapter 3 that the number of ways in which a mere two sequences of only moderate length can be aligned is comparable to current estimates of the number of atoms in the observable universe. The addition of more sequences only increases the number of possibilities. We need both a criterion for evaluating multiple alignments and a computational strategy that will allow us to eliminate large sets of alignments at one stroke.

To describe our evaluation criterion, we will rely on the notion of *projection* of a multiple alignment. To project an alignment of K sequences onto a subset of k sequences, we

- delete the $K - k$ rows of the alignment corresponding to the other $K - k$ sequences, then
- delete every column consisting of k gap symbols.

Although an alignment can be projected onto any subset of its sequences, we will mainly be interested in *pairwise projections* – that is, projections onto pairs of sequences. It makes sense to regard a multiple alignment as appropriate only if its projections are appropriate. Therefore, we will score multiple alignments by a *sum-of-pairs* method: The score assigned to a multiple alignment of k sequences is equal

$$\begin{bmatrix} \text{FSKIGG-HAEEYGAETLERMFIAYPQTKTYFPHF-DLS} \\ \text{VQSSWK-AVSHNEVEILAAVFAAYPDIQNKFSQF-AGK} \\ \text{TNLWGKVNINELGGEALGRLLVVYPWTQRFFEAFGDLS} \end{bmatrix} \quad (60)$$

(a) An optimal alignment of three sequences scored with BLOSUM62 and gap penalty −11.

$$\begin{bmatrix} \text{FSKIGGHAEEYGAETLERMFIAYPQTKTYFPHFDLS} \\ \text{VQSSWKAVSHNEVEILAAVFAAYPDIQNKFSQFAGK} \end{bmatrix} \quad (20)$$

$$\begin{bmatrix} \text{FSKIGG-HAEEYGAETLERMFIAYPQTKTYFPHF-DLS} \\ \text{TNLWGKVNINELGGEALGRLLVVYPWTQRFFEAFGDLS} \end{bmatrix} \quad (46)$$

$$\begin{bmatrix} \text{VQSSWK-AVSHNEVEILAAVFAAYPDIQNKFSQF-AGK} \\ \text{TNLWGKVNINELGGEALGRLLVVYPWTQRFFEAFGDLS} \end{bmatrix} \quad (-6)$$

(b) The three pairwise projections of (a).

$$\begin{bmatrix} \text{VQSSWKAVSHNEV--EILAAVFAAYPDIQNKFSQFAG-K} \\ \text{T-NLWGKVNINELGGEALGRLLVVYPWTQRFFEAFGDLS} \end{bmatrix} \quad (6)$$

(c) An optimal pairwise alignment of the second and third sequences.

Figure 9.1: A multiple alignment and its pairwise projections.

to the sum of the alignment scores of the $k(k-1)/2$ possible projections of the multiple alignment onto pairs of sequences. Of course, each of the pairwise alignments is itself the sum of its column scores; so, if we wish, we can (and normally will) reverse the order of summation and score a multiple alignment column by column. We score each column by summing the scores of the $k(k-1)/2$ pairs of symbols in the column. The score 0 is assigned to a pair of gaps, since the column would be stricken from the projection onto the pair of sequences with the gaps.

As for computational strategy, we will still be able to eliminate large numbers of candidate alignments easily, but we will also have large numbers remaining. Even when the number of sequences is modest (five), identification of the very best multiple alignment is extremely time-consuming. We must rely on the results of fast heuristic methods, which are useful both in and of themselves and for putting time-saving limits on an exact algorithm's search for the best alignment.

In the rest of this chapter, we will first extend the Needleman–Wunsch algorithm to handle three or more sequences and then look at some possible heuristics.

9.1 Extending the Needleman–Wunsch Algorithm

To gain some understanding of the process of constructing multiple alignments, we will first extend Chapter 3's Needleman–Wunsch algorithm to align three sequences so that the sum-of-pairs score is maximized.

Constructing the alignment from the right, we must choose a nonempty subset of the three strings from which we take the last letter to form the column; we will insert

a gap character into this column for the other strings. For example, if the three strings
are HOUSE, HOME, and HOVEL, we have seven options:

$$\left(\begin{array}{l}\text{best align-}\\ \text{ment of}\end{array}\begin{array}{l}\text{HOUS}\\ \text{HOM}\\ \text{HOVE}\end{array}\right)\begin{bmatrix}\text{E}\\ \text{E}\\ \text{L}\end{bmatrix},$$

$$\left(\begin{array}{l}\text{best align-}\\ \text{ment of}\end{array}\begin{array}{l}\text{HOUS}\\ \text{HOM}\\ \text{HOVEL}\end{array}\right)\begin{bmatrix}\text{E}\\ \text{E}\\ \text{-}\end{bmatrix}, \qquad \left(\begin{array}{l}\text{best align-}\\ \text{ment of}\end{array}\begin{array}{l}\text{HOUSE}\\ \text{HOME}\\ \text{HOVE}\end{array}\right)\begin{bmatrix}\text{-}\\ \text{-}\\ \text{L}\end{bmatrix},$$

$$\left(\begin{array}{l}\text{best align-}\\ \text{ment of}\end{array}\begin{array}{l}\text{HOUS}\\ \text{HOME}\\ \text{HOVE}\end{array}\right)\begin{bmatrix}\text{E}\\ \text{-}\\ \text{L}\end{bmatrix}, \qquad \left(\begin{array}{l}\text{best align-}\\ \text{ment of}\end{array}\begin{array}{l}\text{HOUSE}\\ \text{HOM}\\ \text{HOVEL}\end{array}\right)\begin{bmatrix}\text{-}\\ \text{E}\\ \text{-}\end{bmatrix},$$

$$\left(\begin{array}{l}\text{best align-}\\ \text{ment of}\end{array}\begin{array}{l}\text{HOUSE}\\ \text{HOM}\\ \text{HOVE}\end{array}\right)\begin{bmatrix}\text{-}\\ \text{E}\\ \text{L}\end{bmatrix}, \qquad \left(\begin{array}{l}\text{best align-}\\ \text{ment of}\end{array}\begin{array}{l}\text{HOUS}\\ \text{HOME}\\ \text{HOVEL}\end{array}\right)\begin{bmatrix}\text{E}\\ \text{-}\\ \text{-}\end{bmatrix};$$

these correspond to the $2^3 - 1 = 7$ nonempty subsets of the set of three strings.

We can set up a dynamic programming scheme as in Chapter 3, this time with a
cube-shaped table in which $M[\$i1][\$i2][\$i3]$ contains the score of the best possible
alignment of the first $\$i1$ letters of the first string, the first $\$i2$ letters of the second
string, and the first $\$i3$ letters of the third string. Just as we had to fill in edges of our
square table for two strings before proceeding to its interior, we can fill the interior
of the cube only after first filling its edges and then its sides.

Perl code for aligning three sequences is outlined below.

```
sub similarity {
    my($s1,$s2,$s3) = @_;
    ### fill in edges of cube
    foreach my $i1 (0..length($s1)) { $M[$i1][0][0]=$g * $i1 * 2; }
    foreach my $i2 (0..length($s2)) { $M[0][$i2][0]=$g * $i2 * 2; }
    foreach my $i3 (0..length($s3)) { $M[0][0][$i3]=$g * $i3 * 2; }
    ### fill in sides of cube
    ## Side 1
    foreach my $i1 (1..length($s1)) {
        my $aa1 = substr($s1,$i1-1,1);
        foreach my $i2 (1..length($s2)) {
            my $aa2 = substr($s2,$i2-1,1);
            $M[$i1][$i2][0]=max($M[$i1-1][$i2][0]+$g+$g,
                          $M[$i1][$i2-1][0]+$g+$g,
                          $M[$i1-1][$i2-1][0]+$g+p($aa1,$aa2));
        }
    }
}
```

```
## Side 2 ... similar ...
# ... $M[$i1][0][$i3] = max($M[$i1−1][0][$i3]+$g+$g, etc.


## Side 3
# ... $M[0][$i2][$i3] = max($M[0][$i2−1][$i3]+$g+$g, etc.


### fill in interior of cube
foreach my $i1 (1..length($s1)) {
    my $aa1 = substr($s1,$i1−1,1);
    foreach my $i2 (1..length($s2)) {
        my $aa2 = substr($s2,$i2−1,1);
        my $p12 = p($aa1,$aa2);
        foreach my $i3 (1..length($s3)) {
            my $aa3 = substr($s3,$i3−1,1);
            my $p13 = p($aa1,$aa3);
            my $p23 = p($aa2,$aa3);
            $M[$i1][$i2][$i3]
                = max($M[$i1−1][$i2−1][$i3−1]+$p12+$p13+$p23,
                      $M[$i1−1][$i2−1][$i3]+$p12+$g+$g,
                      $M[$i1−1][$i2][$i3−1]+$g+$p13+$g,
                      $M[$i1][$i2−1][$i3−1]+$g+$g+$p23,
                      $M[$i1][$i2][$i3−1]+0+$g+$g,
                      $M[$i1][$i2−1][$i3]+$g+0+$g,
                      $M[$i1−1][$i2][$i3]+$g+$g+0);
        }
    }
}


return ($M[length($s1)][length($s2)][length($s3)]);
}
```

Although this method works reasonably well for three strings of moderate lengths (requiring about two minutes for strings of 100 on your author's laptop), there are several difficulties with extending it to more strings.

- It appears that a different program must be written for each possible number of strings, since the number of nested loops is equal to the number of strings. In typical applications, the number of strings to be aligned may be in the dozens. This appears to require dozens of different programs.
- Furthermore, the size of the program text grows exponentially with the number of strings, since the number of terms over which the maximum is taken in the inner loop is $2^K - 1$ when there are K strings.

Both of these problems can be overcome by skillful changes to the control and data structures of the program, so that a single program can read and count the strings and then fill in the table correctly; we will address this in the next chapter. However, a bigger problem looms:

- The running time of this method increases exponentially in the number of strings.

To see why, suppose we have K strings with a total length of N. For the sake of argument, we can assume that the strings are all roughly of length N/K. Then the table has $(N/K)^K$ entries, and the computation of each entry requires the maximum of $2^K - 1$ distinct terms. Hence the running time is roughly proportional to $(2N/K)^K$.

9.2 NP-Completeness

In fact, the problem of finding a multiple alignment of arbitrary sequences is a member of the class of problems known to computer scientists by the name *NP-complete*.[1] Although a complete treatment of NP-completeness is beyond the scope of this book, its main implications are easily outlined and digested.

Computer scientists regard a problem as *tractable* if some algorithm exists that always solves the problem in time that is proportional to some power of the length of the input. Such problems are said to *solvable in polynomial time*. This definition of tractability has its weaknesses, since a problem that requires time proportional to the 50th power of the input size is not solvable in practice. But the definition has useful mathematical (specifically, closure) properties, and it is indisputable that a problem with *no* polynomial-time algorithm is *intractable*.

In the 1970s, a theory of intractability crystalized around certain problems that had proved resistant to efficient solution by computer. The class of NP-complete problems was defined formally, and a large number of problems ranging from abstruse number-theoretic problems to children's puzzles have since been proven to be members of this class. Its most interesting features may be listed as follows.

- No NP-complete problem has been proven to be solvable in polynomial time.
- No NP-complete problem has been proven to be *un*solvable in polynomial time!
- All NP-complete problems are computationally equivalent in the following sense: If *any* polynomial-time algorithm can be found to solve *any* NP-complete problem, then *every* NP-complete problem can be solved by some polynomial-time algorithm.

---

[1] "NP" stands for "nondeterministic polynomial". Problems in the set NP can be solved in time that grows as a polynomial function of the length of their input by some nondeterministic *Turing machine*. Deterministic Turing machines are the most realistic mathematical models of modern electronic computers. Nondeterministic Turing machines have one very unrealistic feature: they can reliably guess the correct answer to yes–no questions.

- Since so many bright people have tried to solve so many different NP-complete problems for so long with no success, almost no one believes that polynomial-time algorithms exist – but no one can prove this, either!

The good news is that the inputs that typically arise in practice may be much different from the ones contrived to prove NP-completeness. In this case, it may be possible to handle typical inputs satisfactorily. It may also be possible to find a provably good approximation to the best output efficiently even if the problem of finding the exact answer is NP-complete. The bad news is that sometimes even approximating an answer can be shown to be NP-complete.

Biologists sometimes criticize computer scientists for proving that problems are NP-complete and then losing interest. Computer scientists should keep in mind that biologists' data won't disappear just because the analysis required is NP-complete. Laboratory scientists are used to trafficking in approximations, and they may find that an approximation is satisfactory or at least helpful in planning further experiments. However, biologists need to appreciate that some problems are, despite superficial appearances, not amenable to solution by computer. They should be open to possible changes to the design of experiments to give a tractable analysis.

9.3   Alignment Merging: A Building Block for Heuristics

If the direct approach to exact multiple alignment is impractically slow, why is this so? One reason is that there are too many nested loops, one for each of the strings being aligned. The other reason is that there are too many "choices" in the inner loop, a number that is exponential in the number of strings. To be reasonably fast, an alignment program must have only a constant number of nested loops and a constant number of choices. One way to achieve this is by restricting ourselves to aligning only *pairs* of sequences directly – and by using this procedure repeatedly to produce a reasonably good (albeit not necessarily optimal) multiple alignment.

For example, we might adopt an *incremental* approach: aligning the first two sequences; then adding a third without disturbing the residue pairings between the first two; then adding a fourth without disturbing the first three; and so forth. Alternatively, we might employ a *divide-and-conquer* approach, assigning each sequence a "partner" and aligning the pair, then combining aligned pairs into aligned groups of four, then of eight, ..., until only a single multiple alignment of all the sequences remains. In either approach, we need to give some additional thought to the *order* in which the sequences are processed, since some orders may be likely to give better alignments than others. But the two approaches share a common element that we can develop further before exploring these questions – namely, both require that two *alignments* be aligned to each other to form a larger alignment.

We can combine two alignments by a modification of Chapter 3's Needleman–Wunsch algorithm. We will let $M[i][j]$ represent the best multiple alignment score achievable by aligning the first i *columns* of the first alignment with the first j

columns of the second alignment *without disturbing the internal residue pairings of either alignment.* Thus, we will work from the right ends of the small alignments to build the larger. If we are not to disturb the internal structure of the small alignments, then we have only three choices for forming the rightmost column of the new alignment:

1. we can take the rightmost columns from both of the alignments; or
2. we can take the rightmost column from the first alignment and add a gap to *each* sequence in the second alignment; or
3. we can take the rightmost column from the second alignment and add a gap to each sequence in the first alignment.

For example, if we have the two alignments

$$\begin{bmatrix} \text{HOUSE} \\ \text{HO-ME} \end{bmatrix} \quad \text{and} \quad \begin{bmatrix} \text{-HOVEL} \\ \text{ABODE-} \end{bmatrix},$$

then the three options are

$$\left(\text{best alignment of} \begin{array}{c} \begin{bmatrix} \text{HOUS} \\ \text{HO-M} \end{bmatrix} \\ \begin{bmatrix} \text{-HOVE} \\ \text{ABODE} \end{bmatrix} \end{array} \right) \begin{bmatrix} \text{E} \\ \text{E} \\ \text{L} \\ \text{-} \end{bmatrix},$$

$$\left(\text{best alignment of} \begin{array}{c} \begin{bmatrix} \text{HOUSE} \\ \text{HO-ME} \end{bmatrix} \\ \begin{bmatrix} \text{-HOVE} \\ \text{ABODE} \end{bmatrix} \end{array} \right) \begin{bmatrix} \text{-} \\ \text{-} \\ \text{L} \\ \text{-} \end{bmatrix}, \qquad \left(\text{best alignment of} \begin{array}{c} \begin{bmatrix} \text{HOUS} \\ \text{HO-M} \end{bmatrix} \\ \begin{bmatrix} \text{-HOVEL} \\ \text{ABODE-} \end{bmatrix} \end{array} \right) \begin{bmatrix} \text{E} \\ \text{E} \\ \text{-} \\ \text{-} \end{bmatrix}.$$

No matter how many *sequences* we may be aligning, we have only three choices, because we have constrained ourselves to leave the two pre-existing *alignments* undisturbed. Likewise, with only two alignments to work with, we need only two nested loops. Of course, evaluating the scores of the three possible last columns takes longer than for two sequences; in fact, if K sequences are being aligned then the time is proportional to $K(K-1)/2 = O(K^2)$, since this many pairs of symbols must be considered. But $O(K^2)$ is much less than $O(2^K)$.

9.4 Merging Alignments in Perl

In this section, we will explore the details of merging alignments in Perl. Our first step is to develop subroutine scoreColumn, which receives a list of residues and gap symbols as its argument list and returns the sum-of-pairs score of any column containing those symbols. Next we write planAlignmentMerge, which fills in a dynamic programming table for two alignments in a fashion analogous to that of Chapter 3's similarity for two sequences. Last, we write mergeAlignments, which (once the table is filled) can trace through its entries to build the optimal alignment.

One task that will be performed repeatedly is to compute the sum-of-pairs score of a proposed column in the alignment. Subroutine scoreColumn compares every residue in the column to every other and sums the scores. The column will be passed to subroutine scoreColumn as a list of characters. To compute the gap penalty for the column, separate tallies of gap and residue symbols are kept.

```perl
sub scoreColumn {
    my @col = @_;
    my ($gaps,$aas,$score) = (0,0,0);
    while (@col) {
        my $aa = shift @col;
        ($gaps++, next) if $aa eq "-";
        $aas++;
        foreach my $aa1 (@col) {
            next if $aa1 eq "-";
            $score += p($aa,$aa1);
        }
    }
    return $score+($g * $gaps * $aas);
}
```

Subroutine planAlignmentMerge is analogous to Chapter 3's similarity; it fills the table @M so that $M[$i][$j] contains the best multiple alignment score achievable by aligning the first $i columns of the first alignment with the first $j columns of the second alignment – without disturbing their internal residue pairings.

The first issue to be addressed is the representation of alignments. The two alignments will be passed to planAlignmentMerge as two references to lists of sequences with added gap symbols. If either alignment consists of a single sequence, then that sequence must be passed as a reference to a list of one item. In subroutines planAlignmentMerge and mergeAlignments, these references will be stored in the variables $S and $T.

Initialization of row 0 and column 0 of matrix @M is a bit more complicated than in the case of two individual sequences. If the columns of $T have been exhausted, then the cost of aligning the remaining columns of $S is not merely the cost of inserting gaps into $T; the cost of the internal pairings within $S must also be paid. The simplest approach is to use scoreColumn to score each column.

planAlignmentMerge must frequently extract the $ith column from one or the other of the alignments. This can be carried out by mapping the substring operator over the list of strings making up the alignment:

```
      my @Scol = map { substr($_,$i−1,1) } @$S;
```

```perl
sub planAlignmentMerge {
    my($S,$T) = @_;
    my ($sLen,$tLen) = (length($$S[0]),length($$T[0]));
    my @Sgaps = (("-") x @$S);
    my @Tgaps=(("-") x @$T);
    my (@M,@how);
    $M[0][0] = 0;
    foreach my $i (1..$sLen) {
        $how[$i][0] = "|";
        my @Scol = map { substr($_,$i−1,1) } @$S;
        $M[$i][0] = $M[$i−1][0]+scoreColumn(@Scol,@Tgaps);
    }
    foreach my $j (1..$tLen) {
        $how[0][$j] = "-";
        my @Tcol = map { substr($_,$j−1,1) } @$T;
        $M[0][$j] = $M[0][$j−1]+scoreColumn(@Sgaps,@Tcol);
    }
    foreach my $i (1..$sLen) {
        foreach my $j (1..$tLen) {
            my @Scol = map { substr($_,$i−1,1) } @$S;
            my @Tcol = map { substr($_,$j−1,1) } @$T;
            $M[$i][$j] = $M[$i−1][$j−1]+scoreColumn(@Scol,@Tcol);
            $how[$i][$j] = "\\";
            my $left = $M[$i][$j−1]+scoreColumn(@Sgaps,@Tcol);
            if ($left>$M[$i][$j]) {
                $M[$i][$j] = $left;
                $how[$i][$j] = "-";
            }
            my $up = $M[$i−1][$j]+scoreColumn(@Scol,@Tgaps);
            if ($up>$M[$i][$j]) {
                $M[$i][$j] = $up;
                $how[$i][$j] = "|";
            }
        }
    }
    return ($M[$sLen][$tLen],\@how);
}
```

One feature that distinguishes planAlignmentMerge from Chapter 3's similarity is the appearance of the array @how. This array is used to record which of the three alignment options contributes the best score to a particular entry in @M. The entries of @how contain \, |, or -, to indicate the direction of the optimal path in the alignment matrix. Recording each alignment decision explicitly spares us the trouble of "rethinking" each one when returning to the table to reconstruct the best alignment after filling @M. planAlignmentMerge returns both the best score and a reference to the @how array as its result.

As always, we will build the new alignment from right to left. To facilitate this, we have subroutine prependColumnToAlignment to add a new column on the left of an existing alignment. The alignment is passed to this subroutine as a reference to a list of strings; the new column is passed as a list of characters.

```perl
sub prependColumnToAlignment {
    my ($A,@col) = @_;
    foreach (@$A) { $_ = (shift @col).$_ };
}
```

Now we have at hand all the ingredients necessary to write subroutine merge-Alignments, which accepts two alignments as arguments and returns a new multiple alignment of all the sequences in its two arguments. The result is the highest-scoring alignment that doesn't disturb the internal pairings of its two arguments.

The indices $i and $j begin at the "lower right" corner of the table. Each iteration of the **while**-loop adds a column to the alignment and decreases at least one of the indices. The subroutine returns when the indices reach the upper left corner $i = $j = 0.

```perl
sub mergeAlignments {
    my ($S,$T) = @_;
    my ($score,$how) = planAlignmentMerge($S,$T);
    my @result = (("") x (@$S+@$T));
    my ($i,$j) = (length($$S[0]),length($$T[0]));
    my @Sgaps = (("-") x @$S);
    my @Tgaps=(("-") x @$T);
    while ($i>0 || $j>0) {
        if ($$how[$i][$j] eq "\\") {
            my @Scol = map { substr($_,$i−1,1) } @$S;
            my @Tcol = map { substr($_,$j−1,1) } @$T;
            prependColumnToAlignment(\@result,@Scol,@Tcol);
            $i−−; $j−−;
        } elsif ($$how[$i][$j] eq "|") {
```

```
            my @Scol = map { substr($_,$i−1,1) } @$S;
            prependColumnToAlignment(\@result,@Scol,@Tgaps);
            $i−−;
        } elsif ($$how[$i][$j] eq "-") {
            my @Tcol = map { substr($_,$j−1,1) } @$T;
            prependColumnToAlignment(\@result,@Sgaps,@Tcol);
            $j−−;
        }
    }
    return \@result;
}
```

Finally, we may use the following subroutine to rescore our multiple alignment once it has been constructed.

```
sub scoreMultipleAlignment {
    my @alignment = @_;
    my $score;
    foreach my $i (0..length($alignment[0])−1) {
        $score += scoreColumn(map {substr($_,$i,1)} @alignment);
    }
    return $score;
}
```

9.5   Finding a Good Merge Order

There is little theory to guide us in choosing an order for combining our sequences into larger and larger alignments. A useful general rule is to combine the most similar sequences first. This ensures that the contributions of these projections to the overall score are "locked in" and cannot be disrupted by first aligning with relatively dissimilar sequences.

Taking this strategy to its logical conclusion, the subroutine below aligns K strings in $K − 1$ stages. In the ith stage, it chooses from among the $K − i + 1$ existing alignments the two whose merge will make the greatest possible increase (or least possible decrease) to the total score of all alignments in the set. Although such a *greedy strategy* is not guaranteed to find the very best possible multiple alignment, it does at least refer to the desired goal in evaluating its choices.

```
sub greedyStrategyA {      ## argument list is a list of sequences
    my @aligns = map { [$_] } @_;      ## change them to singleton alignments
    while (@aligns>1) {
```

```perl
        ### find the two best ones to merge
        my @alignScores = map { scoreMultipleAlignment($_) } @aligns;
        my ($besti,$bestj,$bestDelta) = (-1,-1,-999999999);
        foreach my $i (0..$#aligns) {
            foreach my $j ($i+1..$#aligns) {
                my ($score,$how) = planAlignmentMerge($aligns[$i],$aligns[$j]);
                my $delta = $score-$alignScores[$i]-$alignScores[$j];
                ($besti,$bestj,$bestDelta) = ($i,$j,$delta)
                    if $delta > $bestDelta;
            }
        }
        ### replace them with their merge
        $aligns[$besti] = mergeAlignments($aligns[$besti],$aligns[$bestj]);
        splice(@aligns,$bestj,1);
    }
    return $aligns[0];
}
```

Another possible greedy approach is to (i) merge the two sequences with the highest similarity score and then (ii) add sequences to this alignment one by one, always adding the sequence that maximizes the increase to the multiple alignment. This strategy, which could be called *incremental*, is implemented by the following subroutine.

```perl
sub greedyStraetgyB {
    my @ss = @_;        ## argument list is a list of sequences
    ### find best pair to merge first
    my ($besti,$bestj,$bestScore)=(-1,-1,-999999999);
    foreach my $i (0..$#ss) {
        foreach my $j ($i+1..$#ss) {
            my ($score,$how) = planAlignmentMerge([$ss[$i]],[$ss[$j]]);
            ($besti,$bestj,$bestScore) = ($i,$j,$score)
                if $score > $bestScore;
        }
    }
    ### merge them
    my $alignment = mergeAlignments([$ss[$besti]],[$ss[$bestj]]);
    splice(@ss,$bestj,1);
    splice(@ss,$besti,1);
    while (@ss) {
        ### find best sequence to add to alignment
        my ($besti,$bestScore)=(-1,-999999999);
        foreach my $i (0..$#ss) {
```

```
            my ($score,$how) = planAlignmentMerge([$ss[$i]],$alignment);
            ($besti,$bestScore) = ($i,$score) if $score > $bestScore;
        }
        ### add best sequence to alignment; remove from @ss
        $alignment = mergeAlignments([$ss[$besti]],$alignment);
        splice(@ss,$besti,1);
    }
    return $alignment;
}
```

9.6   Exercises

1. Complete a dynamic programming table for the alignment of the two alignments

$$\begin{bmatrix} \text{HOUSE} \\ \text{HO-ME} \end{bmatrix} \quad \text{and} \quad \begin{bmatrix} \text{-HOVEL} \\ \text{ABODE-} \end{bmatrix},$$

   showing both the best score and the best alignment for each table entry.

2. Our program spends most of its time in subroutine scoreColumn. If alignments $S and $T have m and n columns respectively, then planAlignmentMerge makes about $3mn + m + n$ calls to scoreColumn. However, closer inspection shows that only about $mn + m + n$ different arguments are passed.

   Show how to alter planAlignmentMerge to make only $mn + m + n$ calls to scoreColumn. The best solution uses no additional memory.

9.7   Complete Program Listings

Directory chap9/ of the software distribution includes file direct.pl, containing our direct extension of the Needleman–Wunsch algorithm to three sequences, as well as file merge.pl, containing our program for constructing multiple sequence alignments by merging smaller alignments.

9.8   Bibliographic Notes

Feng and Doolittle (1987) were first to underline the importance of a "progressive" approach to multiple alignment. The series of CLUSTAL programs (see Thompson, Higgins, and Gibson 1994) have for many years been the standard against which other programs are measured; T-COFFEE (Notredame, Higgins, and Heringa 2000) is an interesting recent alternative.

The PSI-BLAST program of Altschul et al. (1997) develops and uses the profiles mentioned on page 127 for protein database searching. Techniques for speeding up the search for the best alignment using the sum-of-pairs criterion are described in Carillo and Lipman (1988), Gupta, Kececioglu, and Schäffer (1995), and Miller (1993).

Altschul, Carroll, and Lipman (1989) explain why the straightforward sum-of-pairs criterion is inappropriate when the evolutionary distances between the various pairs of sequences are not similar; they also offer rationales for assigning weights to pairs of sequences for a "weighted sum-of-pairs" objective function for multiple alignment.

Comparisons of available programs can be found in Thompson, Plewniak, and Poch (1999), McClure, Vasi, and Fitch (1994), and Gotoh (1996).

## CHAPTER TEN

Multiple Sequence Alignment II

In the previous chapter we saw that finding the very best multiple alignment of a large number of sequences is difficult for two different reasons. First, although the direct method is general, it seems to require a different program text for each different number of sequences. Second, if there are K sequences of roughly equal length then the number of entries in the dynamic programming table increases as fast as the Kth power of the length, and the time required to fill each entry increases as the Kth power of 2.

In this chapter, we will address both of these difficulties to some degree. It is, in fact, possible to create a single program text that works for any number of strings. And, although some inputs seem to require that nearly all of the table entries be filled in, careful analysis of the inputs supplied on a particular run will often help us to avoid filling large sections of the table that have no influence on the final alignment. Part of this analysis can be performed by the heuristic methods described in the previous chapter, since better approximate alignments help our method search more quickly for the best alignment.

10.1   Pushing through the Matrix by Layers

As a comfortable context for learning two of the techniques used in our program, we will begin by modifying Chapter 3's subroutine similarity for computing the similarity of two sequences. Recall that, in that program, we filled a dynamic programming table @M row by row, using the Perl statement

---

$M[$i][$j] = max($M[$i−1][$j]+$g, $M[$i][$j−1]+$g, $M[$i−1][$j−1]+$p);

This statement "pulls" needed information into $M[$i][$j] all at once. Our new strategy for filling the table is slightly different.

- Rather than filling the table row by row, we will fill it by *layers*. The layer of an entry is determined by summing its indices; for example, $M[4][7] is in layer $4+7 = 11$. With two sequences, we find that each table entry in layer ℓ is computed from entries in layers $\ell - 1$ and $\ell - 2$ and is independent of other entries in layer ℓ. For example, $M[4][7] in layer 11 is computed from $M[3][6], $M[3][7], and $M[4][6]; these three lie in layers 9 and 10. In general, with K sequences, we find that an entry in layer ℓ is computed from entries in layers $\ell - 1, \ell - 2, \ldots, \ell - k$. Therefore, filling layer by layer ensures that information is available when needed, and the entries within a single layer can be computed in any order.

- Whereas previously we processed an entry by updating it to reflect the information in the $2^K - 1$ entries upon which it depended, the task will now be to update each of the $2^K - 1$ entries *that depend upon it*. In other words, rather than filling an entry by "pulling" information into higher-numbered layers from lower-numbered layers, we will "push" information from lower-numbered layers to higher-numbered layers. For example, processing $M[4][7] in layer 11 will consist of pushing information into $M[5][7], $M[4][8], and $M[5][8] in layers 12 and 13 rather than pulling from layers 9 and 10.

The new version of similarity shown below can be substituted directly into the program in Chapter 3.

```
sub similarityByPushAndLayers {
    my($s,$t) = @_;
    foreach my $i (0..length($s)) {
        foreach my $j (0..length($t)) {
            $M[$i][$j] = -999999;
        }
    }
    $M[0][0]=0;
    foreach my $layer (0..(length($s)+length($t))) {
        my $i = min($layer,length($s));
        my $j = $layer-$i;
        while ($j<=length($t) && $i>=0) {
            $M[$i+1][$j] = max($M[$i+1][$j], $M[$i][$j]+$g);
            $M[$i][$j+1] = max($M[$i][$j+1], $M[$i][$j]+$g);
            my $p = p(substr($s,$i,1),substr($t,$j,1));
            $M[$i+1][$j+1] = max($M[$i+1][$j+1], $M[$i][$j]+$p);
            $i--; $j++;
        }
    }
    return ($M[length($s)][length($t)]);
}
```

To extend this approach to a larger, undetermined number of sequences, we will replace our multidimensional array @M with a hash %M. The keys of %M will be strings representing lists of indices. For example, $M[30][82][27] will become $M{"30,82,27"}.

This approach has both advantages and disadvantages. The principle advantage is that Perl is indifferent to the number of integers that might be embedded in the string being used as hash key. Furthermore, whereas the appearance of an assignment statement like $M[30][82][27] may cause the allocation of space for hundreds of table entries like $M[30][82][0] through $M[30][82][26], the expression $M{"30,82,27"} never allocates space for more than one hash-table entry. Since we are hoping to avoid filling large numbers of table entries, we should also be happy to save the space they might occupy. The primary disadvantage is that a good deal of time will be spent with **split** and **join** to change strings like "30,82,27" into lists like (30,82,27) and back again.

Our overall strategy will be to work layer by layer, beginning with layer 0, which contains only the origin. Just before processing an entry, we will determine whether or not it is *relevant*. Our definition of relevance varies from strategy to strategy. (See Sections 10.10.2 and 10.10.3.) Ideally, we would like to fill in exactly the set of table entries that must be traced through in order to reconstruct the alignment after the table is filled. Of course, this is impossible. So we deem an entry irrelevant as soon as available information indicates that it cannot possibly be on the reconstruction path. If our strategy is heuristic, we may exclude the entries if it simply seems unlikely that the entry will be needed for the optimal reconstruction. Because we want to minimize the number of entries processed, we will always be on the lookout for excuses to declare an entry irrelevant.

If an entry is relevant, we will use its score to update the score of its *successors* in the table. Two entries p and s are predecessor and successor if and only if each index of s is either the same as or 1 greater than the corresponding index of p. An entry (other than the origin) can be relevant if and only if it has at least one relevant predecessor. We will be able to avoid processing entries with no relevant predecessor at all – they will never even appear on our agenda. We do this by maintaining a data structure in the array @pendingLayers that contains only successors of relevant entries in the current layer. When a relevant entry is processed, we take care that its successors are stored in this structure. When we complete a layer and advance to the next, we test for relevance *only* the entries stored in that structure. We say that an entry is *activated* when added to @pendingLayers and *deactivated* once processed and removed.

@pendingLayers is not a simple list of active entries; instead, active entries are classified by layer into a list of references to lists of entries. An active entry lying three layers beyond the current layer, for example, will be stored in the list pointed to by $pendingLayers[2]. Since there can never be more active layers than there are sequences, @pendingLayers has a fixed length that is independent of the length of the sequences.

The general flow of control for filling the dynamic programming table is shown in the following code excerpt.[1]

---

[1] Note that Perl interprets the expression ($inRange=0,**last**) in the innermost loop as an attempt to construct a list of two items. It successfully determines that the first element is 0 but then is forced to terminate the loop while trying to determine the second element.

```
sub planMultipleAlignment {
    my ($relevant,@ss) = @_;
    ...
    my @pendingLayers = ( );
    foreach (@ss) { push @pendingLayers, [ ]; }      ## build list of empties
    my $origin = join(",", ((0) x (scalar @ss)));
    my $currentLayer = [$origin];      ## prime the pump
    $M{$origin} = 0;
    my $layer = 0;
    my $count = 0;
    while (@$currentLayer) {      ## while something left in current layer
        print "LAYER $layer:";
        my $relevantCount = 0;
        while (my $vec = pop @$currentLayer) {      ## get entry in current layer
            my @vec = split(",", $vec);
            my $inRange = 1;
            foreach (0..$#vec) {
                ($inRange=0, last) if $vec[$_]>$goal[$_];
            }
            if ($inRange && &$relevant($M{$vec},\@vec)) {
                ### this entry is relevant
                ### calculate successors and update their scores
                ...
            }
        }
        $layer++;      ## advance layer number
        $currentLayer = shift @pendingLayers;      ## remove list for next layer
        push @pendingLayers, [ ];      ## add an empty list for new active layer
    }
}
```

The subroutine call &$relevant($M{$vec},\@vec) requires some explanation.
$relevant is one of the arguments to planMultipleAlignment. For this expression
to execute without error, its value must be a *subroutine reference*. A reference to a
subroutine can be created with the operator \&. Later we will define different subroutines for determining relevance:

```
sub tunnelRelevant { ... };
sub carilloRelevant { ... };
sub yes { return 1; }
```

By passing different subroutine references to planMultipleAlignment, we will be
able to use different relevance tests in the same program:

```
planMultipleAlignment(\&tunnelRelevant,@sequences);
planMultipleAlignment(\&carilloRelevant,@sequences);
planMultipleAlignment(\&yes,@sequences);     ## fill entire table
```

To call the subroutine whose reference is stored in $relevant, we use the subroutine dereference operator &. The technique of passing subroutines' references as arguments is very powerful, but many programmers find it confusing and so it should be used sparingly.

We can now turn our attention to how individual relevant entries are handled. As we saw in Chapter 9, each entry has $2^K - 1$ successors that must be updated. The successors' indices in the dynamic programming table can be found by increasing any nonempty subset of the entry's indices by 1. We call the numbers $\{1, 2, \ldots, 2^K - 1\}$ *directions*. We can translate any entry plus direction into the entry's successor in that direction by treating the direction number as a binary number or bit vector and then adding corresponding bits and indices. For example, since 13 is 01101 in binary, the successor of $(4, 23, 6, 19, 11)$ in direction 13 is $(4 + 0, 23 + 1, 6 + 1, 19 + 0, 11 + 1) = (4, 24, 7, 19, 12)$.

To translate a direction number into bits, we will use the bit operations & (bitwise and) and $>>$ (right shift). Specifically, $direction & 1 gives the rightmost bit of $direction, and $direction$>>$1 shifts $direction right one bit. These operations are equivalent to taking the remainder and quotient of $direction by 2. Direction 13 is translated into bits as follows:

```
13 & 1 --> 1;   13 >> 1 --> 6;
 6 & 1 --> 0;    6 >> 1 --> 3;
 3 & 1 --> 1;    3 >> 1 --> 1;
 1 & 1 --> 1;    1 >> 1 --> 0;
 0 & 1 --> 0;    0 >> 1 --> 0;
```

In bits, we have:

```
01101 & 1 --> 1;   01101 >> 1 --> 00110;
00110 & 1 --> 0;   00110 >> 1 --> 00011;
00011 & 1 --> 1;   00011 >> 1 --> 00001;
00001 & 1 --> 1;   00001 >> 1 --> 00000;
00000 & 1 --> 0;   00000 >> 1 --> 00000;
```

In our program, we use $>>=$ instead of $>>$; the statement x>>$=3; is equivalent to $x=$x$>>3;.

At the beginning of the process of generating a successor, the current entry's indices (in @vec) are copied into @succ, and @column is initialized with a gap symbol for each sequence. Each 1-bit extracted from the direction triggers an addition to the appropriate index and the insertion of an amino acid from the appropriate sequence. The 1-bits are also counted to facilitate the successor's insertion into the @pendingLayers structure.

If a score is already recorded for the successor, then it remains only to ensure that that score is updated to the new score if it represents an improvement. If no score is recorded then the new score must be recorded, and the new entry must be activated by inserting it into @pendingLayers.

```perl
if ($inRange && &$relevant($score,\@vec)) {
    $relevantCount++;
    foreach my $direction (1..$numDirections) {
        my @succ = @vec;
        my @column = (("-") x @succ);
        my $layerDelta = 0;     ## count 1-bits
        for (my $i=0; $direction; $i++, $direction>>=1) {
            if ($direction & 1) {     ## this row has an increment
                $succ[$i]++;
                $column[$i] = substr($ss[$i],$succ[$i]-1,1);
                $layerDelta++;     ## count 1-bits
            }
        }
        my $succ = join(",", @succ);
        my $nuscore = $score+scoreColumn(@column);
        if (!defined($M{$succ})) {
            ($M{$succ},$how{$succ}) = ($nuscore, $vec);
            push @{$pendingLayers[$layerDelta-1]}, $succ;
        } elsif ($nuscore>=$M{$succ}) {
            ($M{$succ},$how{$succ}) = ($nuscore,$vec);
        }
    }
} else {
    delete $M{$vec};
    delete $how{$vec};
}
```

Once the dynamic programming table is filled, the actual alignment can be reconstructed. The procedure for doing so is pleasingly short. Each table entry in the $how table, which was filled and returned by planMultipleAlignment, contains the index string of the predecessor that gives the best score. We construct the column to be prepended to the alignment by splitting the predecessor's indices to the current table entry's.

```perl
sub reconstructAlignment {
    my ($how,@ss) = @_;
    my @result = (("") x @ss);
```

```perl
    my $origin = join(",", map {"0"} @ss);
    my $current = join(",", map {length($_)} @ss);
    my @gaps=(("-") x @ss);
    while ($current != $origin) {
        my @current = split(",",$current);
        my $previous = $$how{$current};
        my @previous = split(",", $previous);
        my @column = @gaps;
        for (my $i=0; $i<@ss; $i++) {
            if ($current[$i] != $previous[$i]) {
                $column[$i] = substr($ss[$i],$previous[$i],1);
            }
        }
        prependColumnToAlignment(\@result,@column);
        $current=$previous;
    }
    return \@result;
}
```

The following subroutine constructs the best multiple alignment after filling in the entire dynamic programming table.

```perl
sub yes { return 1; }

sub pushAlign {
    my (@ss) = @_;     ## a list of sequences (strings)
    my ($score,$how) = planMultipleAlignment(\&yes,@ss);
    print "$score\n";
    return ($score,reconstructAlignment($how,@ss));
}
```

10.2   Tunnel Alignments

We now turn our attention to the first of two strategies for determining relevance, a "tunnel digging" strategy. This strategy is simple, and it works especially well if the sequences to be aligned are very similar. In this case, the alignment path through the dynamic programming table is relatively direct and straight. The tunnel strategy "digs" a straight tunnel of predetermined width straight through the dynamic programming table from the origin to the goal.

The width w of the tunnel can be varied to trade off speed for confidence in the optimality of the alignment. If the tunnel is too narrow, the best alignment overall will be missed because it passes through entries outside the tunnel. A wide tunnel takes longer to fill, and the effort may be wasted if the actual alignment path stays too near the center of the tunnel.

The center of the tunnel is the line passing through the dynamic programming table from $M_{0,0,\ldots,0}$ to $M_{\ell_1,\ell_2,\ldots,\ell_K}$, where ℓ_j is the length of the jth sequence. Only entries inside the tunnel are considered relevant. $M_{i_1,i_2,\ldots,i_K}$ lies inside the tunnel if its distance from the center line (i.e., the nearest point on the center line) is less than w. To determine this distance, we must use a little linear algebra to rewrite $(i_1, i_2, \ldots, i_K)$ as the sum of two orthogonal vectors,

$$(i_1, i_2, \ldots, i_K) = \alpha(\ell_1, \ell_2, \ldots, \ell_K) + (x_1, x_2, \ldots, x_K). \tag{10.1}$$

In the language of linear algebra, $\alpha(\ell_1, \ell_2, \ldots, \ell_K)$ is the *projection* of $(i_1, i_2, \ldots, i_K)$ onto $(\ell_1, \ell_2, \ldots, \ell_K)$. This notion of projection is related (though not identical) to the notion of projection of multiple alignments. Observe that $M_{i_1,i_2,\ldots,i_K}$ lies inside the tunnel if its length of $(x_1, x_2, \ldots, x_K)$ is less than w – that is, if

$$x_1^2 + x_2^2 + \cdots + x_K^2 < w^2.$$

The subroutine tunnelRelevant implements this computation. Notice that the definition of tunnelRelevant is *nested inside* the definition of tunnelAlign. This has two effects.

- The name tunnelRelevant has meaning only within the definition of tunnelAlign. It is not possible to call tunnelRelevant by this name in other subroutines or the main program.
- When tunnelRelevant is called, all the names defined inside tunnelAlign have the same meanings inside tunnelRelevant. Thus, tunnelRelevant has access to @goal, $goalSq, and $width.

In this case, the second effect is the desired one. The actual calls to tunnelRelevant take place outside the scope of tunnelAlign, in subroutine planMultipleAlignment.

```
sub tunnelAlign {
    my ($width,@ss) = @_;
    my @goal = map { length($_) } @ss;
    my $goalSq = 0;
    foreach (@goal) { $goalSq += $_*$_; }

    sub tunnelRelevant {
        my ($score,$vecref) = @_;
        my ($i,$veclen2,$proj,$sslen2) = (0,0,0,0);
        foreach $i (0..@ss-1) {
            $veclen2 += $$vecref[$i]*$$vecref[$i];
            $proj += $$vecref[$i]*$goal[$i];
```

```
        }
        return (($veclen2 − ($proj*$proj / $goalSq)) < $width*$width);
    }


    my ($score,$how) = planMultipleAlignment(\&tunnelRelevant,@ss);
    return ($score,reconstructAlignment($how,@ss));
}
```

Many varations on the tunnel strategy are conceivable. We could use a different definition of distance from the center of the tunnel to determine relevance, such as the so-called L_1 metric, which specifies that $M_{i_1,i_2,\ldots,i_K}$ lies inside the tunnel if $(x_1, x_2, \ldots, x_K)$ of equation (10.1) satisfies

$$x_1 + x_2 + \cdots + x_K < w.$$

We could also make the table-filling process adapt to the scores it finds by shifting the tunnel's center slightly as we passed from layer to layer, always setting it adjacent to the highest score seen in the previous layer. All of these strategies are heuristic and may give poor results on some inputs.

10.3   A Branch-and-Bound Method

In this section, we will consider how to use the score of a good multiple alignment to speed up the search for the *best* alignment. For a concrete example, we will use the four protein sequences

```
GVLTDVQVALVKSSFEEFNANIPKNTHRFFTLVLEIAPGAKDLFSFLKGSS
SPLTADEASLVQSSWKAVSHNEVEILAAVFAAYPDIQNKFSQFAGK
VHLSGGEKSAVTNLWGKVNINELGGEALGRLLVVYPWTQRFFEAFGDLS
VLSPADKTNVKAAWGKVGAHAGEYGAEALERMFLSFPTTKTYFPHFDLS
```

These sequences were extracted from four related proteins. Although it would be more appropriate from a biological standpoint to use a nonuniform scoring scheme such as BLOSUM62, the examples will be easier to digest if we stick to our familiar $+1/-1/-2$ scoring scheme.

We assume that an alignment with score -200 is already known; its details are not important. Since we know that the best alignment score is no lower, we call this score the *lower bound*. The better the lower bound, the faster the search for the best alignment. (The approximation methods we have seen so far actually produce much better lower bounds for this input set, but that, too, is unimportant for illustrating the method.)

Now suppose that we have advanced to layer 80 in the construction of the dynamic programming table. We will look at two examples of entries in this layer to see how we might find an excuse to spare ourselves the effort of updating their successors in the table.

Example 1. First, we consider the point at which we process $M\{"19,22,21,18"\}$. If this entry is part of the path of the final solution, then the final solution must have the following form.

best	GVLTDVQVALVKSSFEEFN	best	ANIPKNTHRFFTLVLEIAPGAKDLFSFLKGSS
align-	SPLTADEASLVQSSWKAVSHNE	align-	VEILAAVFAAYPDIQNKFSQFAGK
ment	VHLSGGEKSAVTNLWGKVNIN	ment	ELGGEALGRLLVVYPWTQRFFEAFGDLS
of	VLSPADKTNVKAAWGKVG	of	AHAGEYGAEALERMFLSFPTTKTYFPHFDLS

If the sum of the scores of these two alignments is at least -200, then this entry is on the path of *some* alignment that is better than the best one we already know about. It may not be the best overall, but we can't yet eliminate it as a candidate.

Because the algorithm is processing $M\{"19,22,21,18"\}$, we know that its final value has been fixed and is *exactly* the score of the alignment on the left. In this case, $M\{"19,22,21,18"\} = -75$, and the corresponding alignment is

$$\begin{bmatrix} \text{GVLTDVQVALVKSSFEEF--N-} \\ \text{SPLTADEASLVQSSWKAVSHNE} \\ \text{VHLSGGEKSAVTNLWGKVNIN-} \\ \text{V-LSPADKTNVKAAWGKV--G-} \end{bmatrix}.$$

On the other hand, the score of the right-hand alignment is not known. Fortunately, we can put an upper limit on it. It is less than the sum of the scores of the following six alignments, which are optimal for the pairs of strings given:

$$\begin{bmatrix} \text{ANIPKNTHRFFTLVLEIAPGAKDLFSFLKGSS} \\ \text{VEILAAV--FAAYPD-IQN--K--FSQFAGK-} \end{bmatrix}, \begin{bmatrix} \text{ANIPKNTHRFFTLVLEIAPGAKDLFSFLKGSS} \\ \text{ELGGEALGRL--LVVY--PWTQRFFEAFGDLS} \end{bmatrix},$$

$$\begin{bmatrix} \text{ANIPKNTHRFFTLVLEIAPGAKDLFSFLKGSS} \\ \text{AHAGEYGAEALERMFLSFPTTKTYFPHFDLS-} \end{bmatrix}, \begin{bmatrix} \text{V---EILAAVFAAYPDIQNKFSQFAGK-} \\ \text{ELGGEALGRLLVVYPWTQRFFEAFGDLS} \end{bmatrix},$$

$$\begin{bmatrix} \text{V---EILA-AV---FAAYPDIQNKFSQFAGK} \\ \text{AHAGEYGAEALERMFLSFPTTKTYFPHFDLS} \end{bmatrix}, \begin{bmatrix} \text{EL-G--G-EALGRLLVVYPWTQRFFEAFGDLS} \\ \text{AHAGEYGAEALERMFLSFPTTKTYFPHF-DLS} \end{bmatrix}.$$

The scores of these six alignments are -26, -24, -23, -18, -24, and -11, respectively. So the very best score we could hope for from an alignment passing through this table entry is $(-75) + (-26) + (-24) + (-23) + (-18) + (-24) + (-11) = -201$. We call this number the *upper bound* for entry $M\{"19,22,21,18"\}$. Since the upper bound for this entry is less than the lower bound of -200 for the best alignment, we conclude that this entry is irrelevant and hence do not attempt to push its information on to its successors.

(In fact, the best multiple alignment of the suffixes of the strings is

$$
\begin{bmatrix}
\text{ANIPKNTHRFFTLVLEIAPGAKDLFSFLKGSS} \\
\text{V-------EILAAVFAAYPDIQNKFSQF-AGK} \\
\text{EL-G--G-EALGRLLVVYPWTQRFFEAFGDLS} \\
\text{AHAGEYGAEALERMFLSFPTTKTYFPHF-DLS}
\end{bmatrix},
$$

with a score of -138. So the score of the best alignment passing through this entry is only $(-75) + (-138) = -213$.)

Example 2. Now let's consider the entry $M\{\texttt{"20,20,20,20"}\}$. The upper bound for this entry is the sum (-169) of the scores $(-65, -22, -21, -23, -14, -18, -6)$ of the following seven alignments:

$$
\begin{bmatrix}
\text{GVLTDVQVALVKSSFEEFNA-} \\
\text{SPLTADEASLVQSSWKAVSH-} \\
\text{VHLSGGEKSAVTNLWGKVNI-} \\
\text{V-LSPADKTNVKAAWGKVGAH}
\end{bmatrix},
$$

$$
\begin{bmatrix}
\text{NIPKNTHRFFTLVLEIAPGAKDLFSFLKGSS} \\
\text{NEVEILAAVFAAYPDIQN--K--FSQFAGK-}
\end{bmatrix},
\begin{bmatrix}
\text{NIPKNTHRFFTLVLEIAPGAKDLFSFLKGSS} \\
\text{NELGGEALGRLLVVY--PWTQRFFEAFGDLS}
\end{bmatrix},
$$

$$
\begin{bmatrix}
\text{NIPKNTHRFFTLVLEIAPGAKDLFSFLKGSS} \\
\text{AGEYGAEALERMFLSF-PTTKTYFPHFDLS-}
\end{bmatrix},
\begin{bmatrix}
\text{NEV--EILAAVFAAYPDIQNKFSQFAGK-} \\
\text{NELGGEALGRLLVVYPWTQRFFEAFGDLS}
\end{bmatrix},
$$

$$
\begin{bmatrix}
\text{N-EV--EILAAVFAAYPDIQNKFSQFAGK} \\
\text{AGEYGAEALERMFLSFPTTKTYFPHFDLS}
\end{bmatrix},
\begin{bmatrix}
\text{N-ELGGEALGRLLVVYPWTQRFFEAFGDLS} \\
\text{AGEYGAEALERMFLSFPTTKTYFPHF-DLS}
\end{bmatrix}.
$$

Since the upper bound for this entry (-169) exceeds the lower bound (-200) for the best multiple alignment, we *cannot* safely eliminate this entry from further consideration. We will update its successors as appropriate.

It will eventually turn out that the best alignment of the suffixes in Example 2 is

$$
\begin{bmatrix}
\text{NIPKNTHRFFTLVLEIAPGAKDLFSFLKGSS} \\
\text{N-EV--EILAAVFAAY-PDIQNKFSQF-AGK} \\
\text{N-ELGGEALGRLLVVY-PWTQRFFEAFGDLS} \\
\text{AGEYGAEALERMFLSF-PTTKTYFPHF-DLS}
\end{bmatrix},
$$

so that the score of the best alignment passing through this entry is $(-65) + (-112) = -177$.

The best alignment overall turns out to be

$$
\begin{bmatrix}
\text{GVLTDVQVALVKSSFEEFNANIPKNTHRFFTLVLEIAPGAKDLFSFLKGSS} \\
\text{SPLTADEASLVQSSWKAVSHN--EV--EILAAVFAAYPDIQNKFSQF-AGK} \\
\text{VHLSGGEKSAVTNLWGKVNIN--ELGGEALGRLLVVYPWTQRFFEAFGDLS} \\
\text{V-LSPADKTNVKAAWGKVGAHAGEYGAEALERMFLSFPTTKTYFPHF-DLS}
\end{bmatrix},
$$

with a score of −174. It skips layer 80 by passing directly from entry $M{"20,20, 20,19"} in layer 79 to entry $M{"21,21,21,20"} in layer 83.

10.4   The Branch-and-Bound Method in Perl

It should be clear from Section 10.3 that the method proposed needs to be able to determine – and quickly – the similarity of any two suffixes of any two distinct sequences. This requirement is best met by precomputing a table @c in which entry $c[$p][$q][$i][$j] contains the similarity of the $pth input sequence without its first $i residues and the $qth input sequence without its first $j residues. For fixed $p and $q, values for all $i and $j can be determined in a single run of the Needleman–Wunsch algorithm on reversed versions of the strings. Our code will be shorter and clearer if we simply modify the Needleman–Wunsch algorithm to work from the right ends of the sequences. The following subroutine fills each table entry in constant time.

```
sub computeSuffixPairSimilarities {
    my(@ss) = @_;
    my @c;
    foreach my $p (0..$#ss) {
        my $s = $ss[$p];
        my $m = length $s;
        foreach my $q ($p+1..$#ss) {
            my $t = $ss[$q];
            my $n = length $t;
            $c[$p][$q][$m][$n] = 0;
            for (my $i=$m; $i>=0; $i−−) {
                $c[$p][$q][$i][$n] = $g * ($m−$i);
            }
            for (my $j=$n; $j>=0; $j−−) {
                $c[$p][$q][$m][$j] = $g * ($n−$j);
            }
            for (my $i=$m−1; $i>=0; $i−−) {
                for (my $j=$n−1; $j>=0; $j−−) {
                    my $match = scoreColumn(substr($s,$i,1),substr($t,$j,1));
                    $c[$p][$q][$i][$j]
                        = max($c[$p][$q][$i+1][$j]+$g,
                               $c[$p][$q][$i][$j+1]+$g,
                               $c[$p][$q][$i+1][$j+1]+$match);
            }}}}
    return \@c;
}
```

If the total length of all sequences is N then there are no more than

$$\left(\frac{K(K-1)}{2}\right)\left(\frac{N}{K}\right)^2 \leq \frac{N^2}{2}$$

entries, so only $O(N^2)$ time is required for this precomputation.

Given the table @c, it is a rather simple matter to determine whether a given entry is relevant. As in the previous section, we *nest* the definition of subroutine branchBoundRelevant inside the alignment subroutine so that, when passed as an argument to planMultipleAlignment, it will have access to the local variables $lowerBound and $C.

```
sub branchBoundAlign {
    my ($lowerBound,@ss) = @_;
    my $C = computeSuffixPairSimilarities(@ss);

    sub branchBoundRelevant {
        my ($score,$vecref) = @_;
        my $K = @ss;
        my $upperBound = $score;
        foreach my $p (0..$#ss) {
            foreach my $q ($p+1..$#ss) {
                $upperBound += $$C[$p][$q][$$vecref[$p]][$$vecref[$q]];
            }
        }
        return ($lowerBound<=$upperBound);
    }

    my ($score,$how)
        = planMultipleAlignment(\&branchBoundRelevant,@ss);
    return ($score,reconstructAlignment($how,@ss));
}
```

10.5   Exercises

1. Our program initializes @pendingLayers as follows:

```
    my @pendingLayers = ( );
    foreach (@ss) { push @pendingLayers, [ ]; }
```

Why won't the following work just as well?

```
my @pendingLayers = (([ ]) x @ss);
```

2. As a space-saving measure, modify the program to record directions (integers) in the @how table rather than vectors of indices (strings) as well as to use directions to reconstruct the alignment.

3. Investigate the *A∗* search method (Pearl 1984) and apply it to finding the optimal path through a multiple alignment matrix.

10.6 Complete Program Listings

The software distribution includes this chapter's program in the file chap10/bandbalign.pl. A "push"-based version of Chapter 3's similarity program is given in file chap10/pushsimilarity.pl.

10.7 Bibliographic Notes

See Section 9.8 (pp. 139–40).

## CHAPTER ELEVEN

Phylogeny Reconstruction

The urge to record or to reconstruct "family trees" seems to be a strong one in many different areas of human activity. Animal breeders have an obvious interest in pedigrees, linguists have grouped human languages into families descended from a common (and in many cases unattested) ancestor, and, when several manuscripts of the same text have been recovered, biblical scholars have tried to piece together which ones served as sources for which others. Even before the advent of the theory of evolution, naturalists attempted to discern the "Divine Plan" by assigning each known organism to its correct place in a system of nesting categories known as a taxonomy.

More modern biologists have tried to reconstruct the course of evolution by building trees reflecting similarities and differences in relevant features or *characters* of various species. Whereas early work relied upon morphological features such as shape of leaf or fruit for classification, most recent efforts focus on less subjective molecular features. In this chapter, we develop a program for constructing the evolutionary tree – or *phylogeny* – that best accounts for the differences observed in a multiple sequence alignment.

11.1 Parsimonious Phylogenies

Broadly speaking, there are two approaches to reconstructing the phylogeny of a group of species.[1] One approach first reduces the similarities and differences among the n species to $n(n-1)/2$ numerical scores; it then finds the phylogeny that optimizes a certain mathematical function of those scores.

However, the approach we follow focuses on a finite set of *characters,* each taking on one of a finite number of *states* in each species. The characters we will deal with in our examples are *columns in a multiple alignment* of homologous protein sequences, each taking on one of 21 possible states: an amino acid residue or a gap.

---

[1] Although we consistently speak of a group of species, we might also be dealing with a number of organisms of the same species, such as the HIV viruses found among a group of persons possibly infected, directly or indirectly, by a Florida dentist (see Ou et al. 1992).

In a reconstructed phylogeny describing the evolution of species, the species supplied as input will be modern species and will appear without descendents. This is because protein sequences of extinct ancestral organisms are unavailable. In the language of the mathematical theory of graphs, the modern species are the *leaves, tips,*[2] or *external nodes* of the tree, and the reconstructed ancestral species are the *branch nodes* or *internal nodes*. If one node is the immediate descendent of another, they are *joined* by an *edge* or *branch* of the graph, usually indicated diagrammatically by a line. Two nodes joined by a single edge are also said to be *adjacent* or *neighbors*. The word *connected* is reserved to indicate the existence of a *path* of zero or more edges running between the nodes. The edges, taken together, are sometimes said to define the *topology* of the tree. The reconstruction algorithm must not only infer the topology; it must concurrently assign a protein sequence, or at least a statistical model of a protein sequence, to the ancestral species at each branch node.

A common simplifying assumption is that the tree is *bifurcating*, meaning that each branch node has exactly two descendents. This reduces both the sheer number of possible topologies to be considered and also the number of different program cases needed to deal with a single branch node. If the available evidence does not lend strong support to the exact pattern of emergence of the descendents of a particular ancestor, this can normally be detected in a postprocessing step and dealt with by collapsing together two or more branches.

Graph theory distinguishes between *rooted* and *unrooted* trees. Rooted trees have a clearly identified ancestor from which all other nodes descend, and each edge is *directed* from parent to child. In *unrooted* trees, pairs of nodes are identified as immediately related without specifying the direction of descent. Obviously, evolutionary trees are rooted. Unrooted trees, however, are somewhat more tractable algorithmically. For this reason, it is common to construct rooted phylogenies by adding the sequence of a distantly related species known as an *outgroup* to the multiple alignment, constructing the optimal unrooted phylogeny of the result, and then rooting the resulting tree at the point of attachment of the outgroup to the rest of the tree. For example, a phylogeny of reptilian sequences might use a single avian sequence as an outgroup (Figure 11.1).

How will we determine which is best of the many possible bifurcating branching patterns that can be constructed with a given set of sequences at the leaves? *Parsimony* offers one method. Given two topologies, we prefer the one that can be explained more simply – that is, by the *smaller number of mutations.*

Since mutations are counted by comparing sequences at adjacent nodes in the tree, we can evaluate a given topology effectively only if we can quickly assign sequences to the branch nodes so as to minimize the total number of mutations. Fortunately, the time required for this is proportional to the size of the input: the number of species times the number of characters.

---

[2] A term favored by some biologists but unfamiliar to computer scientists.

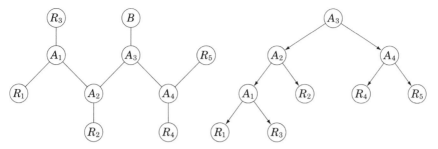

Figure 11.1: An unrooted reptilian phylogeny with an avian outgroup and the corresponding rooted phylogeny. The R_i represent modern reptiles; the A_i, inferred ancestors; and the B a bird.

Unfortunately, the number of possible topologies to be evaluated is very large. To see this, let us begin with three species. There is only one possible bifurcating topology, with three edges emanating from a single ancestor. Now, there are three possible points of attachment for a fourth species, one in the middle of each of the three existing edges. Each gives a distinct topology, and each increases the number of edges from three to five. Thus, wherever we choose to attach the fourth leaf, we have five choices for the fifth leaf, and the resulting tree has seven edges. It is not hard to see that the nth leaf can be attached in any of $(2n - 5)$ places. For example, the unrooted tree in Figure 11.1 has six leaves, and the next leaf could be attached in the middle of any of $(2 \times 6) - 3 = 9$ edges. The total number of distinct topologies for n leaves is therefore

$$(2n - 5)(2n - 7)(2n - 9) \cdots (7)(5)(3) = \prod_{i=4}^{n}(2i - 5).$$

This number, sometimes written $(2n - 5)!!$, satisfies the inequalities

$$2^{n-3}(n - 3)! = \prod_{i=4}^{n} 2(i - 3) < (2n - 5)!! < \prod_{i=4}^{n} 2(i - 2) = 2^{n-2}(n - 2)!.$$

When $n = 10$, it exceeds 2×10^6; when $n = 20$, it is about 2.2×10^{20}. Although we will find ways to avoid evaluating every possible topology, phylogeny reconstruction has been shown to be NP-complete, so prospects for a polynomial-time algorithm are dim.

11.2 Assigning Sequences to Branch Nodes

Our algorithms will mimic the process described for counting topologies. A basic step will be to attach a new leaf (species) to an existing tree by creating a new branch node in the middle of an existing edge. When this happens, we must decide what

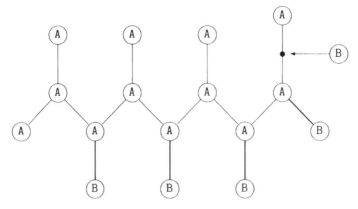

Figure 11.2

sequence to assign to the new branch node, as well as whether the total number of mutations can be decreased by changes to the sequences of other branch nodes. Since there is no interaction between the various characters (alignment columns), we can work this out for each character separately.

We are building the tree incrementally and updating branch node sequences after the addition of each leaf, so we can assume that states were assigned to minimize the number of mutations for the tree that existed before the new leaf was added. It follows that the only changes that could be beneficial are those that set branch nodes to the same state as the new leaf. Furthermore, there can be no advantage to changing a branch node unless at least one of its neighbors is also changed (or is the new leaf). Changes fan outward from the new leaf, and the changed nodes form a connected subgraph of the tree. For example, consider the tree in Figure 11.2, where the As and Bs represent two states of some character and where a new B leaf is to be inserted by creating a new branch node in the middle of the edge indicated.

The fact that the new branch node has two A neighbors might suggest that it, too, should be labeled A as in Figure 11.3. But this tree has five mutations (with darker branches), whereas four suffice (see Figure 11.4).

In order to determine which nodes to change, it is helpful to imagine that the new tree is temporarily rooted at the new leaf. In this rooted tree, no node's state can change unless its parent's state changes; otherwise, the optimality of the state assignments prior to the addition of the new node is contradicted. Questions to be answered for each node include the following.

- If the parent's state changes, what is the best effect that it can have on the number of mutations in the subtree rooted at this node, including the edge joining this node to its parent? We call this number best(n) for node n.
- If the parent's state changes, should this node also change to achieve the best effect? We call this change(n).

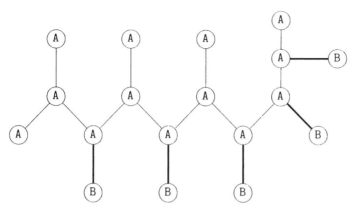

Figure 11.3

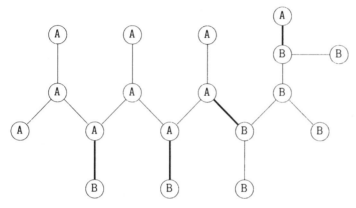

Figure 11.4

Of course, these questions cannot be answered for node n without first answering them for its children c_1 and c_2. Therefore, best(n) and change(n) are defined recursively:

$$\text{best}(n) = \begin{cases} -1 & \text{if } n\text{'s state already matches the new leaf's,} \\ +1 & \text{if } n \text{ is a leaf and currently matches its parent but} \\ & \quad \text{not the new leaf,} \\ 0 & \text{if } n \text{ is a leaf and currently matches neither its} \\ & \quad \text{parent nor the new leaf,} \\ \min(\text{best}_0(n), & \text{if } n \text{ is a branch node not matching the new leaf.} \\ \quad \text{best}_1(n)) \end{cases}$$

Here best$_0(n)$, the best result achievable without changing node n's state, is defined by

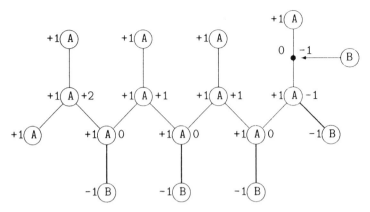

Figure 11.5

$$\text{best}_0(n) = \begin{cases} +1 & \text{if } n \text{ currently matches its parent,} \\ 0 & \text{if } n \text{ currently does not match its parent;} \end{cases}$$

$\text{best}_1(n)$, the best result achievable by changing node n's state, is defined (for internal nodes only) by

$$\text{best}_1(n) = \begin{cases} \text{best}(c_1) + \text{best}(c_2) & \text{if } n \text{ currently matches its parent,} \\ \text{best}(c_1) + \text{best}(c_2) - 1 & \text{if } n \text{ does not currently match its parent.} \end{cases}$$

We thus have

$$\text{change}(n) = \begin{cases} \textbf{false} & \text{if } n\text{'s state already matches the new leaf's,} \\ \textbf{false} & \text{if } n \text{ is a leaf,} \\ \textbf{false} & \text{if } n \text{ is a branch node and } \text{best}_0(n) \leq \text{best}_1(n), \\ \textbf{true} & \text{if } n \text{ is a branch node and } \text{best}_0(n) > \text{best}_1(n). \end{cases}$$

In Figure 11.5, each node has been labeled with values of best_0 (on left) and best_1 (on right).

11.3   Pruning the Trees

The simplest (but slowest) algorithm to find the simplest phylogeny is to mimic the process described for counting topologies. For five species, we proceed as follows.

1. To the single topology possible for the first three species, we attach the fourth species in the first of its three possible locations.
2. To this tree, we attach the fifth species in the first of its five possible positions. We record the number of mutations in the resulting tree.
3. Now move the fifth species to its second possible position. If the resulting tree has fewer mutations than the first, we record the number.
4. Now we try the fifth species in its third, fourth, and fifth possible positions. We record the number of mutations if any tree has fewer than the best so far.

5. We remove the fifth species, and return our attention to the fourth. We move it to its second possible position.

6. Now we return to the fifth species and try it in each of its five possible positions. We continue to record the number of mutations in the best tree.

7. We remove the fifth species, and return our attention to the fourth. We move it to its third position, then try the fifth species in all five possible positions.

This procedure, called *exhaustive search,* will create and evaluate each and every one of the $(2n - 5)!!$ possible phylogenies for the n species.

Under certain circumstances, we may be able to avoid evaluating each and every tree individually. For example, suppose that putting the fourth species in its first position and the fifth in its third position gives a phylogeny with six mutations. If later on we put the fourth species in its second position and discover that we have six or more mutations even before adding the fifth species, it is pointless to evaluate the five possible positions for the fifth species. We can *prune the search* at this point by simply moving on to the next position for the fourth species.

In our small example with only 15 trees to consider *in toto,* little is gained by pruning. But if, for example, we have ten species and can prune out a position for the sixth species, we save ourselves the evaluation of $15 \times 13 \times 11 \times 9 = 19,305$ complete trees – well worth the single comparison required to determine whether pruning is possible.

This method is another example of the *branch-and-bound* technique first seen in Chapter 10. The goal of this technique is to quickly eliminate from further consideration large portions of the problem's *search space,* the set of possible solutions. In Chapter 10, partial multiple alignments were compared to a single approximate solution computed before the start of the branch-and-bound search. The better the approximation, the more quickly the search for the exact answer came to an end.

In the present case, the standard for comparison is constantly updated and improved as more and more complete trees are evaluated. The sooner good approximations to the final solution are found, the fewer trees will ultimately have to be evaluated, because good approximations raise the standard against which all candidate trees must be compared. We could simply leave this to chance and hope that the correct solution or a good approximation will be seen early on in the search, or we could proceed more proactively as in Chapter 10 by applying an approximation technique to the problem to create a relatively high standard of comparison at the beginning of the branch-and-bound search.

One useful approximation method is the *greedy method,* in which we evaluate choices based on local criteria that are similar to the global optimization criterion yet whose optimization does not guarantee the optimization of the global criterion. A natural greedy strategy for constructing a topology with few mutations looks like this.

• To the single topology possible for the first three species, we try attaching the fourth species in each of its three possible locations. Among these three partial phylogenies, we choose the one with the fewest mutations and fix the fourth species in that position.

- To this tree, we attach the fifth species in each of its five possible positions. We fix the fifth species in the position that yields the phylogeny with the smallest number of mutations.

- Now we repeat this process with the sixth through nth species in turn, fixing each in a single position before adding the next.

The resulting tree is not guaranteed to be the best possible; in fact, we might obtain a better one simply by adding the species to the tree in a different order. But the optimization criterion applied at each step does bear an obvious relationship to the overall goal of producing a phylogeny with the fewest possible number of mutations, and we can reasonably expect it to be better than the random tree that is considered first by the plain branch-and-bound algorithm. Furthermore, it is produced by considering a mere $\sum_{i=4}^{n}(2i - 5) = O(n^2)$ partial trees.

11.4   Implementing Phylogenies in Perl

We will use Perl's object-oriented features to implement phylogenies. The package Phylogeny defines a class of the same name, and each interesting phylogeny that we find will be saved as an object of this class. Each phylogeny includes species names, species protein sequences, and a tree structure. The phylogeny package provides several subroutines for creating and copying phylogenies, altering their tree structure, and updating the sequences of branch nodes to minimize mutations. We have chosen to leave the subroutines that implement the exhaustive, greedy, and branch-and-bound algorithms outside the package.

Subroutine Phylogeny—>new, which creates a new phylogeny object, accepts two arguments. The first is a reference to a list of strings, each string representing the protein sequence of one species. The second argument is a reference to a list of strings giving species names or identifiers. Since the algorithm to update the sequences of branch nodes operates on one alignment column at a time, it is advantageous to split the sequence strings into characters and to build for each column a separate list containing its states for the various species. Then, (a reference to) a list of references to the columns is stored in the phylogeny hash. Names are stored as passed, and an empty tree structure is created. Explicitly saving the number of species and number of characters will make some coding less tedious later on.

```
sub new {
    my ($this,$seqs,$names) = @_;
    my @characters;
    for (my $i=0; $i<@$seqs; $i++) {
        my @states = split(//,$$seqs[$i]);
        for (my $j=0; $j<@states; $j++) {
```

```
            $characters[$j][$i] = $states[$j];
        }
    }
    $this =
        bless {characters => \@characters,
                names => ($names||[ ]),
                numTraits => scalar @characters,
                numLeaves => scalar @{$seqs},
                nextBranch => scalar @$seqs,
                tree=>[ ]},
            (ref($this) || $this);
        return $this;
}
```

Trees are graphs, and graphs are typically represented in computer programs by *adjacency lists*. An adjacency list is an array of lists, with one array entry for each node in the graph. We will assign indices to the nodes beginning from 0. Each array entry is a (reference to a) list of the indices of that node's neighbors. A reference to our adjacency list structure is stored under the key tree in the phylogeny hash. Here are two basic subroutines that manipulate the adjacency list:

```
sub neighbors {      ## returns the list of neighbors of v
    my ($this,$v) = @_;
    return @{${$this->{tree}}[$v]};
}


sub joinNewNeighbors {      ## adds an edge between v1 and v2
    my ($this,$v1,$v2) = @_;
    push @{$this->{tree}[$v1]}, $v2;      ## put v2 in v1's list
    push @{$this->{tree}[$v2]}, $v1;      ## put v1 in v2's list
}
```

If we are interested in a list of all pairs of nodes joined by an edge then we can use allEdges, a subroutine that walks down the adjacency list of every node. To avoid including the same edge twice in different orientations, it is added to the result list only when the first index is smaller.

```
sub allEdges {      ## returns a list of all edges in the tree
    my ($this) = @_;
    my $tree = $this->{tree};
    my @result;
```

```perl
for (my $i=0; $i<@$tree; $i++) {
    next unless defined($$tree[$i]);
    foreach (@{$$tree[$i]}) {
        push @result, [$i,$_] if $i<$_;
    }
}
return @result;
}
```

When building a phylogeny, we will sometimes want to know if a node is a leaf or a branch node. The number of leaf nodes is fixed by the number of sequences passed in when the phylogeny object is created, and leaves occupy the lowest positions in the node list. Hence subroutine isLeaf simply compares its argument to $this−>{numLeaves}.

When a good phylogeny is found for a particular set of sequences, it is often useful to be able to save a copy for future reference. We need to make a "deep copy", so that the copy contains no references to any pieces of the original phylogeny. This ensures that further modifications to the original will leave the copy unaffected and vice versa. For example, the slightly mysterious statement

```perl
my @tree = map(\@$_, @{$this−>{tree}});
```

of subroutine copy makes a deep copy of the adjacency list. Let us consider it in detail. Since $this is a reference to a hash, $this−>{tree} is a reference to the adjacency list stored under the key tree, and @{$this−>{tree}} is a copy of the adjacency list. But this is not good enough, since the adjacency list itself is a list of references to lists of neighbors. We need descend deeper and make *copies* of these lists of neighbors. So we apply the **map** operator to the adjacency list. Now what do we do with each item $_ in the adjacency list? First, we make a copy of the list: @$_; and then, we make a reference to that copy: \@$_. The new adjacency list is a list of these references. Subroutine copy makes a deep copy of protein sequences in a similar fashion.

Although we might normally expect a graph package to support manipulation of the adjacency lists only through subroutines to add a single edge or delete a single edge, we gain some opportunities for slightly faster code if we add some subroutines specialized to the task of building phylogenies. Subroutine attachLeafBetween breaks an existing edge by putting a new branch node in its middle and then attaches a new leaf to the branch node. Then it calls updateSeqs to alter branch node sequences as necessary to minimize mutations. Subroutine detachLeafBetween, shown next, reverses the process by removing the branch node and restoring the previous branch node sequences. The elements of $changeList are pairs of reference plus value; the references are scalar references pointing directly to elements of the characters array.

This list is assembled by subroutine updateSeqs and the subroutines it calls when updating the branch node sequences for a new addition. detachLeafBetween reverses these changes.

```
## removes the branch node with neighbors leaf, old1, and old2
## restores the edge between old1 and old2
sub detachLeafBetween {
    my ($this,$leaf,$old1,$old2,$changeList) = @_;
    foreach (@$changeList) {    ## undo changes to characters of branch nodes
        my ($ref,$state) = @$_;
        $$ref = $state;
    }
    my $oldBranch = ($this->removeNeighbors($leaf))[0];
    $this->removeNeighbors($oldBranch);
    $this->{nextBranch}--;
    $this->replaceNeighbor($old1,$oldBranch,$old2);
    $this->replaceNeighbor($old2,$oldBranch,$old1);
}
```

Subroutine replaceNeighbor typifies the performance optimizations possible with more complicated adjacency list manipulations. Rather than using **splice** to remove an old neighbor and then using **push** to add a new one, we can simply assign a new value to the adjacency list location containing the index of the old neighbor. This saves not just a subroutine call but also memory allocation.

```
sub replaceNeighbor {
    my ($this,$v,$oldNeighbor,$newNeighbor) = @_;
    my $r = ${$this->{tree}}[$v];
    for (my $j=0; $j<@$r; $j++) {
        $$r[$j] = $newNeighbor if $$r[$j] == $oldNeighbor;
    }
}
```

Although we normally keep track of mutations incrementally, countMutations – a subroutine that counts all mutations present in the current phylogeny by comparing the sequences of all adjacent pairs of species – is useful as a final check on the incremental updates.

```
sub countMutations {
    my ($this) = @_;
    my $chars = $this->{characters};
```

```perl
    my $n = $this->{numTraits};
    my $mutations = 0;

    foreach ($this->allEdges()) {
        my ($v1,$v2) = @$_;
        for (my $j=0; $j<$n; $j++) {
            $mutations++ if $$chars[$j][$v1] ne $$chars[$j][$v2];
        }
    }
    return $mutations;
}
```

When a new leaf is attached, the sequences of branch nodes are updated to minimize mutations by subroutines updateSeqs, findBestChange, and makeBest-Change. To begin, updateSeqs copies the new branch node's sequence from one of its neighbors in the previously existing tree.

```perl
sub updateSeqs {
    my ($this,$newLeaf,$changeList) = @_;
    my $totalDelta = 0;
    my $newBranch = ($this->neighbors($newLeaf))[0];
    my ($leaf,$old1,$old2) = $this->neighbors($newBranch);

    foreach my $column (@{$this->{characters}}) {
        $$column[$newBranch] = $$column[$old1];
        my $newState = $$column[$leaf];
        next if $$column[$newBranch] eq $newState;
        my @bestChanges;
        my $traitDelta
            = $this->findBestChange($column,$leaf,$newBranch,
                                        $newState,\@bestChanges);
        $this->makeBestChange($column,$leaf,$newBranch,
                                   $newState,\@bestChanges,$changeList);
        $totalDelta += ($traitDelta+1);
    }
    return $totalDelta;
}
```

The recursive function findBestChange computes the function best($curr) and returns its value. It also simultaneously computes change($curr) and saves its value in the array @$bestChanges.

```perl
sub findBestChange {
    my ($this,            ## [in] ref to current phylogeny
        $column,          ## [in] ref to list of states of current character
        $prev,            ## [in] last tree node considered (parent of curr)
        $curr,            ## [in] current tree node
        $newState,        ## [in] state of the character in the new leaf
        $bestChanges      ## [out] list of nodes to change (if parent changes)
    ) = @_;

    my $currState = $$column[$curr];
    my $mutBefore = ($currState eq $$column[$prev]) ? 0 : 1;
    $$bestChanges[$curr] = 0;

    ## Option 0: don't change this vertex's state in subtree
    my $mutAfter = ($currState eq $newState) ? 0 : 1;
    my $delta0 = $mutAfter-$mutBefore;

    ## changing is not always an option ...
    return $delta0 if ($this->isLeaf($curr) || ($currState eq $newState));

    ## Option 1: change this vertex's state in subtree
    my ($next1,$next2) = grep($_ ne $prev, $this->neighbors($curr));
    my $b1
        = $this->findBestChange($column,$curr,$next1,
                                        $newState,$bestChanges);
    my $b2
        = $this->findBestChange($column,$curr,$next2,
                                        $newState,$bestChanges);
    my $delta1 = $b1+$b2-$mutBefore;

    ## return, save best option
    return $delta0 if $delta0 <= $delta1;
    $$bestChanges[$curr] = 1;
    return $delta1;
}
```

Since change(x) indicates only whether x's state should change *if its parent's does*, changes cannot be carried out on the first pass through the tree, which is responsible for propagating information upward from descendent to ancestor. So then make-BestChange makes a second pass through the tree, propagating the information saved in @$bestChanges downward from the root and making state changes as indicated.

```perl
sub makeBestChange {
    my ($this,              ## [in] ref to current phylogeny
        $column,            ## [in] ref to list of states of current character
        $prev,              ## [in] last tree node considered (parent of curr)
        $curr,              ## [in] current tree node
        $newState,          ## [in] state of the character in the new leaf
        $bestChanges,       ## [in] list of nodes to change (if parent changes)
        $changeList) = @_;  ## [out] list of all changes to character states
    my ($next1,$next2) = grep($_ ne $prev, $this->neighbors($curr));
    return unless $$bestChanges[$curr];
    my $traitRef = \$$column[$curr];
    push @$changeList, [$traitRef,$$traitRef];    ## save current state
    $$traitRef = $newState;
    $this->makeBestChange($column,$curr,$next1,$newState,
                          $bestChanges,$changeList);
    $this->makeBestChange($column,$curr,$next2,$newState,
                          $bestChanges,$changeList);
}
```

11.5 Building the Trees in Perl

Our subroutines for building parsimonious phylogenies rather directly implement the strategies we have described. Subroutine buildTreesExhaustively builds every possible phylogeny and saves a copy of the best. It begins by joining the first two species by an edge and then calls the recursive subroutine buildAllTrees to attach the rest of the species in every possible way. Each call to subroutine buildAllTrees is devoted to attaching a particular species to the tree. To do this, a list of all edges is collected, and the species is attached to and then detached from each in turn. While the species is attached, a recursive call is made to attach the remaining species. As outlined in the previous section, attaching and detaching involve more than just changes to the tree topology; modification to the sequences of the branch nodes must be made and subsequently reversed.

```perl
sub buildTreesExhaustively {
    my ($sequences) = @_;
    my $phyl = Phylogeny->new($sequences);
    $phyl->joinNewNeighbors(0,1);
    my $bound = 1E99;
    $count = 0;
    my $best;
    buildAllTrees($phyl,2,$phyl->countMutations(),\$bound,\$best);
    warn "$count calls to buildAllTrees.\n";
    return $best;
}
```

```perl
sub buildAllTrees {
    my ($phyl,$newLeaf,$mutations,$bound,$best) = @_;
    $count++;
    if (!($phyl->isLeaf($newLeaf))) {
        if ($mutations<=$$bound) {
            $$bound = $mutations;
            warn "Call no. $count: Tree with $mutations mutations.\n";
            $$best = $phyl->copy();
        }
        return;
    }
    foreach ($phyl->allEdges()) {
        (my $v1, my $v2) = @$_;
        my $changeList = [];
        my $deltaMut
            = $phyl->attachLeafBetween($newLeaf,$v1,$v2,$changeList);
        buildAllTrees($phyl,$newLeaf+1,
                      $mutations+$deltaMut,$bound,$best);
        $phyl->detachLeafBetween($newLeaf,$v1,$v2,$changeList);
    }
}
```

```perl
sub buildTreeGreedily {
    my ($sequences) = @_;
    my $phyl = Phylogeny->new($sequences);
    $phyl->joinNewNeighbors(0,1);
    my $mutations = $phyl->countMutations();
    for (my $newLeaf=2; $phyl->isLeaf($newLeaf); $newLeaf++) {
        my ($bestDelta,$bestV1,$bestV2) = (1E99,-1,-1);
        foreach ($phyl->allEdges()) {
            (my $v1, my $v2) = @$_;
            my $changeList = [];
            my $deltaMutations
                = $phyl->attachLeafBetween($newLeaf,$v1,$v2,$changeList);
            ($bestDelta,$bestV1,$bestV2) = ($deltaMutations,$v1,$v2)
                if $deltaMutations < $bestDelta;
            $phyl->detachLeafBetween($newLeaf,$v1,$v2,$changeList);
        }
        $phyl->attachLeafBetween($newLeaf,$bestV1,$bestV2,[]);
        $mutations += $bestDelta;
    }
    return $phyl;
}
```

Subroutine buildTreeGreedily implements our greedy strategy and builds a single tree that is not guaranteed to be optimal. Each iteration of its inner loop is responsible for finding the best position for its species relative to the existing partial tree. The new species is attached to and detached from each existing edge without an intervening recursive call, then the species is reattached to the position that minimized mutations. Once every species has been attached in this way, the resulting phylogeny is returned.

Subroutine buildTreesEfficiently implements our branch-and-bound strategy. It begins like buildTreesExhaustively but calls the recursive subroutine buildGoodTrees. Subroutine buildGoodTrees is similar to buildAllTrees, but – before attempting to attach its species – each call compares the number of mutations in the current partial phylogeny to the number in the best known phylogeny. If the latter number has been exceeded, the subroutine simply returns, since the trees it could generate are guaranteed to be suboptimal. Subroutine buildTreesEfficiently takes an upper bound on number of mutations as an optional second argument. This number could come from buildTreeGreedily or another approximation method.

```perl
sub buildTreesEfficiently {
    my ($sequences,$upperBound) = @_;
    my $phyl = Phylogeny−>new($sequences);
    $phyl−>joinNewNeighbors(0,1);
    $upperBound ||= 1E99;    ## the second argument is optional
    $count = 0;
    my $best;
    buildGoodTrees($phyl,2,
                    $phyl−>countMutations( ),\$upperBound,\$best);
    warn "$count calls to buildGoodTrees.\n";
    return $best;
}

sub buildGoodTrees {
    my ($phyl,$newLeaf,$mutations,$bound,$best) = @_;
    $count++;
    if ($mutations>$$bound) {
        warn "Pruning at Leaf $newLeaf...\n";
        return;
    }
    if (!($phyl−>isLeaf($newLeaf))) {
        warn "Call no. $count: Tree with $mutations mutations.\n";
        $$bound = $mutations;
        $$best = $phyl−>copy( );
        return;
    }
```

```
    foreach ($phyl->allEdges()) {
        (my $v1, my $v2) = @$_;
        my $changeList = [ ];
        my $deltaMut
            = $phyl->attachLeafBetween($newLeaf,$v1,$v2,$changeList);
        buildGoodTrees($phyl,$newLeaf+1,
                       $mutations+$deltaMut,$bound,$best);
        $phyl->detachLeafBetween($newLeaf,$v1,$v2,$changeList);
    }
}
```

11.6   Exercise

1. A *perfect phylogeny* is one in which every state of every character arose by a single mutation. In other words, if the ith character assumes s_i different states, then the total number of mutations in the phylogeny is $M = \sum_i (s_i - 1)$.

   Modify the program to find a perfect phylogeny if one exists; if none exists, the revised program should indicate this but need not give the best possible phylogeny. *Hints:* One way to solve this problem is to compute M from the formula given and supply this to build_trees_efficiently as the upper bound. However, it should be possible to make the program significantly faster by exploiting the fact that every partial phylogeny must also be perfect.

11.7   Complete Program Listings

The software distribution includes our phylogeny module chap11/Phylogeny.pm and our branch-and-bound phylogeny program in file chap11/bandbphyl.pl.

11.8   Bibliographic Notes

PAUP (Swofford 1990) is a popular program for constructing phylogenies by the criterion of parsimony. The computational complexity of the problem is analyzed in Day, Johnson, and Sankoff (1986). Felsenstein's PHYLIP program (Felsenstein 1982, 1989) presents an alternative approach based on maximum likelihood.

The outlines of the area in general can be found in Lake and Moore (1998), Swofford et al. (1996), and Nei and Kumar (2000). Sneath and Sokal (1973) is a classic treatment of the distance-based approach (in contrast to the character-based approach pursued in this chapter).

In recent years, the interrelatedness of the phylogeny reconstruction problem and the multiple sequence alignment problem has been recognized: to find a good sequence alignment – "good" in the sense that it aligns analogous residues of homologous proteins – one must be guided by a good phylogeny, and vice versa. Gotoh's (1996) "double nested randomized iterative technique" exemplifies this approach: A phylogenetic tree is estimated, from this a set of weights is produced for scoring alignments with a weighted-sum-of-pairs criterion, and finally the multiple sequence alignment that maximizes the scoring function is found; the whole process is repeated (basing the new tree on the alignment) until the tree or alignment fails to change significantly on some cycle.

## CHAPTER TWELVE

Protein Motifs and PROSITE

In Chapters 7 and 8 we considered one method of determining the function of a new protein sequence: by searching for homologous sequences with known function in a large database like GenBank. Another method is to search for known *motifs* (also called *blocks,* or *segments*) in the protein sequence. A motif is a short (typically 5–15 residues), highly conserved region with a recognized biochemical function. PROSITE is a database of about 1400 motifs to which a new sequence can be compared.

PROSITE has three types of entries for describing motifs.

1. *Patterns* are simply sets of short sequences, expressed in a concise notation. For example, the pattern [ASV]-S-C-[NT]-T-x(2)-[LIM] . represents a set of $3 \cdot 1 \cdot 1 \cdot 2 \cdot 1 \cdot 20^2 \cdot 3 = 7200$ different sequences of residues (including, e.g., VSCTTQPM and SSCTTAAL). If any one of these is a substring of a new, uncharacterized protein sequence, then the new sequence *matches* the pattern, and the matching substring is likely to perform the function described in the database – in this case, the substring is probably a "glyceraldehyde 3-phosphate dehydrogenase active site". The result of pattern matching is dichotomous; the protein either matches or fails to match the pattern. However, the PROSITE database entry does contain statistics on the number of false positives and false negatives observed when the pattern is applied to the entire SWISS-PROT database of protein sequences. PROSITE[1] includes 1314 patterns. We will look at patterns in detail in Section 12.2.

2. *Matrices,* also known as *profiles,* provide statistical models of motifs. Successive columns of the matrix give scores for the occurrence of the different amino acids in successive positions of the protein sequence. A simple profile for planet Threamino (p. 114) might look like this:

A	5	1	−6	0	4
B	0	2	3	−1	0
C	−4	−2	3	3	−6

---

[1] Release 16.0 of July 1999.

To apply the profile to a new sequence, its substrings would be aligned to the columns of the profile and the resulting scores analyzed statistically. For example, the substring ABCBA would be assigned score $5 + 2 + 3 + (-1) + 4 = 13$. More sophisticated profiles include penalties for deleting columns of the profile or inserting residues between profile columns. PROSITE presently includes 56 profiles.

3. *Rules* are English prose descriptions of motifs that do not fit well into the preceeding frameworks. PROSITE includes only six rules.

In this chapter, we will concentrate on methods for matching PROSITE patterns and target sequences and reporting on the location of the matches. In Section 12.1, we will look at the format of the PROSITE database and sketch a module for reading its entries. In Section 12.2, we will learn more about Perl's built-in pattern-matching features and learn how to translate PROSITE patterns into Perl regular expressions for quick matching. Section 12.3 describes suffix trees, an extremely useful data structure for searching for the presence of many small strings in a single large string. In Section 12.4, we apply suffix trees to the task of speeding up the search for all PROSITE patterns in a single new protein sequence.

12.1   The PROSITE Database Format

Relative to the public DNA databases, PROSITE is small, consisting of two files with a total of about 5.4 Mb describing about 1400 motifs. The main file, `prosite.dat`, contains the information most suitable for automatic processing: the motif specification as well as accession numbers, an identifier, a one-line description, time stamps, and annotation of the precise location within the motif of the chemical bonds that typify its interactions with other molecules. The second file, `prosite.doc`, contains a free-form summary and bibliography for each motif. We will concentrate on reading the first file and then develop a package (similar to Chapter 6's sequence readers) for attaching to a file of PROSITE-style entries and retrieving patterns one by one.

We will create a new object for reading a PROSITE database by passing a string giving the filename of the database to the method Prosite—>new. Since this method is much simpler than SeqReader—>new in Chapter 6, we will not describe it in detail here. Method Prosite—>close is nearly identical to its analog in Chapter 6.

The interesting work is performed by Prosite—>readMotif, which reads the next entry from the database and collects various labeled parts of the entry into a hash. Each and every line in a PROSITE file is labeled with a two-letter code in columns 1 and 2, and these codes are the keys of the hash. The value attached to a given key is the concatenation of the lines labeled with that key. There are thirteen different codes, but we will pay attention to only four in this chapter.

1. An ID line begins each entry. Columns 3, 4, and 5 are always empty. Columns 6 and beyond contain a unique alphanumeric identifier for the entry followed by PATTERN, MATRIX, or RULE.

2. DE begins lines containing a one-line description of the entry.
3. PA lines contain patterns. A single motif entry may have more than one PA line if the pattern is too long to fit on a single line.
4. // marks the last line of each entry.

The first several lines in the database are usually a series of comment (CC) lines describing the entire database. The first entry begins with the first ID line. We can search ahead for this line and save its contents with the statements:

```
sub readMotif {
    my ($this) = @_;
    my $fh = $this->{fh};
    my $line;
    while (($line=<$fh>) && ($line !~ /^ID/)) { };
    return undef unless $line;
    my %entry;
    $entry{ID} = substr($line,5);
    ...
```

The caret ˆ in the pattern /ˆID/ is a special symbol for requiring that ID be the *first* two characters in the line for the pattern to succeed. If the file is exhausted before such a line is found, we return the undefined value to signal that no motif entries remain.

The following code reads the file line by line and then appends each line to the hash entry corresponding to its two-letter code. Since columns 3–5 are always empty, they are removed before storage. Every entry should end with //, but we also accept the end of the file as the end of an entry. A reference to the newly filled hash is the return value of the method.

```
sub readMotif {
    ...
    while (($line=<$fh>) && ($line !~ m|^//|)) {
        $line =~ s/(..).../;
        $entry{$1} .= $line;
    }
    return \%entry;
}
```

12.2   Patterns in PROSITE and Perl

PROSITE patterns consist of a series of elements separated by hyphens. Elements consist of a residue specification, possibly followed by a repetition specifier. The residue specification may take any of the following forms.

- A single upper-case letter from the abbreviations given in Figure 1.2. The letter specifies that exactly this residue must appear in the target string.
- The wildcard x (or X) matching any single residue.
- Square brackets enclosing two or more upper-case letters. The specifier [LIVM], for example, matches a single occurrence of any of the four residues L, I, V, or M.
- Curly braces enclosing two or more upper-case letters. The specifier {LIVM} matches a single occurrence of any residue *except* L, I, V, or M.

The repetition modifier may take either of two forms:

- if (x), the element matches exactly x successive residues meeting the residue specification;
- if (x,y), the element matches at least x and at most y successive residues.

The repetitions of a specification are independent in the sense that [LIVM](3) matches $4^3 = 64$ strings, including VML and ILM, not just the four strings LLL, III, VVV, and MMM. If no modifier is present, the element must match exactly one residue.

Finally, the entire series of elements may be preceded by < or followed by >, which respectively indicate that the motif must appear at the very beginning or very end of the protein.

In Perl, patterns are also called *regular expressions*. We have already seen a few simple Perl regular expressions. The simplest way for us to tackle PROSITE patterns in Perl is to (a) use Perl's string-handling facilities to translate the PROSITE pattern into a Perl pattern and then (b) use the Perl version in a Perl pattern-matching operation. In other words, we will use Perl's pattern matcher two ways: first to build a piece of Perl code – a Perl pattern – and then to execute that new piece of code.

Our overall attack will be to split the PROSITE pattern on – after stripping off < and >. Then we will translate each element in turn. Finally, we will re-join the translated elements to form the final Perl pattern.

```
sub perlizePattern {
    my ($this,$pattern) = @_;
    $pattern =~ s/\.$//;
    my ($left,$right);
    $left = '^' if ($pattern =~ s/^\<//);
    $right = '$' if ($pattern =~ s/\>$//);
    my $newpattern = join(", map(perlizeElement($_), split('-',$pattern)));
    return $left . $newpattern . $right;
}
```

We include the parameter $this because this subroutine will be included as a method in our package Prosite.pm.

To translate individual elements into Perl, we first separate the components specifying residue and repetition with the statement

---

my ($residue,$repetition) = ($elem =~ /^([^\(]*)(\(.*\))?$/);

When evaluated in a list context like this, a pattern-matching expression returns a list of the matched strings ($1,$2,$3,...). The statement above is therefore equivalent to

---

$elem =~ /^([^\(]*)(\(.*\))?$/;
my ($residue,$repetition) = ($1,$2);

Let's look at the pattern in some detail. [^\(] matches every character except the left parenthesis. The backslash \ signals that the following parenthesis loses its usual special meaning as a grouping device within patterns; it is "just another character". The asterisk * signals that the preceeding item can be repeated zero or more times. The parentheses wrapped around ([^\(]*) have their familiar function; they cause the substring matched to be saved in $1. So everything in $elem up to the first left parenthesis will be assigned to $1 and eventually $residue.

The second half of the pattern is devoted to the repetition. \(.*\) matches a left parenthesis, any number of any character, and a right parenthesis. The surrounding parentheses in (\(.*\))? cause the repetition specification to be saved in $2, and the question mark indicates that the preceding (grouped) item is optional. The initial caret and final dollar sign force the pattern to match the *entire* string representing the PROSITE pattern element.

Once these two components have been separated, it is a simple matter to change Prosite's x, { }, and ( ) to the ., [^], and {} required in the equivalent Perl pattern.

```
sub perlizeElement {
    my ($elem) = @_;
    my ($residue,$repetition) = ($elem =~ /^([^\(]*)(\(.*\))?$/);
    $repetition =~ tr |()|{}|;
    $residue =~ tr/xX/../;
    return "[^$1]$repetition" if $residue =~ /\{(.*)\}/;
    return "$residue$repetition";
}
```

A short driver program named `prosite.pl` is included at the end of this chapter. It uses `Prosite.pm` to open a PROSITE database, read its patterns, translate them into Perl, and search for each of them in a sequence file.

12.3   Suffix Trees

If there is anything Perl does well, it is string searches. Its string-matching engine is highly optimized. All the same, the time it takes to search for a pattern in a target string depends on the length of the target; doubling the target doubles the search time. And, in the case of a PROSITE search, the entire target string must be re-examined for each of PROSITE's 1400 or so patterns. In this section, we develop a method that allows a string to be analyzed once so that the presence or absence of a query string can be determined in time proportional to the length of the *query* string rather than the length of the much longer target string. Of course, the preprocessing requires time enough to look at the entire target string. But once preprocessing is completed, the resulting analysis can be used over and over to search for as many patterns as desired. In the following section, we will extend this method to search for PROSITE patterns instead of simple strings.[2]

The *suffix tree* is an ideal data structure for quick identification of substrings of a fixed target string. A suffix tree can be constructed in space and time proportional to the target string's length. Once constructed, it can be used repeatedly to determine whether query strings are suffixes or substrings of the target string in time proportional to the length of the query and independent of the target string.

Suffix trees provide an example of *digital* searching. Despite their name, digital techniques are not limited to numbers; they exploit the binary nature of all computer data representations by dividing search keys into "digits" consisting of one or more bits and using individual digits as indices to access arrays. In this way, digital techniques are able to make many-way decisions in constant time, in contrast to *comparative* techniques (such as binary search trees) with decisions based only on trichotomous comparisons.[3]

Our exposition and programming are simplified if we assume that each string is terminated by a special "end-of-string" character that cannot appear elsewhere in any string. In our examples, we will assume that strings are composed of lower-case letters a through z and terminated by a semicolon ;. Figure 12.1 depicts a suffix tree for the string banana and illustrates the following general properties of suffix trees.

1. Each node of the tree contains a nonempty substring of the target string. (The root is the single exception; it always contains the empty string.)
2. If we consider any path from the root to a leaf, the concatenation of the substrings contained in the nodes on the path form a suffix of the target string. Conversely, every suffix can be formed by concatenating the substrings on some root-to-leaf path. If we say that a node *represents* the substring formed by concatenating the substrings on the path from the root to the node, then every leaf represents a suffix and every suffix is represented by some leaf.
3. Any two children of the same parent have substrings that begin with different letters.
4. Every internal node has at least two children.

---

[2] Experienced Perl programmers may be thinking that this is just what Perl's **study** operator does. In a few casual experiments, however, using **study** had no measurable effect on the speed of PROSITE searching, while in others it sent the pattern-matcher into an apparent infinite loop.

[3] Comparisons of entire keys with one of three results: "less", "equal", or "greater".

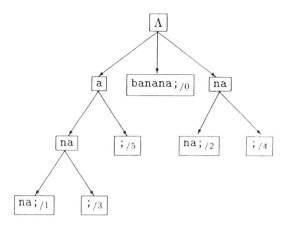

Figure 12.1: Suffix tree for banana.

5. Every leaf node is labeled with the starting position of the suffix it represents, and the node's substring ends in ; .

Properties 1 and 2 make it possible to determine whether a particular query string is a substring of the target string by searching for a matching path from the root. Property 3 makes it possible to do so in time linear in the query's length by eliminating the need to explore more than one alternative path. Property 4 ensures that the memory used to store the suffix tree is linear in the target string's length. Property 5 means that the set of all starting points of a given substring can be collected from the set of leaves descended from the node representing the substring in time linear in the number of starting points.

To search for a query string, we start at the root and then repeatedly descend from parent to child. At each step of the descent, we advance from the current node (if possible) to the single child beginning with the same letter as the query string, then delete from the query string (if possible) the new node's substring. The descent halts when any one of the following occurs.

- We simultaneously reach end-of-string characters in the query and the target, signaling that the query is a suffix of the target.
- We reach the end of the query string but not the target string, signaling that the query is a substring (prefix of a suffix) of the target but not a suffix.
- We find no child beginning with the same letter as the query string, signaling that the query is not a substring of the target.
- The substring in the node is not a prefix of the query string and therefore cannot be deleted. This, too, signals that the query is not a substring of the target.

Using the object-oriented techniques outlined in Chapter 6, we represent the suffix tree in its entirety by a hash reference blessed into the class SuffixTree. Suffix tree methods will therefore be implemented in the file SuffixTree.pm. To avoid a proliferation of files, individual nodes in the tree will be represented by unblessed hash references, although it would be possible to create another module for them.

The blessed hashes representing suffix tree objects will have two hash keys, `target` and `root`. The value corresponding to `target` is the entire target string whose suffixes are stored in the tree. The hash reference representing the node at the root of the tree will be stored under key `root`.

The unblessed hashes for individual nodes will all have the two hash keys `off` and `len` identifying the substring of the target that the node "stores" by starting point (offset) in the target and length. This technique requires only constant space to record any substring of any length and linear space to record all substrings in all nodes; if substrings were stored explicitly in each node, the resulting redundancy could consume quadratic space. Leaf nodes have the key `startAt` giving the starting position of the suffix represented by the path from the root to the leaf. The hash also contains a key–value pair for each child of the node; the key is the first letter of the substring in the child node, and the associated value is the hash reference representing the node. Thus, a suffix tree for `banana` is constructed explicitly by the following expression:

```
bless
{target=>'banana;',
root=>
    {off=>0, len=>0,                                    # empty substring
    'a'=>{off=>1, len=>1,                               # substring 'a'
        'n'=>{off=>2, len=>2,                           # substring 'na'
            'n'=>{off=>4, len=>3, startAt=>1},          # substring 'na;'
            ';'=>{off=>6, len=>1, startAt=>3}           # substring ';'
            },
        ';'=>{off=>6, len=>1, startAt=>5}               # substring ';'
    },
    'b'=>{off=>0, len=>7, startAt=>0},                  # substring 'banana;'
    'n'=>{off=>2, len=>2,                               # substring 'na'
        'n'=>{off=>4, len=>3, startAt=>2},              # substring 'na;'
        ';'=>{off=>6, len=>1, startAt=>4}               # substring ';'
    },
    ';'=>{off=>6, len=>1, startAt=>6}                   # substring ';'
    }
},
"SuffixTree";
```

Under this scheme of representation, it is not too difficult to implement the search procedure described above.

```
sub isSubstring {
    my ($this,$query) = @_;
    $query = lc $query;
    my $p = $this->{root};
```

```
    while ($query) {
        $p = $$p{substr($query,0,1)};
        return 0 if !$p;
        my $patLen = min($$p{len}, length $query);
        my $pStr = substr($this->{target},$$p{off},$patLen);
        return 0 if $query !~ s/^$pStr//;
    }
    return $p;
}
```

As usual, the parameter $this is the hash reference blessed into the class SuffixTree. The string to be searched for is stored in $query. The **while**-loop begins with the current node $p at the root and continues until the query string is totally consumed. In each iteration of the **while**-loop, the first remaining character of the query is used as a hash key in $p to find the correct child to descend to. If no child is found, 0 is returned. The substring stored implicitly in node $p is extracted from the target and stored in $pStr. If the query is too short to match the entire target substring, then a shorter version is extracted. If $pStr matches the beginning (or all) of $query, it is deleted from $query; otherwise, 0 is returned. Finally, if the descent consumes the entire query then the loop halts and the node $p is returned.

The advantage of returning $p instead of just 1 is that we can, if desired, collect the set of starting points of the query substring. An easy way to do this is by maintaining a queue of descendents of the node representing the query and to repeatedly replace one internal node in the queue with its children.

```
sub gatherStartPoints {
    my ($this,$p) = @_;
    my @results;
    my @pending = ($p);
    while (@pending) {
        $p = shift @pending;
        push @results, $$p{startAt} if defined($$p{startAt});
        push @pending, map($$p{$_}, grep(/^.$/, keys %$p));
    }
    return @results;
}
```

Of course, we do not generally want to construct suffix trees by hand and then key them in as a set of nested hashes. Instead, we want to create suffix trees by statements like

```
my $t = SuffixTree->new("banana");
```

A naive algorithm for constructing suffix trees works from left to right in the target string. To process the string banana, it first adds the one suffix of b to the tree, then the two suffixes of ba (i.e., ba and a), then the three suffixes of ban, and so forth. In the jth stage, $j - 1$ suffixes are formed by adding the jth letter to the suffixes of the previous stage, and another suffix is formed of the jth letter all by itself. The process is expressed very roughly by the following Perl-like pseudocode:

```
foreach my $j (0..length($target)−1) {
    foreach my $i (0..$j) {
        add substr($target,$i,$j−$i+1) to tree;
    }
}
```

The difficulty with this code is that its running time is proportional to the *square* of the length of the target, *even if* a single substring can be added in constant time. Fortunately, a surprising series of shortcuts allows us to achieve linear running time for the insertion of all suffixes combined – even though a single substring *cannot* be added in constant time!

The first shortcut arises when we discover that a path for the current substring already exists in the tree. For example, suppose we are building a suffix tree for the string hodgepodge and have advanced to suffixes ending at the second g (i.e., to a value of 8 for j). When i is 6, our task is to add the substring odg. We discover, however, that no change to the tree is necessary because odg has already been entered into the tree – it happened back when j was 3 and i was 1. This discovery has even broader implications: if odg was added when j was 3, then dg and g (the next two substrings on our agenda) were also added back when j was 3. We can thus skip dg and g and go on to the suffixes ending with the final e. In fact, this is generally true: If any substring is found to be already present in the tree, then the order of insertion guarantees that all suffixes of the substring are also present; the inner loop can be short-circuited and j can be advanced. So we can modify our pseudocode to read:

```
foreach my $j (0..length($target)−1) {
    foreach my $i (0..$j) {
        add substr($target,$i,$j−$i+1) to tree;
        last if substr($target,$i,$j−$i+1) was already present;
    }
}
```

The next shortcut comes about when change *is* required in the tree to accommodate a new substring. Although details vary, we will soon see that these changes always involve the creation of a new child containing exactly the last letter of the substring being inserted, that is, substr($target,$j−1,1). But at this point, it is very easy – and consistent with our ultimate goal – to add the *entire* suffix beginning at position $i to the new child node. If we add the entire suffix immediately then we can remove starting position $i from further consideration, like this:

```
my $first = 0;
foreach my $j (0..length($target)−1) {
    foreach my $i ($first..$j) {
        add substr($target,$i,$j−$i+1) to tree;
        if (substr($target,$i,$j−$i+1) was already present) {
            last;    ## exit inner loop
        } else {
            add substr($target, $i, length($target)−$i) to tree;
            $first = $i+1;
        }
    }
}
```

Adding the whole suffix to the new node takes only constant time, because we store the suffix implicitly in the node by setting two indices rather than by copying its characters! If we had to add k letters one by one, we would need time proportional to k.

Now every execution of the body of the inner loop eliminates one value of either $i or $j (and sometimes both) from further consideration. The quantity $first+$j is 0 just before the first execution of the body of the inner loop, and it increases by at least 1 (sometimes 2) during each execution. The body cannot execute when either $j or $first is greater than the length of the target string. Therefore, the number of times the body can execute is no more than twice the number of letters in the target string. For the seven-character string banana, the body of the loop executes only ten times, as summarized in the following table:

$j	$i	substring	action
0	0	b	insert banana; and increase $first
1	1	a	insert anana; and increase $first
2	2	n	insert nana; and increase $first
3	3	a	increase $j
4	3	an	increase $j
5	3	ana	increase $j
6	3	ana;	insert ana; and increase $first
6	4	na;	insert na; and increase $first
6	5	a;	insert a; and increase $first
6	6	;	insert ; and increase $first

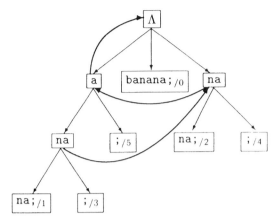

Figure 12.2: Suffix tree for banana with suffix links.

It might now seem that we can achieve linear time overall only if each substring can be added in constant time. Actually, we are able to achieve linear time overall without such a tight bound on the individual additions. To achieve this we add *suffix links,* which allow us to avoid beginning our search at the root for every addition.

12.3.1 Suffix Links

Suffix links satisfy the following properties.

1. Only internal nodes (i.e., nodes with children) have suffix links, and every internal node except the root has exactly one suffix link.
2. If an internal node represents the substring $a_1a_2a_3 \ldots a_k$, then the node at the other end of its suffix link represents $a_2a_3 \ldots a_k$.
3. Suffix links always point to other internal nodes. For suppose that a node represents $a_1a_2a_3 \ldots a_k$ and is internal. Then there must be at least two letters b_1 and b_2 for which $a_1a_2a_3 \ldots a_kb_1$ and $a_1a_2a_3 \ldots a_kb_2$ are in the tree. But then their suffixes $a_2a_3 \ldots a_kb_1$ and $a_2a_3 \ldots a_kb_2$ must also be in the tree, and the node representing $a_2a_3 \ldots a_k$ must also be internal. As an example, consider the string household. The node representing ho must be internal, since both household and hold are suffixes. It follows that the node representing o must also be internal in order to accommodate the suffixes ousehold and old.
4. The "mapping" defined by suffix links "preserves ancestry". In less highfalutin terms, if p is an ancestor of q, then p's suffix link is an ancestor of q's suffix link. This holds because p's substring is a prefix of q's string, and this continues to be true when the first letter is dropped from each string. household again provides an example. The node for ho is an ancestor of the nodes for hold and household, so it follows that the node for o is an ancestor of those for old and ousehold.

Figure 12.2 shows a suffix tree with suffix links.

In our implementation, the suffix link of a node is stored in its hash under the key `sfx`. Suffix links are required for fast construction of suffix trees, but they make no contribution to subsequent searches. Our implementation allows them to be removed easily after construction if space is at a premium.

Our tree-building method, SuffixTree−>new, takes the following form:

```
sub new {
    my ($this,$target) = @_;
    my $root = {off=>0,len=>0};
    $$root{sfx} = $root;
    $target = lc $target . ";";
    my ($i,$j,$p,$pDepth,$needSfx) = (0,0,$root,0,0);
    while ($j<length $target) {
        ($i,$j,$p,$pDepth,$needSfx)
            = add($target,$i,$j,$p,$pDepth,$needSfx);
        $pDepth = 0 if $pDepth < 0;
        $needSfx = undef if $needSfx == $root;
        $j++ if $i > $j;
    }
    return bless {target=>$target, root=>$root};
}
```

The loop calls subroutine add to make sure that each added suffix is included in the tree. This subroutine is invoked with the target string and five other values; it returns a list of five new values to be passed to the next invocation.

The first two of these values, $i and $j, are simply the starting and ending points of the substring within the target. Although the loop structure looks different than in our earlier pseudocode, add is in fact invoked with precisely the same values of $i and $j. Subroutine add itself decides whether to increase $i (if the tree had to be modified) or $j (if the substring was already present) and includes the updated values in the list it returns. The last statement in the **while**-loop updates $j in situations corresponding to the termination of the inner loop in the pseudocode.

The third and fourth values, $p and $pDepth, describe the location in the tree where the search for the place to add the next substring should commence. Subroutine add is able to determine this quickly using suffix links. Node $p is where the search begins, and integer $pDepth is the number of letters of the substring already accounted for by the path from the root to $p.

Property 3 of suffix links states that, in a complete suffix tree, every internal node's suffix link points to another internal node. However, during construction, we may find that a newly created internal node represents a substring whose suffix is not yet represented by an internal node. In this case, we are guaranteed that the required node will be created in the algorithm's very next step. The fifth and final value in the list passed to and returned by add is $needSfx, a pointer to the single node in the tree lacking its proper suffix link. This node's suffix link is always filled in on the very next call to add. $needSfx is undefined when no node lacks a suffix link.

Subroutine add begins by picking up the search for the substring at the node where it was left off on the last call to add. Nodes $p, $q, and $r are (respectively) the current node, its parent, and its grandparent, if all exist. If $p is advanced off the tree, then the other two variables can be used to regain a foothold. $sLeft gives the length of the portion of the suffix that is not accounted for by the substrings on the path from the root to $p. At the end of this stage, we have descended as far as possible into the tree on the current string.

```
sub add {
    my ($target,      ## the target string
        $i,           ## the position in $target at which the substring starts
        $j,           ## the position in $target at which the substring ends
        $p,           ## the suffix tree node from which searching is to begin
        $pDepth,      ## number of letters in path from root through $p's parent
        $needSfx)     ## new internal node waiting for a suffix pointer, if any
        = @_;
    my $sLeft = ($j+1−$i)−$pDepth;       ## length of unmatched portion
    my $sLetter = substr($target,$j,1);   ## new letter to append
    my ($q,$r,$pLen);
    while ($p && ($sLeft>($pLen=$$p{len}))){     ## go down as far as possible
        ($pDepth,$sLeft) = ($pDepth+$pLen,$sLeft−$pLen);
        ($r,$q,$p) = ($q,$p,$p−>{substr($target,$i+$pDepth,1)});
    }
    my $qDepth = $pDepth−$$q{len};       ## no. of letters, root to $q's parent
    ...
}
```

It is important to note that we need not check every letter during this search. The order of insertion guarantees that the current suffix minus its last letter is already in the tree somewhere; we only need to look at the one letter per node that determines the correct path from parent to child.

Next, subroutine add examines both node $p and the substring to be added to determine how the tree must be modified, if at all. A skeleton of the conditional statements that distinguish among the three cases is shown below; our next step is to deal separately with each of these cases.

```
sub add {
    ...
    if (!$p) {     ## substring is absent; we fell off the tree
        ...
```

```
    } else {       ## we're still on the tree, but the last letter may mismatch
        my ($pOff,$pLen) = ($$p{off},$$p{len});
        my $pLetter = substr($target,$pOff+$sLeft−1,1);
        if ($sLetter eq $pLetter) {       ## substring is already present
            ...
        } else {       ## substring is absent; we must split an existing node
            ...
        }
    }
}
}
```

The first case occurs when an existing node, $q, precisely represents all but the very last letter of the suffix. In this case, the search pushes $p off the tree because $p−>{substr($target,$i+$pDepth,1)} does not exist. Node $q receives a new child to represent the suffix, extending all the way to the end of the target string. We are finished with the current starting point, so we return an increased value of $i but do not change $j. If $q already has a suffix link, we can use it to start the next search. The depth of the suffix link is one less than $q's depth $qDepth, because its string omits the first letter of $q's string. If the new node is $q's first child then $q has no suffix link and we must move up two nodes to $q's parent $r, with depth $qDepth−$$r{len}, to find a suffix link for beginning the next search. We include $q in the return list to have its suffix link assigned on the next call to add.

```
    $$q{$sLetter} = {off=>$j, len=>(length($target)−$j), startAt=>$i};
    $$needSfx{sfx} = $q if ref($needSfx);
    if ($$q{sfx}) {       ## parent already internal node
        return ($i+1, $j, $$q{sfx}, $qDepth−1, undef);
    } else {       ## parent becomes internal node
        return ($i+1, $j, $$r{sfx}, $qDepth−$$r{len}−1, $q);
    }
```

The second case occurs when the suffix is already present at node $p. In this case, we increase $j and report that no suffix links are currently lacking. The next search starts right where this one left off, since we are not dropping a letter from the left end of the search string:

```
    $$needSfx{sfx} = $q if ref($needSfx);
    return ($i, $j+1, $q, $qDepth, undef);
```

The last case is the most complicated. It occurs when the string represented by node $p is longer than the suffix and the last letter of the suffix disagrees with $p's string. In this case, $p's string must be broken into two pieces assigned to two separate nodes: one for the part matching the suffix and one for the unmatched remainder. The node for the common part has two children, one for the old string and one for the new suffix. Naturally, we extend the suffix to the end of the target string; correspondingly, we increase $i to signal that we are finished with this starting point. The next search begins from the suffix link of the parent $q of the split node.

```
## must split node to add suffix;
## one new node contains the rest of this suffix
my $new1 = {off=>$j, len=>(length($target)−$j), startAt=>$i};

## the old node retains the latter part of its original key
$$p{off} = $pOff+$sLeft−1;
$$p{len} = $pLen−$sLeft+1;

## a second new node contains common start of key and substring;
## it points to the two differing endings
my $new2 = {off=>$pOff, len=>$sLeft−1};
$$new2{$pLetter} = $p;
$$new2{$sLetter} = $new1;

## parent $q of original node $p must now point to new one
$q−>{substr($target,$i+$pDepth,1)} = $new2;
$$needSfx{sfx} = $new2 if ref($needSfx);

## we've put whole suffix in a new node; we're done
return ($i+1, $j, $$q{sfx}, $qDepth−1, $new2);
```

In all three cases, we take care to update suffix links if $needSfx is defined.

12.3.2   The Efficiency of Adding

We must now convince ourselves that the time required for all calls to add together is only linear in the length of the target string. This point is not obvious, because the **while**-loop in add makes it clear that individual calls may take more than constant time. We must therefore find a way to show that the total number of steps taken through the tree along child and suffix links is at most linear.

Let n be the length of the target string. We have already shown that the number of calls to add is no more than $2n$. We have previously used the term *depth* to refer to the length of the string along the path from the root to a given node. Now, we will refer to the number of *nodes* along that path as the *level*. Level n is achieved only if the n letters of the target are identical, and so it should be clear that no node has level greater than n. The first seach begins at level 0. Suppose that, on the ith call to add, $p begins at level L_i and ends at level L_i'. Then the ith call takes $(L_i' - L_i)$ steps following child links plus 1 step following a suffix link. If add is called m times, the total number of steps is

$$(L_1' - L_1 + 1) + (L_2' - L_2 + 1) + \cdots + (L_{m-1}' - L_{m-1} + 1) + (L_m' - L_m + 1)$$

$$= (m \cdot 1) + (L_1' - L_1) + (L_2' - L_2) + \cdots + (L_{m-1}' - L_{m-1}) + (L_m' - L_m)$$

$$= m - L_1 + (L_1' - L_2) + (L_2' - L_3) + \cdots + (L_{m-1}' - L_m) + L_m'$$

$$= m - L_1 + L_m' + \sum_{i=1}^{m-1} (L_i' - L_{i+1})$$

$$\leq m - 0 + n + \sum_{i=1}^{m-1} (L_i' - L_{i+1}).$$

Now $(L_i' - L_{i+1})$ is just the difference in level between where one search ends and the next begins. Since we find the starting point of a search by following a suffix link from either the parent or the grandparent of the node where the previous search ends, this difference is at most 2 plus the decrease from following the suffix link. The next surprise is that following a suffix link decreases level by at most 1. The comment accompanying Property 3 of suffix links provides the explanation. If the string $a_1 a_2 a_3 \ldots a_k$ must be split among several nodes because some of its prefixes are shared with other strings in the tree, then string $a_2 a_3 \ldots a_k$ must be split in the same places to accommodate the suffixes of those prefixes. For example, since hotdoghotrod must be divided as hot + doghotrod to accommodate hotrod in the tree, its suffix otdoghotrod must be similarly divided into ot + doghotrod to accommodate otrod. There is only one place that $a_1 a_2 a_3 \ldots a_k$ could be split that $a_2 a_3 \ldots a_k$ is not: between a_1 and a_2. On the other hand, $a_2 a_3 \ldots a_k$ could be broken in many places that $a_1 a_2 a_3 \ldots a_k$ is not. For example, house and ouse are suffixes of outhouse; house is at level 1, but ouse = ou + se is at level 2 to accomodate ou + thouse.

We have just shown that $(L_i' - L_{i+1}) \leq 2 + 1 = 3$. So the total number of steps is at most

$$3n + \sum_{i=1}^{m-1} 3 \leq 3(m + n) \leq 3(2n + n) \leq 9n,$$

which is linear in n as we had wished.

12.4 Suffix Trees for PROSITE Searching

In the previous section, we developed a package for a data structure that allows us to search quickly for substrings of a fixed target string. For our PROSITE application, however, we need to be able to search not for substrings given explicitly but rather for substrings specified implicitly by a PROSITE pattern. Fortunately, it is not too hard to modify our searching procedure to explore multiple paths that partially match our pattern.

First, we break the pattern into its pieces and form a linked list. Next, we write a recursive subroutine for searching the suffix tree for substrings matching the pattern. We use a recursive control structure because we may need to continue searching on more than one descendent of a given node.

Since suffix trees have a large number of applications that do not involve PROSITE patterns, we will not add the new search methods to SuffixTree.pm. Instead, we create a new package PrositeSuffixTree.pm that inherits all the methods of SuffixTree.pm. The header of this file is:

```
package PrositeSuffixTree;
require SuffixTree;
@ISA = qw(SuffixTree);
use strict;
```

The first step in the search process is to break the pattern into its pieces and form a linked list. For simplicity, we will not handle variable-length elements like [LIVM](5,9); only about one sixth of PROSITE patterns include these. We will treat < and > as just another letter, so we separate them from their neighbors with -. We split the elements from each other on -. Then we break the individual element into residue and repetition by calling pairForm. The pairs are put into a linked list. For example, when searchForPattern is called for the PROSITE "N-myristoylation site" pattern

G-{EDRKHPFYW}-x(2)-[STAGCN]-{P},

it is transformed into the Perl linked list

['g',1, ['[^edrkhpfyw]',1, ['.',2, ['[stagcn]',1, ['[^p]',1, []]]]]].

This linked list is formed out of five three-element lists. The first two entries in each are residue and repetition specifications. The third entry is a reference to another three-element list. (The last three-element list has a reference to an empty list.) This representation was chosen to avoid repeated copying of parts of the pattern; instead, a single reference can be included in more than one place.[4]

---

[4] The notation Prosite::perlizeElement is new; it means to call the subroutine perlizeElement in the package Prosite. It differs from Prosite->perlizeElement in that it does *not* shift 'Prosite' into @_ before calling perlizeElement. We could have omitted Prosite:: altogther by adding use Prosite; to the beginning of PrositeSuffixTree.pm, as we have frequently done with Util.pm.

```perl
sub searchForPattern {
    my ($this,$pattern) = @_;
    $pattern =~ s/\n//gs;      ## remove newlines
    $pattern =~ s/\.$//;
    $pattern =~ s/^</<-/;
    $pattern =~ s/>$/->/;
    my $patLinks=[];
    foreach (reverse split("-",$pattern)) {
        $patLinks = [pairForm(Prosite::perlizeElement($_)),$patLinks];
    }
    return ($this->searchHelp($this->{root},$patLinks));
}

sub pairForm {
    my ($perlPatElem) = @_;
    my ($spec,$rep) = ($perlPatElem =~ /^([^\{]*)(\{.*\})?$/);
    $rep =~ s/[\{\}]//g;
    $rep ||= 1;
    return ($spec,$rep);
}
```

Method searchForPattern calls searchHelp with a reference to the root node and the preprocessed pattern. searchHelp copes with the need to explore more than one path by calling itself recursively on some number of children of the current node $node.

The first task of searchHelp, however, is to determine whether the substring stored in the current node conforms to the pattern. The target substring is retrieved and matched against successive elements of the pattern list until either the match fails, the substring is exhausted, or the pattern is exhausted. In the case of a mismatch, searchHelp returns immediately, aborting search on this path. If the pattern is exhausted then the search has been successful, and the target indices corresponding to this node are gathered and returned as the value of searchHelp. If the substring is exhausted without exhausting the pattern, further exploration among the children of the current node is required.

To match a pattern element of $rep repetitions of residues satisfying $spec, we construct a Perl pattern with {}. If the match succeeds, we can proceed to the next pattern element. However, there are two different reasons it could fail. If the substring contains residues that did not match the pattern, we should abort the search. On the other hand, it could be that the residues present matched $spec but that the substring was too short to match $rep times. To check for this latter case, we keep track of the length of the substring in $ssLen and construct a pattern that matches $ssLen repetitions of $spec when the first match fails. If the second match fails, we abort the search. If it succeeds, we will need to look for ($rep−$ssLen) repetitions of $spec in the children of the current node. So we add this requirement to $patLinks in a new link at the beginning of the linked list.

```perl
sub searchHelp {
    my ($this,$node,$patLinks) = @_;
    my $ssLen = $$node{len};
    my $ss = substr($this->{target},$$node{off},$ssLen);
#   print "At node $node representing $ss \n";
    my ($spec,$rep);
    while ($ss && @$patLinks) {
        ($spec,$rep,$patLinks) = @$patLinks;
        my $pat = "$spec\{$rep\}";
#       print "Comparing $ss to $pat \n";
        if ($ss !~ s/^$pat//) {
            return () if ($ssLen>$rep);
            my $pat = "$spec\{$ssLen\}";
            return () if ($ss !~ s/^$pat//);
        }
        $ssLen -= $rep;
    }
    ## reached end of substring for this node
    $patLinks = [$spec,-$ssLen,$patLinks] if $ssLen < 0;
    if (@$patLinks==0) { return ($this->gatherStartPoints($node)); }
    ...
```

Next, we must sort through the children of the current node to determine which ones warrant further exploration by a recursive call. If the next residue specification is a positive one listing the satisfactory residues, then the easiest approach is to split this specification into individual residues and to make recursive calls for any residue for which the current node has children. However, if the specification is negative or a wildcard, then it is easier to begin with a list of the keys of the current node and winnow out those that do not meet the specification.

```perl
sub searchHelp {
    ...
    $spec = $$patLinks[0];
    if ($spec =~ /[\^\.]/) {
        return map(($this->searchHelp($$node{$_},$patLinks)),
                   grep(/^$spec$/,(keys %$node)));
    }
```

```
   else {
      return map(($this->searchHelp($$node{$_},$patLinks)),
                      grep($$node{$_},(split //,$spec));
   }
}
```

To print a list of all the occurrences of substrings matching a PROSITE motif entry from the database, we can use the following driver:

```
sub listMotifs {
    my ($this,$motif) = @_;
    my $pattern = lc $$motif{PA};
    return unless $pattern;
    my $description = $$motif{DE};
    $description =~ s/\n//gs;    ## remove newlines
    return "Can't handle $description\n" if $pattern =~ /\,/;
    foreach ($this->searchForPattern($pattern)) {
        print "$description seen at position $_\n";
    }
}
```

Our strategy for conducting a PROSITE analysis of a protein sequence is to construct a suffix tree for the sequence and then search the suffix tree for each of the patterns. The search may take time greater than the length of the pattern (e.g., if every possible string matching the pattern occurs somewhere in the target!), but the depth of search in the tree is no greater than the maximum length of the pattern, and deep searches occur only when the pattern is partially matched.

If we have a *set* of protein sequences to be analyzed, then we could repeat the procedure of the last paragraph for each sequence. But we could also form a single suffix tree containing all the sequences! The easiest way to do this is simply to build a suffix tree of the concatenation of the individual sequences. Then we can search for a single motif in *all* of the sequences with a single call to searchForPattern. It is this approach that is taken in the driver program `protree.pl`.

12.5   Exercises

1. The International Union of Pure and Applied Chemistry (IUPAC) has assigned the following interpretations to four of the five letters that remained after one-letter codes were assigned to the 21 natural amino acids:

```
B:   D or N
X:   any
J:   I or L
Z:   E or Q
```

At present, only X is found in PROSITE. Modify perlizePattern to permit the use of the other four letters in PROSITE patterns.

2. Write a subroutine patternCardinality that determines the number of different strings that match a given PROSITE pattern. On the first pass, you may ignore factors contributed by variable repetition modifiers like (2, 9). When you include these, remember that a single substring may match a pattern in two or more ways; ABABA matches the pattern x(0,10)-A-x(0,10) in three different ways.

3. The motif searching program based on suffix trees gives the position where the pattern matches in the concatenation of the protein sequences. Improve the program to report the name of the individual protein matched and the position relative to the beginning of the individual sequence. To accomplish this efficiently, it may help to build two lists as the proteins are read in: one containing the names of the sequences, and one containing their lengths.

4. Can the suffix tree package be extended to allow more strings to be added after the initial string is constructed? Remember that we are not really interested in substrings of the long, concatenated string involving pieces of more than one of the individual input sequences (i.e., substrings containing "><").

5. Extend the motif searching program based on suffix trees to print out the substrings it has matched (like the original motif searching program) rather than simply giving the starting position.

6. Extend the motif searching program based on suffix trees to handle PROSITE patterns with variable-length elements like "{LIVM}(8,11)".

7. Since every child of the current node meets the wildcard specification, couldn't we save a little time by changing the end of searchHelp to read as follows?

```
sub searchHelp {
    ...
    $spec = $$patLinks[0];
    if ($spec eq '.') {
        return map(($this->searchHelp($$node{$_},$patLinks)),
                (keys %$node));
    }
    elsif ($spec =~ /[\^\.]/) {
        return map(($this->searchHelp($$node{$_},$patLinks)),
                grep(/^$spec$/,(keys %$node));
```

```
    }
    else {
        return map(($this−>searchHelp($$node{$_},$patLinks)),
                    grep($$node{$_},(split //,$spec)));
    }
}
```

8. Draw suffix trees for the following strings:
   (a) `biostatisticians`
   (b) `hocuspocus`
   (c) `hobsonjobson`[5]
   (d) `mississippi`
   (e) `wooloomooloo`
   (f) `householder`

9. Method SuffixTree−>isSubstring returns true if and only if its argument is a
   substring of the target string. Modify it to return true if and only if its argument
   is a *suffix* of the target.

12.6 Complete Program Listings

The software distribution includes files `Prosite.pm` and `prosite.pl` for motif
matching with Perl patterns, as well as files `SuffixTree.pm`, `SuffixTree.pm`,
`PrositeSuffixTree.pm`, and `protree.pl` for motif matching with suffix trees.

12.7 Bibliographic Notes

Extensive coverage of suffix trees and the various algorithms for their construction
(Wiener's, McCreight's, and Ukkonen's) can be found in Gusfield's (1997) excellent
book.

A recent description of the PROSITE database is found in Hoffmann et al. (1999).

---

[5] Both "hocus-pocus" and "hobson-jobson" are deformations of phrases from religious ceremonies.

## CHAPTER THIRTEEN

Fragment Assembly

In Chapter 3, we introduced the sequence assembly problem and discussed how to determine whether two fragments of a DNA sequence "overlap" when each may contain errors. In this chapter, we will consider how to use overlap information for a set of fragments to reassemble the original sequence. Of course, there is an infinite number of sequences that could give rise to a particular set of fragments. Our goal is to produce a reconstruction of the unknown source sequence that is reasonably likely to approximate the actual source well.

In Section 13.1 we will consider a model known as the *shortest common superstring problem*. This simple model assumes that our fragments contain no errors and that a short source sequence is more likely than a longer one. If the input set reflects the source exactly, then only exact overlaps between fragments are relevant to reconstruction. This approach's focus on exact matching allows us to use once more that extremely versatile data structure from Chapter 12, the suffix tree.

Beginning in Section 13.2, we consider another solution, a simplified version of the commonly used PHRAP program.[1] PHRAP assumes that errors are present in the sequence files and furthermore that each base of each sequence is annotated with a *quality value* describing the level of confidence the sequencing machinery had in its determination of the base. It will not be possible to present all of the issues and options dealt with by PHRAP, but in Sections 13.3–13.7 we will present several important aspects of its approach in a Perl "mini-PHRAP".

13.1 Shortest Common Superstrings

The shortest common superstring problem has as input an arbitrary number of strings called *fragments*. The fragments are assumed to be exact substrings of some unknown larger string, the *target*. The goal is to reconstruct the target by finding the *shortest* string that contains each of the fragments as a substring. The problem is easy if no

---

[1] "PHRAP" stands for *Phil [Green]'s Revised Assembly Program.*

two input strings overlap; then any concatenation of the fragments is an acceptable solution. Two strings *overlap* if some nonempty prefix of one is identical to a suffix of the other; `dollhouse` and `household` have a five-letter overlap (`house`) in one "direction" and a one-letter overlap (d) in the other. However, if many different pairs of fragments overlap, then it is more difficult to decide which overlaps to exploit in forming a short superstring.

For example, suppose the input set is

`{maryhadalit, ittlelam, lelambits, itsfleecewaswhit, ecewaswhite, iteassnow}`

The reconstruction

`itsfleecewaswhittlelambitsmaryhadaliteassnowecewaswhite`

exploits three different overlaps in the input set. But a closer look (and some knowledge of *Mother Goose*) helps us to find a shorter superstring,

`maryhadalittlelambitsfleecewaswhiteassnow`

If one of the fragments is a substring of another, then the presence of the smaller string in the input set will have no influence on the final solution. On the other hand, allowing such fragments makes for a number of special cases when programming a solution to the problem. Therefore, we will make the assumption that the input set is "substring free". It is not difficult to filter out unwanted substrings at the beginning by using suffix trees.

Let the input set be $\{f_1, f_2, \ldots, f_k\}$. One way to describe a candidate solution is to give for each fragment f_i the range of indices $[l_i, r_i]$ at which it can be found in the superstring. The substring-free condition means that neither $l_i = l_j$ nor $r_i = r_j$ unless $i = j$; furthermore, $l_i < l_j$ if and only if $r_i < r_j$. This means that we can say "f_i lies to the left of f_j" as a shorthand for "$l_i < l_j$ and $r_i < r_j$".

Given a fixed left-to-right ordering of the f_i – for example, $(f_1, f_2, f_3, \ldots, f_k)$ – we can easily find the shortest common superstring in which the fragments appear in that order. We begin by overlapping f_2 as much as possible with f_1. We can accomplish this by building a suffix tree for f_1 and searching for the longest prefix of f_2 that we can find in the suffix tree. We then use a suffix tree of f_2 to overlap f_3 as much as possible with f_2, and so on. It is not hard to see that this greedy approach is reliable; there is no possible future gain to be had from accepting less than the maximum overlap at any stage. If the length of the longest overlap of f_i and f_j is denoted by $o(f_i, f_j)$, then the length of the solution is

$$\sum_{i=1}^{k} |f_i| - \sum_{i=1}^{k-1} o(f_i, f_{i+1}).$$

Since $\sum_{i=1}^{k} |f_i|$, the total length of all the fragments, is independent of their ordering, it follows that the best ordering is the one that maximizes $\sum_{i=1}^{k-1} o(f_i, f_{i+1})$. To find this ordering, we can precompute $o(f_i, f_j)$ for all ordered pairs of fragments by constructing a suffix tree for each fragment and searching for a prefix of every other fragment in each tree. Then we can use this information to evaluate each of the $k!$ possible orderings of the fragments.

Our problem is equivalent to the graph-theoretic problem of finding a *Hamiltonian path* of maximum weight in a directed graph.[2] A Hamiltonian path is a path along the edges of the graph that visits every vertex exactly once. The relevant graph in our case has k vertices $\{v_1, v_2, \ldots, v_k\}$, one for each of the fragments $\{f_1, f_2, \ldots, f_k\}$. There are $k(k-1)$ edges, one for each ordered pair of distinct fragments, and the weight of the edge from v_i to v_j is $o(f_i, f_j)$. This problem is similar to the well-known *traveling salesperson problem*.[3] Both are famous examples of the difficult class of *NP-complete* problems described in Section 9.2.

Although we cannot expect to solve large instances of any NP-complete problem, we can take certain steps to reduce running time. Here, as in Chapter 10, we adopt a branch-and-bound approach to avoid evaluating orderings of the fragments that cannot possibly yield the shortest superstring. First, we write code to read in fragments and compute overlaps. Next, we develop a program that exhaustively considers all possible orderings. Finally, we will modify the program to prune obviously suboptimal orderings from our search as soon as possible.

We will represent our graph by a hash table whose keys are the fragments and whose values are lists of (fragment, overlap) pairs. (This data structure is usually called an *adjacency list*.) For the example given previously, the hash could be entered literally as

---

```
{"maryhadalit" => [["ittlelam",2],["lelambits",0],
                   ["itsfleecewaswhit",2],["ecewaswhite",0],["iteassnow",2]],
 "ittlelam" => [["maryhadalit",1],["lelambits",5],
                   ["itsfleecewaswhit",0],["ecewaswhite",0],["iteassnow",0]],
 "lelambits" => [["maryhadalit",0],["ittlelam",0],
                   ["itsfleecewaswhit",3],["ecewaswhite",0],["iteassnow",0]],
 "itsfleecewaswhit" => [["maryhadalit",0],["ittlelam",2],
                   ["lelambits",0],["ecewaswhite",10],["iteassnow",2]],
 "ecewaswhite" => [["maryhadalit",0],["ittlelam",0],
                   ["lelambits",0],["itsfleecewaswhit",0],["iteassnow",3]],
 "iteassnow" => [["maryhadalit",0],["ittlelam",0],
                   ["lelambits",0],["itsfleecewaswhit",0],["ecewaswhite",0]]
}
```

---

[2] A graph in which every edge has a source and a destination; for example, a rooted phylogeny.

[3] A traveling salesperson seeks to plan a business trip visiting every city in a sales territory once before returning home, simultaneously seeking to minimize total travel time (or expense) on the road. Our problem differs slightly because our traveler never returns home.

Close adherence to the principles of software engineering would lead us to create a package defining a class of graphs that would be available for reuse to solve other problems. But our program involves just a single graph, and our path to a working program is shortened if we simply define a global variable %graph and subroutine addWeightedEdge to add edges to it. Its arguments are the left fragment, the right fragment, and the length of their longest overlap. The subroutine creates a new pair in the adjacency list of the left fragment.

```
sub addWeightedEdge {
    my ($v1,$v2,$weight) = @_;
    push @{$graph{$v1}}, [$v2,$weight];
}
```

The first task of the program is to read in the fragments, compute their overlaps, and save the results in the graph. Suffix trees are used to compute overlaps. The hash %tree holds a suffix tree value for each of its fragment keys. As each new fragment is read from input (one per line), a suffix tree is created for it. Then, the new fragment is searched for in the existing suffix trees and the older fragments are similarly searched for in the new suffix tree. The results are saved in the graph. At the end of the process, the suffix trees are no longer needed. By removing the value %tree, we allow Perl to reclaim the memory they occupy for reuse.

```
while (my $in=<STDIN>) {
    chomp($in);
    $graph{$in} = [ ];
    my $inTree = SuffixTree−>new($in);
    foreach (keys %tree) {
        addWeightedEdge($in, $_, $inTree−>maxSuffix($_));
        addWeightedEdge($_, $in, $tree{$_}−>maxSuffix($in));
    }
    $tree{$in} = $inTree;
}
%tree = undef;
```

As we enumerate all possible orderings of the fragments, we use the global hash %visited to avoid including a single fragment more than once. Globals $bestString and $bestWeight record the total weight (total overlap) of the string of highest weight seen so far by the algorithm. We iterate through the fragments, invoking search to enumerate all orderings with a given fragment first.

```
my $bestWeight = 0;
my $bestString = "";
my %visited;
foreach (keys %graph) { search($_,0,$_); }
print "\n\nShortest Common Superstring is:\n$bestString\n";
```

Subroutine search extends the superstring by one fragment – its argument $v –
then calls itself recursively to make another extension. If fragment $v has already
been included, then search returns immediately; otherwise, the current superstring
is compared to the best so far and is recorded if better. Fragment $v is "marked" in
%visited before iteration through the other fragments for recursive calls; it is then
unmarked before the subroutine returns, removing $v from the incomplete ordering.
The arguments of search are the vertex corresponding to the fragment most recently
added, the weight of the path so far, and the superstring constructed for the first
through most recent fragments.

```
sub search{
    my ($v,$pathWeight,$pathString) = @_;
    return if $visited{$v};
    if ($pathWeight>$bestWeight) {
        ($bestWeight,$bestString) = ($pathWeight,$pathString);
        print "New Best ($pathWeight): $pathString\n";
    }
    $visited{$v} = 1;
    foreach my $edge (@{$graph{$v}}) {
        my ($u,$edgeWeight) = @$edge;
        search($u,
                $pathWeight+$edgeWeight,
                $pathString . substr($u,$edgeWeight));
    }
    $visited{$v} = 0;
}
```

The method just programmed will correctly find the shortest common superstring
of the fragments – given enough time. We have already argued that we cannot expect to find a truly "efficient" (polynomial-time) solution to this problem, but there
are some steps we can take to speed things up. In particular, we can adopt a branch-
and-bound approach that allows us to return immediately from a recursive call that
has no hope of producing a superstring shorter than the best already known.

There are two aspects to the branch-and-bound strategy. One is to organize the
search so that the lower bound – the value of the best known solution – is driven up

quickly. This means that the most promising candidates should be investigated first. In this case, sorting the adjacency lists into decreasing order by weight is productive, since this causes search to try to extend the superstring with fragments that give large overlap first. The adjacency lists can be sorted with the following Perl statements:

```perl
foreach (keys %graph) {
    my @temp = sort { $$b[1] <=> $$a[1]} @{$graph{$_} };
    $graph{$_} = \@temp;
}
```

The other aspect is to devise a fast-to-compute but reasonably accurate upper bound on the improvement possible by extending the current partial solution to its optimal completion. For our problem, we must analyze the most overlap we can possibly hope to obtain in adding the remaining unordered fragments to the current incomplete ordering. Suppose that we have established the first p fragments of the ordering and that $k - p$ remain. Extending the reconstruction to completion means adding an overlap onto the left end of each of the $k - p$ remaining fragments. Furthermore, each of these overlaps must be with the right end of either the pth fragment or with one of the other $k - p$ remaining fragments. Let f be one of the $k - p$ remaining fragments. We find the largest remaining overlap involving the left end of f and the right end of either the pth fragment or the other $k - p - 1$ unordered fragments. We repeat this process for each of the $k - p$ remaining fragments. The sum of these $k - p$ largest left overlaps is an upper bound on what we can hope to achieve by extending the current incomplete ordering.

We can find these $k - p$ weights by traversing the $k - p + 1$ adjacency lists of the sources and maintaining $k - p$ running maxima for the different destinations. This involves at most $n(k - p)$ comparisons of weights. This number is small compared to the $(k - p)!$ orderings that can be eliminated if the upper bound shows that further extension is pointless. Actually, since the adjacency lists are sorted, we can short-circuit the traversal of any list as soon as an edge with weight 0 is found.

The upper bound on improvement to the current incomplete ordering is calculated by subroutine prospects, shown below. Its arguments are $newV, the last vertex on the current path, and $viz, a reference to the array recording which vertices have and have not been visited. The hash entry $maxIn{$s} records the largest among the weights of the edges coming from an unvisited vertex or $newV into the vertex representing the fragment $s.

```perl
sub prospects {
    my ($newV,$viz) = @_;
    my @unviz = grep { !$$viz{$_} } (keys %graph);
    my %maxIn;
```

```
    foreach my $v (@unviz) {
        foreach my $edge (@{$graph{$v}}) {
            my ($u,$edgeWeight) = @$edge;
            last if $edgeWeight == 0;
            $maxIn{$u} = $edgeWeight if $edgeWeight > $maxIn{$u};
        }
    }
    my $result = 0;
    foreach my $v (@unviz) { $result += $maxIn{$v}; }
    return ($result−$maxIn{$newV});
}
```

Subroutine prospects introduces the useful Perl operator **grep**. Its operands are (i) an expression involving $_ and (ii) a list. The result is a filtered version of the list, containing only the items for which the expression returned a value regarded as "true" by Perl.

The upper bounds computed by prospects can be used to cut off pointless searches by adding the following as the third line of subroutine search:

```
    return if ($pathWeight+prospects($v,\%visited)) <= $bestWeight;
```

13.2   Practical Issues and the PHRAP Program

Finding shortest common superstrings is an interesting problem, but as a model of sequence assembly it is so greatly oversimplified as to be entirely impractical. Among the issues that must be addressed by a practical program are the following.

1. *The uncertainty of the base-calling process.* For a number of reasons, the "light show" produced by laser stimulation of the DNA in the sequencing gel is not an unambiguous progression of flashes of exactly one of four colors, as implied by Chapter 3. Instead, the sequencing machinery records a *trace* of the sequence consisting of measurements of the strength of each of the four colors at each of thousands of points in time. Base-calling software (like PHRAP's partner PHRED) then analyzes the trace and produces a sequence of bases, attaching an evaluation of the probability of correctness to each base. This combination of sequence and *base qualities* is called a *sequencing read,* or simply a *read.*

2. *Input set size.* The number of reads involved in a sequence assembly may approach or surpass 100,000, so even a cleverly designed branch-and-bound program will be impossibly slow. Any practical approach must apply fast algorithms to optimize the objective function approximately.

3. *Vector contamination.* The DNA sequences being studied have often been preserved and replicated by introducing them into bacterial or yeast artificial chromosomes (BACs and YACs), plasmids, or other hosts. Often, some of the host sequence, or the sequence of the vector used to introduce the object of study into the host, will remain when the DNA is sequenced. This is especially likely near the ends of reads. Practical software must be prepared to deal with pairs of sequences whose overlap has been obscured by a vector sequence attached to the overlapping end.

4. *The double-stranded nature of DNA.* Real sequencing data includes reads from both strands of the target. Assembly requires that each read be assigned not just a position but also an *orientation* ($5'$–$3'$ or $3'$–$5'$) on the reconstructed target.

5. *Long repeated substrings within the target.* These can lead to confusion about which similarities in fragments reflect common origin in the target and which do not.

6. *Chimerical fragments.* Sometimes two pieces of DNA from disconnected regions of the target become artificially joined during the cloning or sequencing process. The presence in input of a read of the resulting sequence – called a *chimera* – presents an obvious problem. Unfortunately, the percentage of chimerical reads in some sequencing projects can run as high as 30% or more.

7. *Sequencing chemistry.* Two chemical processes, the *dye-terminator* process (described in Chapter 3) and the *dye-primer* process, are commonly used to prepare samples for the agarose gel. Each is prone to a characteristic set of errors. For example, the dye-terminator chemistry sometimes fails to reveal G following A.

PHRAP attempts to address all of these issues. Space permits our simplified version to address only the first two with any degree of trueness to PHRAP. Our program proceeds in six stages.

1. *Reading inputs.* The files produced by the base-calling software are read in, and a simple data structure is created for each sequencing read.

2. *Aligning reads.* The output of this stage is a collection of high-scoring local alignments between pairs of reads. It would be far too slow to align each read to every other with the Smith–Waterman algorithm. This stage therefore includes a filter for determining which pairs (and even which regions within pairs) are likely to yield high-scoring alignments. Of course, we are not guaranteed to find every single high-scoring alignment.

3. *Adjusting qualities.* If two reads have a high-scoring alignment with a region of identity but one read's qualities are low over this region while the other's are high, then it makes sense to boost the qualities in the first read. On the other hand, in a region of the alignment where there is disagreement, it makes sense to reduce the qualities of one or both reads. The adjustments made to a particular read should depend, of course, on the totality of the information provided by *all* alignments to the read.

4. *Assigning reads to contigs.* It is unusual for a sequence assembly to result in a single uninterrupted reconstruction of the target; normally several disconnected regions of the target are identified, reconstructed, and presented separately in the output. The term *contig* refers to a set of reads that appear to form an uninterrupted stretch of contiguous bases in the target sequence. PHRAP employs a *greedy method* to assign reads to contigs. First, each read is assigned to a contig by itself. Then, the alignments are considered one by one, from highest- to lowest-scoring, and the contigs containing the two aligned reads are merged into a single contig unless this would impose an unsatisfiable condition on the relative positions of all reads in the alignment.

5. *Readjusting qualities.* In this stage, we discard the adjusted qualities produced in the third stage and recompute adjusted qualities following a similar process. In the third stage, we had to assume that all high-scoring alignments reflected common origin in the target sequence. After reads have been assigned to contigs, however, we can assume that an alignment between reads assigned to different contigs do *not* reflect common origin, no matter how high the score. Therefore, this stage ignores such alignments.

6. *Developing consensus sequences for contigs.* In this stage, we use the lists of reads and alignments assigned to each contig to construct a sequence with quality annotations. After grouping together positions on the individual reads that the alignments suggest have a common origin on the target, we use a *dynamic programming method* to find, for each such group, the best sequence beginning at that point in the contig and extending to the right.

We will expand on each stage in turn in the following sections.

13.3   Reading Inputs for Assembly

PHRAP's companion base-calling program, PHRED, produces files in FASTA format (described in Chapter 6) for the nucleotides of its reads. For each sequence in the FASTA file, it creates a separate file of quality information. PHRED assigns integer quality values between 0 and 99.

To simplify, we will use *single-digit* values 0 through 9 in a nonstandard file format we call "FASTQ". Just as FASTA entries begin with >, FASTQ entries begin with Q>. The quality values may either be interspersed with the bases, like this:

```
a3a4c3g7t8t8g8c7 ...
```

or follow them, like this:

```
aacgttgc ...
34378887 ...
```

A FastqReader can be created by small modifications to Chapter 6's FastaReader.

Our program creates an object of class DnaRead for each of the sequences in the FASTQ file. As usual, these objects are blessed references. In this program, we will need to be able to use objects as hash keys.

This poses a bit of a problem, because Perl requires that all hash keys be strings; for example, in the code

```
my %hash;
foreach my $i (1..5) { $hash{$i}++; }
```

the integer values of $i are implicitly converted to the strings "1", "2", "3", "4", and "5" before being used as hash keys. This is easily overlooked by the programmer, since Perl converts strings (in the proper format) to integers with equal facility. The following code assigns the integer 15 to $x:

```
my $x;
foreach (keys %hash) { $x += $_ }
```

Similarly, in the statement $hash2{bless {}, "Bob"}++, the blessed reference to an empty hash will be converted to a string like Bob=HASH(0x8102038). However, Perl provides *no* operator to convert strings representing references back to the original references, and the following code will terminate with an error:

```
foreach (keys %hash2) { $_->dump(); }
```

One way for us to solve this problem is to attach an integer "serial number" to each DnaRead as it is created and use this number as the hash key rather than the reference. To allow access to the original object based on the serial number alone, we add a *package-level variable* named @AllReads to the DnaRead package:

```
package DnaRead;
my @AllReads;
```

This is a *single* variable shared by all objects and subroutines of the package. Whenever a new DnaRead is created, a reference to it is pushed onto @AllReads:

```
sub new {
    my $this = bless { ...
                    SerialNumber => scalar(@AllReads)},
                    ($this || ref $this);
```

```
    push @AllReads, $this;
    return $this;
}
```

Thus, when necessary, a serial number $i can be converted back into a blessed reference to the corresponding DnaRead by the call DnaRead−>GetBySerial($i), where GetBySerial is defined in **package** DnaRead by:

```
sub GetBySerial {
    my ($this,$num) = @_;
    return $AllReads[$num];
}
```

Although not required, the name of this subroutine begins with an upper-case letter. This is a useful convention for subroutines that refer only to package-level variables rather than to any particular object.[4]

With package-level variables, we can use a few other shortcuts. We can organize the keys (field names) of the objects' hash tables into two lists: @ImmutableFields, set once and for all when the object is created; and @MutableFields, changeable at any time during execution.

```
my @ImmutableFields =
    ("Bases",          ## the nucleotide sequence of the read; a string
     "Length",         ## number of nucleotides in the read
     "Name",           ## the name of the sequence; a string
     "OrigQual",       ## ref to list of qualities caller assigned to bases
     "SerialNumber");  ## this read's index in @AllReads

my @MutableFields =
    ("Alignments",     ## ref to list of alignments to this read
     "AdjQual",        ## ref to list of current adjusted quality estimates
     "Contig",         ## ref to contig to which read is currently assigned
     "ContigStart",    ## approx. interval of contig covered by this read
     "ContigEnd");
```

Then we can concisely define methods to access and set each field (Bases, OrigQual, AdjQual, SetAdjQual, SetContigEnd, etc.) by the following lines, which are executed when another file executes **use** DnaRead.

----

[4] Package-level variables and subroutines correspond to the "static" variables and methods of C++ and Java.

```
foreach (@MutableFields) {
    eval "sub set$_ { \$_[0]−>{$_} = \$_[1] } ;" ;
}


foreach (@MutableFields,@ImmutableFields) {
    eval "sub $_ { \$_[0]−>{$_} } ;" ;
}
```

Lists @ImmutableFields and @MutableFields are also conveniently available for use by debugging and dumping subroutines.

To create a new DnaRead, we simple invoke DnaRead−>new, which initializes all immutable and some mutable fields.

```
sub new
{
    my ($this,      ## literal "DnaRead" or ref to an existing DnaRead
        $name,      ## identifier of this sequences
        $bases,     ## the bases
        $qualities  ## reference to list of qualities corresponding to bases
        ) = @_;

    my $this
        = bless {Bases => $bases,
                 Length => length($bases),
                 Name => $name,
                 OrigQual => $qualities,
                 AdjQual => \@$qualities,     ## makes a copy of the array
                 SerialNumber => scalar(@AllReads)}, ($this || ref $this);
    push @AllReads, $this;
    return $this;
}
```

13.4 Aligning Reads

Our alignment routines are collected in **package** JudiciousAligner. The output of this stage consists of a collection of Alignment objects – each representing a high-scoring local alignment between two reads – which is organized in two data structures as follows.

1. A hash, %alignmentsByPair, each key of which is a pair of serial numbers of reads. The corresponding value is a reference to a list of blessed references to Alignments involving that pair of reads.

2. Lists attached to each DnaRead object. The Alignments field of each DnaRead object holds a reference to a list of blessed references to all Alignments involving that particular read and any other read.

Thus, each Alignment appears in three places: once in the hash, and twice in the lists attached to reads.

As previously mentioned, it would be impractically slow to use the Smith–Waterman algorithm to align every pair of reads. PHRAP solves this problem by performing an alignment only if two reads share a common 14-base-long substring or *14-mer*. Although it is possible for two reads to have a high-scoring alignment without 14 identical bases, it is extremely unlikely. The first step in building alignments is therefore to construct a hash whose keys are 14-mers and whose values are lists of all locations where the key 14-mer occurs. Each location is represented as a pair consisting of read serial number and index of the first base of the 14-mer. This hash is constructed by subroutine build14merHash.

```
sub build14merHash {
    my ($readList) = @_;
    my %hash14;
    foreach my $read (@$readList) {
        my $seq = $read->Bases();
        my $readIndex = $read->SerialNumber();
        my $lasti = (length $seq)-15;
        for (my $i=0; $i<$lasti; $i++) {
            my $word = substr($seq,$i,14);
            $hash14{$word} ||= [];
            push @{$hash14{$word}}, [$readIndex,$i];
        }
    }
    return \%hash14;
}
```

The choice of the number 14 is driven by two factors. If the number is too small, then the number of pairs that pass through the filter to alignment by the Smith–Waterman algorithm becomes too large. On the other hand, if it is too large then pairs of reads with common origin may be excluded by the filter because of sequencing errors.

The next step is to use this hash to generate pairs to be aligned. In fact, PHRAP does not even fully align reads sharing a 14-mer; it only fills selected ranges of diagonals, or *bands,* of the alignment matrix near the shared 14-mers. The subroutine findCandidateBands uses the hash of 14-mers to produce (i) a hash mapping pairs of reads into the index of the diagonals upon which shared 14-mers occurs and (ii) a hash mapping pairs to lists of promising bands.

To produce the first hash, the list attached to each 14-mer is examined, and every (read, index) pair is paired with every other in the list, so that a list of k elements produces $k(k-1)/2$ pairs. The indices are subtracted to give the index of the diagonal (or *offset*) and the offset is added to the list of offsets for the pair of reads. (Perl's value operator returns a list of the values stored in a hash; **foreach** (**values** %h) { ... } is equivalent to **foreach my** $key (**keys** %h){ $_ = $h{$key};... }.)

```
sub findCandidateBands {
    my ($hash14mers) = @_;
    ## build hash mapping pairs of reads to offsets with matching 14-mers
    my %pairHash;
    foreach (values %$hash14mers) {
        my @wordLocs = sort { $$a[0]<=>$$b[0] || $$a[1]<=>$$b[1] } @$_;
        for (my $i=0; $i<@wordLocs; $i++) {
            my ($read1,$loc1) = @{$wordLocs[$i]};
            for (my $j=$i+1; $j<@wordLocs; $j++) {
                my ($read2,$loc2) = @{$wordLocs[$j]};
                $pairHash{$read1,$read2} ||= [ ];
                push @{$pairHash{$read1,$read2}}, ($loc2−$loc1);
    }}}
    ...
```

The second hash is created by converting each offset into a band of width 42 centered at the center of the 14-mer. Overlapping bands are then merged by creating a sorted list of all bands' left and right endpoints, sorting it, and scanning from left to right. As we scan, we need to keep track of the number of bands we are currently within. To help, we mark left endpoints with +1 and right endpoints with −1. When the current number of bands drops to zero, the processing of a set of overlapping bands has been completed.

```
sub findCandidateBands {
    ...
    foreach my $pair (keys %pairHash) {
        my @endPoints;
        foreach my $offset (@{$pairHash{$pair}}) {
            push @endPoints, [$offset−14,1], [$offset+28,−1];
        }
        $pairHash{$pair} = bandsFromEndpoints (@endPoints);
    }
    return \%pairHash;
}
```

```
sub bandsFromEndpoints {
    my @endpoints = sort { $$a[0]<=>$$b[0] || $$b[1]<=>$$a[1] } @_;

    my $nesting = 0;
    my $currLeft;
    foreach (@endPoints) {
        my ($endPoint,$deltaNesting) = @$_;
        $currLeft = $endPoint
            if ($nesting==0 && $deltaNesting==1);
        push @bands, [$currLeft,$endPoint]
            if ($nesting==1 && $deltaNesting==-1);
        $nesting += $deltaNesting;
    }
    return \@bands;
}
```

Next, the candidate bands are aligned by subroutine alignBand to see if they give a high-scoring alignment. The algorithm used is a slightly modified version of the Smith–Waterman algorithm of Chapter 7. If the band to be considered extends from diagonal d_1 to diagonal d_2, we want to fill only those elements $M_{i,j}$ of the alignment matrix that satisfy $d_1 \leq j - i \leq d_2$. Since the matrix is sparsely filled, we can save space by representing it by a hash table rather than a multidimensional array. The reward for a pair of matching nucleotides is 1; mismatches cost -2; and columns with gaps cost -8 each.[5] After the band of the matrix is filled, the highest-scoring alignment is saved as a string of Ms for matches, Ss for mismatches or *substitutions*; Is for insertions in the second string with respect to the first (gap in first); and Ds for deletions in the second string with respect to the first (gap in second). These operations are by now routine, so we omit further discussion of the code. If successful, alignBand creates and returns an Alignment object describing the result. The fields of an Alignment object are described by these commented lines from `Alignment.pm`:

```
package Alignment;
...
my @ImmutableFields =
    ("Read1","Read2",     ## refs to objects describing the two reads aligned
     "Start1","Start2",   ## starting positions of alignment in each of above
     "End1","End2",       ## ending positions of alignment in each of above
```

---

[5] PHRAP uses a system of affine gap penalties as described in Exercise 3.7, where gap initiation costs -12 and gap extension costs -2.

    "RawScore",   ## *raw Smith–Waterman score of the local alignment*
    "Path",   ## *a string (MSDI)∗ for path through alignment matrix*
    "MeanOffset",   ## *mean offset of aligned positions*
    "SerialNumber");

```
my @MutableFields =
    ("WeightedScore",      ## S–W score weighted by column quality
     "SupportedContig");   ## contig whose construction alignment supports
```

The work of alignBand is overseen by alignBands, which iterates first through the pairs of reads that have candidate bands for alignment and next through the bands themselves. Once all the bands for a particular pair have been aligned, the results are sorted by score and stored in the hash %alignmentsByPair. Ultimately, alignBands returns a reference to this hash.

```
sub alignBands {
    my ($readList,$candidateBands) = @_;
    my $MINSCORE = 30;
    my %alignmentsByPair;
    ## try to align each band for each pair
    foreach my $pair (keys %$candidateBands) {
        my ($r1,$r2) = @$readList[split($;,$pair)];
        my ($sn1,$sn2) = ($r1->SerialNumber(),$r2->SerialNumber());
        my @alignments;
        foreach my $band (@{$$candidateBands{$pair}}) {
            my $alignment = alignBand($r1,$r2,$band);
            push @alignments, $alignment
                if $alignment->RawScore() > $MINSCORE
        }
        if (@alignments) {      ## did any alignments meet MINSCORE criterion?
            @alignments =       ## mark best and save all
                sort {$b->RawScore() <=> $a->RawScore()} @alignments;
            $alignmentsByPair{$sn1,$sn2} = \@alignments;
        }
    }
    return \%alignmentsByPair;
}
```

At the highest level, the alignment process is managed by makeAlignmentMaps, the only subroutine exported by **package** JudiciousAligner. After invoking the subroutine already described to create hashes of 14-mer locations, candidate alignment bands, and then high-scoring alignments, it iterates through the alignments in

%alignmentsByPair and adds each to the Alignments list of each of its reads. As a return value, it passes on a reference to %alignmentsByPair.

```perl
sub makeAlignmentMaps {
    my ($readList) = @_;
    my $fourteenmerHash = build14merHash($readList);
    my $candidateBands = findCandidateBands($fourteenmerHash);
    my $alignmentsByPair = alignBands($readList,$candidateBands);
    ## returns a hash mapping pairs of reads to their best alignments
    ## index the alignments by read instead of just by pair
    my @alignmentsByRead;
    foreach my $pair (keys %$alignmentsByPair) {
        my ($r1,$r2) = split($;,$pair);
        my $alignments = $$alignmentsByPair{$pair};
        $alignmentsByRead[$r1] ||= [ ];
        push @{$alignmentsByRead[$r1]}, @$alignments;
        $alignmentsByRead[$r2] ||= [ ];
        push @{$alignmentsByRead[$r2]}, @$alignments;
    }
    foreach my $read (@$readList) {
        $read->setAlignments
            ($alignmentsByRead[$read->SerialNumber( )] || [ ]);
    }
    return $alignmentsByPair;
}
```

13.5   Adjusting Qualities

We now turn our attention to the issue of errors and varying confidence in the nucleotides received from the base caller.

In Chapter 3, we described the process of reading the bases of a sequence from the gel as a simple matter of shining a laser on a certain part of the gel and seeing which of four colors results. In practice, we may find combinations of colors in various strengths that make the correct base less than obvious. There can be many reasons for this. Some involve local inconsistencies in the density of the gel itself. Some relate to the fact that certain DNA sequences may have sections capable of forming RNA-like secondary structures, and these can impede the progress of some copies through the gel. And, of course, DNA polymerase is prone to make occasional errors in replication; this is especially likely to happen when the template contains several consecutive identical nucleotides.

The actual output of the sequencing machinery is not a sequence but rather a *trace* consisting of a set of four color intensity levels at each of several thousand

```
qual2 LLLL   LLHHHHHHHHHHHHHHHHHH   HHHHHHHHHHHHHHHHHHHHHHLLLLLLLLL
read2 xxxy--xxxxxxxxxxxxxxxxxyy--xyyyxxxxxyyyxxxxxxxxxxxxxyyyyy
read1 xxxxxx-x-xxxxxxxxxxxxxxxxxxxxxxx---xxxxxxxxxxxxxxxxxxxxx
qual1 LLLLLL L LLLLLLLLLLLLLLHHHHHHHHHH   HHHHHHHHHHHHHHHHHHHHHHHH
               ||||||||||||||           |||
adjustments    HHHHHHHHHHHHHH            LLL
```

Figure 13.1: Some possible quality adjustments to `read1` suggested by an alignment with `read2`, where L and H indicate low and high quality.

points along the gel. The process of turning the trace into a predicted sequence is known as *base calling*. PHRED, the base-calling program most frequently teamed with PHRAP, outputs a predicted sequence of nucleotides in one file and a sequence of corresponding *quality values* in another. Each quality value represents PHRED's estimate of the probability that the corresponding nucleotide is incorrect. If p is this probability, then PHRED outputs the quality value

$$q_{\text{PHRED}} = -10 \cdot \log_{10}(p).$$

For example, if PHRED is 99.9% sure of a particular base call then its quality value is

$$q_{\text{PHRED}} = -10 \cdot \log_{10}(1 - 0.999) = 30.$$

If we invert the formula, we have

$$p = (0.1)^{0.1 q_{\text{PHRED}}}.$$

PHRED's quality values are rounded to integers and restricted to the range $[0, 99]$.

In our simplified program, we will use single-digit qualities. With p the probability of error just described, we define our quality values by

$$q = -\log_{10}(p), \quad \text{and} \quad p = (0.1)^q.$$

The base-calling program must assign qualities to a read using only the information available from a single trace. When we process the read in this stage, however, we have other sources of information: alignments between reads suggesting a common origin, and the qualities of the reads aligned to the read under scrutiny. In this stage, we exploit this information to revise the qualities assigned to the bases of all reads, one by one. We always retain the original qualities assigned by the base caller in the field OrigQual of DnaRead objects; adjusted qualities are saved in AdjQual.

If two reads have a high-scoring alignment with a region of identity yet one read's qualities are low over this region while the other's are high, then it makes sense to boost the qualities in the first read. On the other hand, in a region of the alignment where there is disagreement, it makes sense to reduce the qualities of one or both reads. (See Figure 13.1.) Of course, different alignments to the same region may suggest contradictory revisions; therefore, any adjustments should depend on the *totality* of the information provided by all alignments to the read, not just a single alignment. These observations, in turn, suggest the outlines of PHRAP's strategy: Consider each read in turn for quality adjustment; then fill two arrays, @bestMatch and @bestAlternative, with entries for each base of the read.

1. $bestMatch[$i] contains the highest quality assigned to any base aligned to and identical to the read's $ith base in any alignment.
2. $bestAlternative[$i] contains the highest quality assigned to any base or gap aligned to but *not* identical to the read's $ith base in any alignment. (To assign a quality to a gap, we look at the quality of one of its neighboring nucleotides.)

To fill these arrays, we simply need to consider the alignments for the current read one by one, walking through its columns to update the arrays. Once the arrays are filled, we can iterate through the bases of the read from left to right, updating each quality value by referring to corresponding positions of the arrays.

Our program implements this strategy in a package named QualityReviser, which exports a single subroutine, reviseQualities. This subroutine supervises calls for each read to collectMatchedQualities to fill arrays @bestMatch and @bestAlternative, and then to adjustQuality, to update the read's AdjQual field. reviseQualities has an additional function: it also oversees the computation of a weighted alignment score in which each column's score is multiplied by the lower quality value for the column.[6]

```
sub reviseQualities {
    my ($readList,$alignmentsByPair,$pass) = @_;
    foreach my $read (@$readList) {
        adjustQuality($read,collectMatchedQualities($read,$pass));
    }
    foreach (values %$alignmentsByPair) {
        foreach (@$_) { computeWeightedScore($alignment); }
    }
}
```

Subroutine collectMatchedQualities iterates over all alignments relating to its argument $read as recorded in its Alignments field. In the first pass (pass 0), all alignments are used; in the second, only those with positive weighted scores; and in the final pass, only those used in the construction of contigs. The passes are otherwise identical. The first step with each alignment is to gather information from the alignment and its two reads into local variables. This is made a bit tedious by the fact that the read being adjusted may be recorded as either Read1 or Read2 in the alignment. The next step is to establish the position in each read corresponding to the first

---

[6] PHRAP estimates a log likelihood score for each alignment based on more solid statistical foundations.

column of the alignment in siteIn1 and siteIn2. Finally, we begin to advance through the alignment column by column.

```perl
sub collectMatchedQualities {
    my ($read,      ## the read to be revised
        $pass       ## 0, 1, or 2
        ) = @_;
    my @bestAlternative;
    my @bestMatch = @{$read->OrigQual()};     ## COPIES!!
    foreach my $alignment (@{$read->Alignments()}) {
        next if $pass==1 && $alignment->WeightedScore()<0;
        next if $pass==2 && !($alignment->SupportedAlignment());

        ## we want "read1" to refer to the read being adjusted
        my $swap = ($read==$alignment->Read2());
        my $read1 = $read;     ## just to give it a name with a number
        my $read2 = $swap ? $alignment->Read1() : $alignment->Read2();
        my $path = $alignment->Path();
        $path =~ tr/ID/DI/ if $swap;
        my $seq1 = $read1->Bases();
        my $seq2 = $read2->Bases();
        my $origQual2 = $read2->OrigQual();

        ## work through the local alignment column by column;
        ## compute indices of first bases aligned
        my $siteIn1 = $alignment->Start1()-1;
        my $siteIn2 = $alignment->Start2()-1;
        ($siteIn1,$siteIn2) = ($siteIn2,$siteIn1) if swap;
        foreach my $column (split(//,$path)) {
            ...
            ...
        }     ## foreach my $column
    }     ## foreach my $alignment
    ## finished looking at each alignment for current sequence
    return (\@bestMatch,\@bestAlternative);
}
```

When we advance to the next column, we advance our position in at least one of the reads. Match and substitution columns cause a simple update to one entry in @bestMatch and @bestAlternative, respectively. Columns with a gap require somewhat more work.

```perl
foreach my $column (split(//,$path)) {
    $siteIn1++ unless $column eq "I";
    $siteIn2++ unless $column eq "D";

    if ($column eq "M") {     ## match
        $bestMatch[$siteIn1]
            = max($bestMatch[$siteIn1],$$origQual2[$siteIn2]);
    } elsif ($column eq "S") {     ## substitution
        $bestAlternative[$siteIn1]
            = max($bestAlternative[$siteIn1],$$origQual2[$siteIn2]);
    } else {     ## insertion or deletion
        my $base = ($column eq "D") ?
            substr($seq1,$siteIn1,1) : substr($seq2,$siteIn2,1);
        my ($lo2,$hi2) = findBaseRange($seq2,$base,$siteIn2);
        next if ($hi2<$lo2);     ## no matching base in $seq2

        my $q2 = min(@$origQual2[$lo2..$hi2]);
        my ($lo1,$hi1) = findBaseRange($seq1,$base,$siteIn1);

        foreach ($lo1..$hi1) {
            $bestAlternative[$_] = max($bestAlternative[$_],$q2);
        }
    }
}
```

The general approach to gaps in either read is to (i) determine which of the four nucleotides is aligned against the gap and then (ii) treat runs of that nucleotide adjacent to the column as a group. (The limits of the runs themselves are found by findBaseRange, a straightforward subroutine omitted here.)

For example, we may have

```
... gctaaaaccat ...
... ggt--aaccgt ...
```

No matter which read is being adjusted, we will update @bestAlternative's entry for each position in the run of as – that is, four entries if the upper read is being adjusted or two if it is the lower. The value used to update @bestAlternative is the *lowest* quality found among the as in the other read. If the alignment shown were the only one dealing with these bases and the initial qualities were

```
    5421
... gctaaaaccat ...
... ggt--aaccgt ...
    33
```

then the revision process would turn these into

```
          4320
... gctaaaaccat ...
... ggt--aaccgt ...
           22
```

pushing the qualities toward agreement about the length of the run.

13.6 Assigning Reads to Contigs

The assignment of reads to contigs is performed by package ContigBuilder. The package exports a single subroutine, assignReadsToContigs, which returns a reference to a list of Contig objects. Like the DnaRead package, package Contig provides objects with a number of fields along with access and update methods. Contig objects have a serial number that we use mainly for human-readable dumps and traces.

```
package Contig;
...
my @ImmutableFields =
    ("SerialNumber");     ## records order in which contig objects are created

my @MutableFields =
    ("Bases",           ## the consensus sequence of the contig; a string
     "Quals",           ## ref to list of qualities assigned to consensus
     "Reads",           ## ref to list of included reads
     "AlignedPairs",    ## ref to list of alignments pertaining to this contig
     "Length",          ## estimated length of contig
     "FirstStart",      ## where leftmost read starts, wrt contig's 0; often <0
     "LastEnd",         ## where rightmost read ends, relative to contig's 0
     "Score");          ## sum of nucleotide qualities for this contig
```

The first step performed by assignReadsToContigs is to create one contig object for each read and to assign the read to that contig. The next step is to filter the list of alignments to ensure that no two overlap in either of their two reads. This is carried out by subroutine findBestAlignments. Finally, the alignments are passed to mergeByAlignment in descending order by score. mergeByAlignment uses the information in the alignment to decide whether to merge the two contigs containing the aligned reads.

```
package ContigBuilder;
sub assignReadsToContigs {
```

```perl
    my ($readList,$alignmentsByPair) = @_;
    my @allContigs;
    foreach my $read (@$readList) {
        my $contig = Contig->new([$read],1,0,$read->Length()-1);
        $read->setContig($contig);
        $read->setContigStart(0);
        $read->setContigEnd($read->Length()-1);
        push @allContigs, $contig;
    }
    ## make read layout, using greedy algorithm
    my $sortedAlignments = findBestAlignments($alignmentsByPair);

    ## process each pair from highest- to lowest-weighted score
    foreach my $alignment (@$sortedAlignments) {
        mergeByAlignment($alignment);
    }
    return [grep { @{$_->Reads()}>0 } @allContigs];
}
```

Subroutine findBestAlignments iterates in its outermost loop over the set of all pairs of reads that have alignments. The alignments for each pair are sorted by score (ascending). Then each alignment is compared to every higher-scoring alignment. If an overlap is found, the lower-scoring alignment is removed from further consideration. If no overlap is found with any higher-scoring alignment, then the alignment is added to the list bestAlignments. The code uses a *labeled loop* and *labeled **next**-statement*. Normally, a **next**-statement advances to the beginning of the next iteration of its innermost enclosing loop. When labeled, it advances to the next iteration of the loop bearing the same label. (Of course, the label must be attached to a loop that *encloses* the **next**-statement.) Thus, in the code below, the candidate alignment to be included in @bestAlignments is $alignments[$i], but the loop searching for reasons to reject the candidate is indexed by $j. When a reason is found, we must increase the value of $i, not $j, to reject the candidate.

```perl
sub findBestAlignments {
    my ($alignmentsByPair) = @_;
    my @bestAlignments;
    foreach my $pair (keys %$alignmentsByPair) {
        my ($r1,$r2) = split($;,$pair);
        next if $r1>$r2;     ## don't look at the same list twice
        my @alignments =
            sort { $a->WeightedScore()<=>$b->WeightedScore() ||
                   $a->RawScore()<=>$b->RawScore() }
                @{$$alignmentsByPair{$pair}};
```

```
ILOOP:
    foreach my $i (0..$#alignments) {
        my $cand = $alignments[$i];
        my ($cStart1,$cEnd1) = ($cand−>Start1( ),$cand−>End1( ));
        my ($cStart2,$cEnd2) = ($cand−>Start2( ),$cand−>End2( ));
        foreach my $j ($i+1..$#alignments) {
            my $other = $alignments[$j];
            my ($oStart1,$oEnd1) = ($other−>Start1( ),$other−>End1( ));
            my ($oStart2,$oEnd2) = ($other−>Start2( ),$other−>End2( ));
            ## dismiss candidate if it is overlapped by higher-scoring
            ## alignment in either read
            next ILOOP unless ($cStart1>$oEnd1) || ($cEnd1<$oStart1);
            next ILOOP unless ($cStart2>$oEnd2) || ($cEnd2<$oStart2);
        }
        push @bestAlignments, $cand;
    }
}
## sort out the survivors by score for greedy algorithm
@bestAlignments =
    sort { $b−>WeightedScore( )<=>$a−>WeightedScore( ) ||
            $b−>RawScore( )<=>$a−>RawScore( ) }
        @bestAlignments;
    return \@bestAlignments;
}
```

Subroutine mergeByAlignment receives an Alignment as its only argument. If the aligned reads are already in the same contig, the alignment is analyzed to see whether it is consistent with the current approximate layout of the contig. Consistency is defined in terms of possible offsets between the two reads: the offset implied by the alignment versus the offset implied by the (higher-scoring) alignments already supporting the contig.

For example, if the alignment pairs the ith base of one read and the jth of the other, does the current approximate layout of the contig as defined by the ContigStart and ContigEnd fields of the two reads also put these two bases in nearby positions on the contig? If the answer to this question is "yes", then the alignment is added to the contig's list of supporting alignments. If not, it is dropped from further consideration.

If the alignments are in different contigs, then testContigMerge is invoked to decide whether the alignment provides sufficient justification to merge the two contigs. If so, mergeContigs is called to seal the deal.

findAlignmentOffset's computation is relatively self-explanatory arithmetic involving the Length, ContigStart, and ContigEnd fields of the two Reads and the MeanOffset field of the Alignment. Likewise, mergeContigs's task is routine: update the Contig, ContigStart, and ContigEnd fields of the reads in one of the two contigs; combine the Reads and AlignedPairs fields; and update the FirstStart and LastEnd fields in one Contig object while erasing them in the other.

```perl
sub mergeByAlignment {
    my ($alignment) = @_;
    my ($read1,$read2) = ($alignment->Read1(),$alignment->Read2());
    return if $read1 == $read2;

    ## find the contigs whereof the two reads are currently members
    my ($contig1,$contig2) = ($read1->Contig(),$read2->Contig());

    ## find the offset between the two contigs implied by the alignment
    my $offset = findAlignmentOffset($alignment);

    if ($contig1==$contig2) {      ## already in same contig
        if (abs($offset)<25) {       ## further support for this contig
            $alignment->setSupportedContig($contig1);
            push @{$contig1->AlignedPairs()}, $alignment;
        }
    }
    else {    ## different contigs; try to merge
        if (testContigMerge($alignment,$offset)) {
            mergeContigs($contig1,$contig2,$offset);
            $alignment->setSupportedContig($contig1);
            push @{$contig1->AlignedPairs()}, $alignment;
        }
    }
}
```

Subroutine testContigMerge is somewhat more complicated. Generally, it is preferable to fail to combine two contigs that *are* contiguous in the target than to combine contigs that are not really contiguous. In order to avoid erroneous merges, testContigMerge considers all alignments between pairs of reads involving one read from each contig of the proposed merge. It seeks to determine whether any credible alignments suggest significantly different offsets than the alignment being tested. If so, the merge is rejected.

The missing element is a definition of "credible". PHRAP makes several passes through the alignments, becoming somewhat more credulous on each pass. For simplicity, we are making a single pass through the alignments. We regard an alignment as providing a "credible" contradiction to the tested alignment if its score is no more than 20% less.

To iterate over the alignments joining the two contigs, we iterate over the Alignments field of each read contained in the second alignment and check each read for membership in the first contig.

```perl
sub testContigMerge {
    my ($alignment,      ## alignment on which to base proposed merge
        $offset)         ## offset between contigs implied by alignment
        = @_;

    my ($aRead1,$aRead2) = ($alignment->Read1(),$alignment->Read2());

    ## find the contigs whereof the two reads are currently members
    my ($contig1,$contig2) = ($aRead1->Contig(),$aRead2->Contig());
    my $alignmentScore = $alignment->WeightedScore();
    my $rejectCutoff = $alignmentScore-0.20 * abs($alignmentScore);

    ## for each read in contig2
    foreach my $c2Read (@{$contig2->Reads()}) {
        ## for each alignment to the read
        foreach my $align2 (@{$c2Read->Alignments()}) {
            ## $c1Read is a read in contig 1
            my $c1Read = $align2->Read2();
            my $swap = ($c1Read==$c2Read);
            $c1Read = $align2->Read1() if $swap;
            ## we need alignments between pairs in the contigs being merged
            next if $c1Read->Contig() != $contig1;

            ## What offset between contigs is implied by this alignment?
            ## How different is this from previous estimates?
            ## If a lot, including both $alignment and $align2 in the contig
            ## will put a "strain" on the contig.
            my $alignOffset = ($swap?-1:1)*findAlignmentOffset($align2);
            my $strain = abs($offset+$alignOffset);
            my $align2score = $align2->WeightedScore();

            ## If the strain is large and $align2 is not grossly less reliable
            ## than $alignment, then we cannot safely merge.
            return 0 if ($strain>25) && ($align2score>=$rejectCutoff);
        }    ## for my $align2
    }    ## for my $read2
    return 1;    ## merge is OK
}
```

13.7   Developing Consensus Sequences

Subroutines for deriving consensus sequences for contigs from their reads and supporting alignments are encapsulated in the ConsensusBuilder package. It exports a single subroutine, constructConsensusSequences, for use by the main program.

```
sub constructConsensusSequences {
    my ($contigs) = @_;
    foreach (@$contigs) {
        if (@{$_->getReads()}>1) {
            constructConsensusFromGraph($_);
        } else {    ## singleton contig
            my ($read) = @{$_->Reads()};
            $_->setLength($read->Length());
            $_->setBases($read->Bases());
            $_->setQuals(join("",@{$read->AdjQual()}));
            $_->setScore(sum(@{$read->AdjQual()}));
        }
    }
}
```

This subroutine's only argument is a reference to a list of Contigs. It considers each Contig in turn, calling subroutine constructConsensusFromGraph for those having more than one read while simply copying information from the single DnaRead object to the Object for those having only one.

Subroutine constructConsensusFromGraph works in three stages.

1. First, the alignments of the contig are processed to assemble sets of positions on its reads that correspond to the same position in the contig's sequence. The sets consist of (read, position) pairs that we call *read-sites*. We call a position on the contig a *contig-site*. Read-sites corresponding to the same contig-site are *equivalent*. Therefore, we name the subroutine to which this task is delegated findEquivalentReadSites. It has three return values. The first is a reference to a hash mapping each contig-site to a list of equivalent read-sites. The second is a reference to a hash mapping each read to a list of positions that have been found to be equivalent to some position in another read. The last is a reference to a union-find structure (see Appendix C) mapping each read-site to its contig-site.

2. Next, the natural left-to-right orderings of the read-sites within the individual reads are found and saved explicitly in hashes named %rsPredecessors and %rsSuccessors. This information is important because it establishes an approximate left-to-right ordering on the contig-sites in the guise of a partial order $\prec$. Specifically, if (r, i) and (r, j) are read-sites with $i < j$ and if S_1 and S_2 are

contig-sites containing (r, i) and (r, j) respectively, then $S_1 \prec S_2$; in this case, we call the substring $r_i r_{i+1} r_{i+2} \ldots r_{j-1}$ a *snippet*. The task of filling the hashes %rsPredecessors and %rsSuccessors is delegated to findReadSiteOrder.

3. Finally, constructConsensusFromGraph itself uses the information gathered in the first two steps to assign a best-possible suffix, or *completion,* to each contig-site. It works approximately right to left by dynamic programming. The "possible" suffixes from contig-site S are derived from chains of successors

$$S = S_0 \prec S_1 \prec S_2 \prec \cdots \prec S_k,$$

where each link $S_i \prec S_{i+1}$ in the chain contributes to the suffix a snippet of one of the reads. The "best" suffix is the one in which the sum of the quality values on all snippets is maximized.

findEquivalentReadSites identifies equivalent read-sites by scanning alignment paths for long runs of matches. When a run of eight or more matching nucleotides is found, read-sites are established on both reads at four-base intervals and recorded as equivalent in the UnionFind structure $uf. Read-sites are encoded as strings consisting of read serial number and position on read, separated by Perl's standard separator $;.

```
sub findEquivalentReadSites {
    my ($contig) = @_;
    my %readSitePairs;
    my $uf = UnionFind->new();      ## maps readSites to contigSites
    foreach my $alignment (@{$contig->AlignedPairs()}) {
        my ($site1,$site2) = ($alignment->Start1(),$alignment->Start2());
        my $serial1 = $alignment->Read1()->SerialNumber();
        my $serial2 = $alignment->Read2()->SerialNumber();
        my $path = $alignment->Path();
        while ($path) {
            my $column = substr($path,0,1);
            $site1++ unless $column eq "I";
            $site2++ unless $column eq "D";
            if ($path =~ /^M{8}/) {
                while ($path =~ s/^M{4}//) {
                    my $rsPair1 = $serial1 . $; . $site1;
                    my $rsPair2 = $serial2 . $; . $site2;
                    $uf->union($rsPair1,$rsPair2);
                    $readSitePairs{$rsPair1} = 1;
                    $readSitePairs{$rsPair2} = 1;
                    $site1 += 4; $site2 += 4;
                }
```

```
            } else { $path =~ s/^.//; }
        }
    }
    ...
}
```

Once the equivalences are established, it is easy to fill the hashes that map (a) contig-sites to lists of read-sites and (b) reads to lists of sites. Iteration through all read-sites is easy because each has been made a key of the hash readSitePairs.

```
sub findEquivalentReadSites {
    ...
    my (%equivClasses,%sitesOnRead);
    foreach my $rsPair (keys %readSitePairs) {
        my $root = $uf->find($rsPair);
        $equivClasses{$root} ||= [ ];
        push @{$equivClasses{$root}}, $rsPair;
        my ($read,$site) = split($;,$rsPair);
        $sitesOnRead{$read} ||= [ ];
        push @{$sitesOnRead{$read}}, $site;
    }
    return (\%equivClasses,\%sitesOnRead,$uf);
}
```

Subroutine findReadSiteOrder fills the hashes %rsPredecessors and %rsSuccessors that indirectly define the $\prec$ relation between contig-sites. To make explicit the left-to-right orderings among the read-sites, it is necessary only to sort the lists of integers that have been created for each read in %sitesOnRead and then traverse them left to right. findReadSiteOrder also adds artificial read-sites $(r, 0)$ and $(r, |r|)$ at the two ends of each read r, as well as a single dummy read-site identified by the string SOURCE at the left end of each read. The left dummy is the predecessor of every $(r, 0)$ read-site.

```
sub findReadSiteOrder {
    my ($equivClasses,$sitesOnRead) = @_;
    ## construct a hash of successors for each read-site (RS) pair
    my %rsSuccessors;       ## maps each RS pair to next pair on same read
    my %rsPredecessors;     ## maps RS pair to previous pair on same read
    ## dummy equiv. class; contains dummy site for each read in contig
    $$equivClasses{"SOURCE"} ||= [ ];
```

```perl
foreach my $read (keys %$sitesOnRead) {
    my @sites = sort {$a<=>$b} @{$$sitesOnRead{$read}};
    $$sitesOnRead{$read} = \@sites;
    my $lastPair = $read . $; . 0;
    push @{$$equivClasses{"SOURCE"}}, $lastPair;
    foreach (@sites) {
        my $rsPair = $read . $; . $_;
        $rsSuccessors{$lastPair} = $rsPair;
        $rsPredecessors{$rsPair} = $lastPair;
        $lastPair = $rsPair;
    }
    my $rsPair = $read . $; . DnaRead->GetBySerial($read)->Length();
    $rsSuccessors{$lastPair} = $rsPair;
    $rsPredecessors{$rsPair} = $lastPair;
    $$equivClasses{$rsPair} = [$rsPair];
}
return (\%rsSuccessors,\%rsPredecessors);
}
```

With this information at hand, constructConsensusFromGraph can begin to find completions for each contig-site as follows. Let

$$S = \{(r_1, i_i), (r_2, i_2), \ldots, (r_k, i_k)\}$$

and

$$S' = \{(r_1', i_i'), (r_2', i_2'), \ldots, (r_{k'}', i_{k'}')\}$$

be two contig-sites, and let $R = \{r_1, r_2, \ldots, r_k\}$ and $R' = \{r_1', r_2', \ldots, r_k'\}$ be the corresponding sets of reads alone. Suppose that read $r \in R \cap R'$ so that $r = r_m = r_n'$ for some m and n. Then let $b(S, S', r)$ be the sum of the quality values for bases i_m through $i_n' - 1$ of read r, and let

$$b(S, S') = \max_{r \in R \cap R'} b(S, S', r)$$

be the largest sum of qualities to be had by advancing from any read-site in S to a site on the same read in S'. Then best(S), the score of the completion of S, can be defined recursively by

$$\text{best}(S) = \max_{S': S \prec S'} (b(S, S') + \text{best}(S')).$$

The recursive definition is reminiscent of others to which we have applied dynamic programming, but with one important difference: the set of values of "best" that must

be computed before computing best(S) is defined not by some simple arithmetic relationships between array indices but instead by the $\prec$ relation stored in the hashes
%$rsSuccessors and %$rsPredecessors. To make sure that the values of successors are available as needed for computing the values of their predecessors, we will
perform a *topological sort* of the contig-sites. The hash %pending will contain, for
each contig-site, the number of successor sites still without values for "best". When
a value is computed for any contig-site, %$rsPredecessors will be used to find and
decrement the count in %pending of each of its predecessors. Whenever one of these
values drops to 0, the contig-site will be added to the @ready list. Each step through
the outer loop will remove an item from @ready, compute its best suffix and score,
and update %pending and perhaps @ready.

To get the process started, constructConsensusFromGraph must count the number of successors of each contig-site in %pending and then collect contig-sites with
0 counts in @ready. Of course, the initial 0 counts will correspond only to the artifi-
cial $(r, |r|)$ read-sites introduced by findReadSiteOrder.

```
sub constructConsensusFromGraph {
    my ($contig) = @_;
    my ($equivClasses,$sitesOnRead,$classNames)
        = findEquivalentReadSites($contig);
    my ($rsSuccessors,$rsPredecessors)
        = findReadSiteOrder($equivClasses,$sitesOnRead);
    my %completion;      ## gives the best possible nucleotide sequence
    my %quality;         ## gives individual base qualities for above
    my %score;       ## gives total score (sum of qualities) of best sequence
    my %pending;     ## number of successors with still-unknown completion
    my @ready;

    foreach my $eqClass (keys %$equivClasses) {
        $pending{$eqClass} = @{$$equivClasses{$eqClass}};
    }
    foreach (keys %$rsPredecessors) {
        push @ready, $_ unless defined $$rsSuccessors{$_};
    }

    while (@ready) {
        my $eqClass = pop @ready;
        my ($bestString,$bestScore);
        foreach my $readSite (@{$$equivClasses{$eqClass}}) {
            my ($readNum,$site) = split($;,$readSite);
            my $predReadSite = $$rsPredecessors{$readSite};
            my $predContigSite = $classNames->find($predReadSite);
```

```
            $pending{$predContigSite}——;
            push @ready, $predContigSite if $pending{$predContigSite}==0;
            my $succReadSite = $$rsSuccessors{$readSite};
            next unless $succReadSite;    ## dummy sinks have no successor
            my $succContigSite = $classNames->find($succReadSite);
            my $succSite = (split($;,$succReadSite))[1];
            my $readEntry = DnaRead->GetBySerial($readNum);

            my $score = $score{$classNames->find($succReadSite)};
            my @quals = @{$readEntry->tAdjQual()}[$site..($succSite-1)];
            foreach (@quals) { $score += $_; }

            next unless ($score>$score{$eqClass});

            $score{$eqClass} = $score;
            my $snippet = substr($readEntry->Bases(),$site,$succSite-$site);
            $completion{$eqClass}=$snippet . $completion{$succContigSite};
            $quality{$eqClass} = join('',@quals,$quality{$succContigSite});
        }
    }
    my ($bestKey,$bestScore);
    foreach (@{$$equivClasses{SOURCE}}) {
        my $succ = $$rsSuccessors{$_};
        ($bestKey,$bestScore) = ($succ,$score{$succ})
            if $score{$succ}>$bestScore;
    }
    $contig->setLength(length($completion{$bestKey}));
    $contig->setBases($completion{$bestKey});
    $contig->setQuals($quality{$bestKey});
    $contig->setScore($bestScore);
}
```

13.8 Exercises

1. Complete the code for FastqReader.pm, as described in Section 13.3. Change verifyFormat and the line controlling the **while**-loop in readSeq. Collect both letters and digits in $tbuff, and eventually separate the two when the end of an entry is found. Add method seqQuality to return the qualities assigned to the bases of the sequence.

2. Reads that have a high-scoring alignment with themselves will almost certainly cause problems during the consensus-building phase of the practical algorithm.

The most rudimentary solution is to simply throw out these reads during the alignment phase. Implement this solution, or a better one.

3. Explain why the designers of Perl decided not to provide either implicit or explicit conversion of strings to reference.

4. To increment the ContigStart field of a DnaRead object $r object, it is necessary to write

---

$r−>setContigStart($r−>ContigStart( )+1);

Investigate Perl 5's experimental lvalue subroutines. Revise DnaRead.pm to allow the statement above to be replaced by

---

$r−>ContigStart( )++

---

5. Subroutine constructConsensusFromGraph has one serious inefficiency: it explicitly stores the entire suffix computed for each string. This means that bases (and qualities) near the end of reads may be copied and stored many times, wasting both space and time. Correct this problem.

   Hint: Begin by replacing the statements

$completion{$eqClass} = $snippet . $completion{$succContigSite};
$quality{$eqClass} = **join**('',@quals,$quality{$succContigSite});

---

with

$contribution{$eqClass} = $snippet;
$quality{$eqClass}
    = **join**('',@{$readEntry−>getAdjQual( )}[$site..$succ−1]);

Record additional information with each contig-site to help you assemble only a single long string after the score of the dummy SOURCE contig-site has been found.

6. Add a new method SuffixTree−>overlap. This method should return the length of the longest prefix of its argument that is also a suffix of the target string.

7. Remember that looking up a string in a hash table takes time proportional to the length of the string, whereas looking up an array entry takes only constant time. One way to speed up our branch-and-bound program for shortest common superstrings would be to redesign our adjacency lists to replace hash look-ups by array

accesses whenever possible. A typical strategy is to assign consecutive integers 0, 1, 2, ... to the fragments as they are read in and simultaneously construct a hash table for translating fragments to integers and an array for translating fragment numbers back to strings. Then the strings can be replaced by the corresponding numbers in the adjacency lists and the hash tables can be replaced by arrays. For example, the graph for "Mary had a little lamb" might become (depending on the assignment of integers):

([[1,2], [2,0], [3,2], [4,0], [5,2]],
 [[0,1], [2,5], [3,0], [4,0], [5,0]],
 [[0,0], [1,0], [3,3], [4,0], [5,0]],
 [[0,0], [1,2], [2,0], [4,10],[5,2]],
 [[0,0], [1,0], [2,0], [3,0], [5,3]],
 [[0,0], [1,0], [2,0], [3,0], [4,0]]
)

Implement this suggestion. (It may be a good idea to create a `Graph.pm` module at this point.)

8. To be able to solve the shortest common superstring problem more quickly, we attempted "to devise a fast-to-compute but reasonably accurate upper bound on the improvement possible by extending the current partial solution to its optimal completion". In graph-theoretic terms, the "optimal completion" is a maximal-weight simple path beginning at the last visited vertex v and passing through each of the unvisited vertices exactly once. In other words, it is the set of edges of maximum weight among the family of sets satisfying the following conditions:[7]

   (a) it includes a path from v to every unvisited vertex;
   (b) it includes no more than one outgoing edge from v;
   (c) it includes no more than one outgoing edge from each unvisited vertex;
   (d) it includes no more than one incoming edge into each unvisited vertex;
   (e) the number of edges in the set equals the number of unvisited vertices;
   (f) it includes no cycles;
   (g) it includes only edges from v or unvisited vertices to unvisited vertices.

   By eliminating some of the conditions, we create a "superfamily" of the possible completions. Since the superfamily includes the optimal completion, its maximum-weight member has weight no smaller than the optimum. The "fast-to-compute but reasonably accurate upper bound" developed on page 201 was found by dropping conditions (a), (b), (c), and (f) but enforcing (d), (e), and (g).

---

[7] These conditions are not independent; for example, (f) is a necessary consequence of the conjunction of (a) and (e).

Find other reasonably fast but more accurate bounds by enforcing different subsets of the conditions – for example, by dropping (c) but enforcing (a), (b), (d), (e), (f), and (g); or by dropping (a), (d), and (f) but enforcing (b), (c), (e), and (g).

9. Another approach to fragment assembly is to regard it as multiple sequence alignment in which there are no penalties for gap symbols introduced at the beginning or end of a fragment. Modify the algorithms of Chapters 9 and 10 to implement this approach; explain its impracticality.

13.9 Complete Program Listings

Directory `chap13/` of the software distribution includes program file `scs.pl` for our shortest common subsequence program and, for our mini-PHRAP program, the files `miniphrap.pl`, `DnaRead.pm`, `Alignment.pm`, `Contig.pm`, `Judicious-Aligner.pm`, `QualityReviser.pm`, `ContigBuilder.pm`, and `Consensus-Builder.pm`.

13.10 Bibliographic Notes

Computer scientists (see Kosaraju, Park, and Stein 1994) have shown that an approximate solution to the shortest common superstring problem can be found efficiently.[8] At first, a method guaranteeing an output no more than four times as long as the true optimum was found. Subsequently, this constant has been reduced, but not sufficiently to make the answers interesting to biologists.

PHRED's authors (Ewing and Green 1998) describe the base-calling process and the idiosyncrasies of different sequencing chemistries. The internals of the PHRAP program have previously been discussed only in code-release documentation. The system is available from its author, Philip Green of the University of Washington. Dean and Staden (1991) developed one alternative to PHRAP; Huang's (1996) CAP assemblers provide another.

---

[8] That is, in time polynomial in the length of the input.

## CHAPTER FOURTEEN

Coding Sequence Prediction with Dicodons

Once a new segment of DNA is sequenced and assembled, researchers are usually most interested in knowing what proteins, if any, it encodes. We have learned that much DNA does not encode proteins: some encodes catalytic RNAs, some regulates the rate of production of proteins by varying the ease with which transcriptases or ribosomes bind to coding sequences, and much has no known function. If study of proteins is the goal, how can their sequences be extracted from the DNA? This question is the main focus of *gene finding* or *gene prediction*.[1]

One approach is to look for *open reading frames* (ORFs). An open reading frame is simply a sequence of codons beginning with ATG for methionine and ending with one of the stop codons TAA, TGA, or TAG. To gain confidence that an ORF really encodes a gene, we can translate it and search for homologous proteins in a protein database. However, there are several difficulties with this method.

- It is ineffective in eukaryotic DNA, in which coding sequences for a single gene are interrupted by introns.
- It is ineffective when the coding sequence extends beyond either end of the available sequence.
- Random DNA contains many short ORFs that don't code for proteins. This is because one of every 64 random codons codes for M and three of every 64 are stop codons.
- The proteins it detects will probably not be that interesting since they will be very similar to proteins with known functions.

Another approach is to develop a statistical model that exploits both the nonuniform distribution of amino acids in real proteins and *codon bias* – the tendency of different codons for the same amino acid to occur nonuniformly. For example, *Bacillus subtilis* coding sequences exhibit obvious codon bias in coding leucine: CCT, the

---

[1] Gene finding includes the identification of catalytic RNAs and other elements as well, but the extraction of protein-coding sequences dominates the area.

most frequently occurring L codon, occurs 4.7 times as often as CTA, the least frequent.[2] Of course, different amino acids appear with different frequencies, too. In *B. subtilis,* the most common codon overall is AAA for lysine and the least common is CCC for proline, with a ratio of 15 to 1.

Codon bias varies not only between genes within a single species but also from species to species. The first type of variation is apparently related to the relative availability of tRNAs corresponding to the various codons. The coding sequences of the most highly expressed proteins rely on the most readily available tRNAs for fast translation; coding sequences with less common codons put the brakes on translation by forcing the ribosome to wait for a rare tRNA. The second type of variation may relate to the effects of sequence differences among ribosomal RNAs on the ribosome's "grip" on different codons during translation. Although it may eventually be possible to derive estimates of codon biases directly from study of a species's rRNA and tRNA sequences or other elements, at present it is most effective to develop statistical models of codon bias for individual species by analyzing a training set for each species to be considered.

This leaves us with a chicken-and-egg problem: If no genes are known for a particular species, how can we build a model to find genes for that species? One answer is to employ a "boot-strapping" process: We can use a model for a closely (or, if necessary, not-so-closely) related species to predict some genes; then we can use those genes as a training set for a new and improved model. The risk, as with all statistical models, is that the training set is not representative of the (unknown) set of all genes and that we overlook potentially interesting genes that are not similar to the training set.

For now, we assume that a training set exists and concentrate on how to process it in order to predict new genes accurately and efficiently.

14.1 A Simple Trigram Model

We will begin with an unrealistically simple version of the gene-finding problem. In this version, the DNA sequence is known to contain exactly one coding sequence for a single-exon gene. Its reading frame is also known; we only need to find where it starts and ends. Of course, we will also need a training set of known coding sequences.

Our first model for predicting genes will be based on log-of-odds scores (see Chapter 4) for each three-base subsequence of our sequence. We will call these *trigrams,* since *codon* correctly refers only to a trigram that actually codes for an amino acid in a coding region. We base our model of coding regions $\mathcal{M}_C$ on the 64 observed frequencies of codons in the training set, $p_C(\text{AAA})$, $p_C(\text{AAC})$, ..., $p_C(\text{TTT})$.

Although it is possible to develop a training set and empirical model of noncoding sequences, it is more typical to model noncoding sequences as random DNA, in

---

[2] 28,259 : 6038, according to an analysis of annotated coding sequences in the GenBank Accession NC_000964.

which each position is statistically independent of every other and each of the four nucleotides is equally likely to appear in each position. Under this model $\mathcal{M}_N$, the trigram probabilities are defined by

$$p_N(\text{AAA}) = p_N(\text{AAC}) = \cdots = p_N(\text{TTT}) = (1/4)^3 = 1/64.$$

So the lod score in bits of a given trigram $b_1 b_2 b_3$ is

$$
\begin{aligned}
\text{lod}(b_1 b_2 b_3) &= \lg\left(\frac{\Pr(b_1 b_2 b_3 \mid \mathcal{M}_C)}{\Pr(b_1 b_2 b_3 \mid \mathcal{M}_N)}\right) \\
&= \lg(p_C(b_1 b_2 b_3)) - \lg(1/64) \\
&= \lg(p_C(b_1 b_2 b_3)) + 6.
\end{aligned}
$$

For a DNA sequence $b_0 b_1 b_2 \ldots b_{n-1}$ with n divisible by 3, we begin by computing

$$\text{lod}(b_0 b_1 b_2), \text{lod}(b_3 b_4 b_5), \text{lod}(b_6 b_7 b_8), \ldots, \text{lod}(b_{n-3} b_{n-2} b_{n-1}).$$

Then, for any subsequence $b_{3i} b_{3i+1} b_{3i+2} \ldots b_{3j} b_{3j+1} b_{3j+2}$, we can compute the score

$$
\begin{aligned}
s(3i, 3j+2) &= \text{lod}(b_{3i} b_{3i+1} b_{3i+2} \ldots b_{3j} b_{3j+1} b_{3j+2}) \\
&= \lg\left(\frac{\Pr(b_{3i} \ldots b_{3j+2} \mid \mathcal{M}_C)}{\Pr(b_{3i} \ldots b_{3j+2} \mid \mathcal{M}_N)}\right) \\
&= \lg\left(\prod_{k=i}^{j} \frac{\Pr(b_{3k} b_{3k+1} b_{3k+2} \mid \mathcal{M}_C)}{\Pr(b_{3k} b_{3k+1} b_{3k+2} \mid \mathcal{M}_N)}\right) \\
&= \sum_{k=i}^{j} \text{lod}(b_{3k} b_{3k+1} b_{3k+2}).
\end{aligned}
$$

Our prediction for the location of the gene should be the substring with the highest score, that is, the i and j for which

$$s(3i, 3j+2) = \max_{p,q} s(3p, 3q+2).$$

Underlying this calculation is an assumption that adjacent codons are statistically independent; this comes into play when the probability of the substring is expressed as the product of the probabilities of the individual trigrams.

A naive approach to finding i and j would be to compute each of the $\sum_{m=0}^{n/3} m \approx n^2/18$ sums separately, performing $j - i$ additions to find $s(3i, 3j+2)$; this would give an algorithm requiring time proportional to the third power of the length of the DNA sequence. Fortunately, we can reduce the time to be linear in the length of the sequence by defining

$$s_{\max}(3j+2) = \max_p s(3p, 3j+2)$$

and noting that

$$\max_{p,q} s(3p, 3q + 2) = \max_{q} \left(\max_{p} s(3p, 3q + 2) \right) = \max_{q} (s_{\max}(3q + 2)).$$

These equations tell us that, if we know the starting point and score of just a *single best* candidate coding sequence ending at each possible position in the sequence, then we can find the best overall in linear time.

But the best candidate at each position is constructed either by extending the best candidate at the previous position or by starting anew with an empty sequence; thus the recurrence

$$s_{\max}(3j + 2) = \max(0, s_{\max}(3j - 1) + \mathrm{lod}(b_{3j}b_{3j+1}b_{3j+2}))$$

holds, and we can use it in a dynamic programming algorithm to compute all needed values of $s_{\max}$ in linear time by working from left to right.

This method has two peculiar deficiencies in its handling of stop codons. The first is that it could possibly allow a stop codon in the middle of its prediction, even though this is impossible in real genes. We can easily correct this by explicitly checking each codon to see if it is a stop codon.

The second flaw is that the prediction will *never* end precisely at the stop codon that terminates the gene. This is because all three stop codons have *negative* lod scores – they are *less* likely to occur in coding sequences than in random sequences – and a subsequence ending (or beginning) with a negative-scoring codon can never have the maximum score. We know that stop-codon lod scores are negative because otherwise some stop codon would occur on average more than once every 64 codons. This would mean that the average protein length is at most 64 amino acids; in fact, very few proteins are this short.

To compensate for this, we can search forward for a stop codon from the end of the subsequence with maximum lod score and then revise the prediction accordingly. The following program incorporates both of these improvements.

```
my %lod = {AAA=>...};     ## maps codons to log-of-odds scores
my %stopCodon = {TAA=>1,TGA=>1,TAG=>1};

my $dna = <STDIN>;
my ($lastScore,$lastStart);
my ($bestScore,$bestStart,$bestEnd);

for (my $i=0; $i<=(length $dna)−3; $i+=3) {
    if ($stopCodon{substr($dna,$i,3)}) {
        ($lastScore,$lastStart) = (0,$i+3);
        next;
    }
    my $trigramScore = $lod{substr($dna,$i,3)};
```

```
    my $newScore = max(0,$trigramScore+$lastScore);
    my $newStart = $newScore ? $lastStart : $i+3;
    ($bestScore,$bestStart,$bestEnd)
        = ($newScore,$newStart,$i) if ($newScore>$bestScore);
    ($lastScore,$lastStart) = ($newScore,$newStart);
}
while (!($stopCodon{substr($dna,$bestEnd,3)})) { $bestEnd += 3; }
print "Likeliest Gene: ", substr($dna,$bestStart,$bestEnd+3−$bestStart);
```

14.2 A Hexagram Model

Although trigrams – corresponding as they do to codons in coding regions – seem a natural choice for developing statistical models of coding sequences, there are no theoretical obstacles to using substrings of other lengths. This section describes a model based on substrings of length 6, or *hexagrams,* and the corresponding frequencies of *dicodons* in coding sequences. Dicodon models, especially when supplemented with models of exon–intron boundaries and upstream elements such as the Shine–Dalgarno sequence, have been found to be quite effective gene predictors. The strength of this model comes from its exploitation of both codon bias and the statistical dependence of adjacent residues in a protein.[3]

We assume as before that a training set of known genes exists. To collect dicodon frequencies, we extract overlapping substrings of length 6 at intervals of 3. For example, if the training set contains the string

        ATGGATCGTCGCCCAATA...

then the dicodons to be counted are

        ATGGAT, GATCGT, CGTCGC, CGCCCA, CCAATA, ...

From the resulting frequencies, we can compute a log-of-odds score for each of the $4^6 = 4096$ different hexagrams by the formula

$$\text{lod}(b_1b_2b_3b_4b_5b_6) = \lg\left(\frac{\Pr(b_1b_2b_3b_4b_5b_6 \mid \mathcal{M}_C)}{\Pr(b_1b_2b_3b_4b_5b_6 \mid \mathcal{M}_N)}\right)$$

$$= \lg(p_C(b_1b_2b_3b_4b_5b_6)) - \lg(1/4096)$$

$$= \lg(p_C(b_1b_2b_3b_4b_5b_6)) + 12.$$

---

[3] The presence of this dependence is clearly documented, even though we have repeatedly assumed in other statistical models that it does not exist. Fortunately, in most cases the dependence does not seem to destroy the practical effectiveness of many methods built on an explicit assumption of its absence.

To predict genes in DNA sequence $b_0 b_1 b_2 \ldots b_{n-1}$, we begin by computing

$$\mathrm{lod}(b_0 b_1 b_2 b_3 b_4 b_5), \mathrm{lod}(b_3 b_4 b_5 b_6 b_7 b_8), \mathrm{lod}(b_6 b_7 b_8 b_9 b_{10} b_{11}), \ldots.$$

We can simplify our equations somewhat by writing $b_{[i:j]}$ for the the subsequence $b_i b_{i+1} b_{i+2} \ldots b_j$. If adjacent hexagrams are nearly independent and $b_{[i:j]}$ is moderately long, then we can estimate its log-of-odds score by

$$\mathrm{lod}(b_{[i:j]}) = \lg\left(\frac{\Pr(b_{[i:j]} \mid \mathcal{M}_C)}{\Pr(b_{[i:j]} \mid \mathcal{M}_N)}\right)$$

$$\approx \lg\left(\prod_{k:0\le 6k\le i-j-5} \frac{\Pr(b_{[i+6k:i+6k+5]} \mid \mathcal{M}_C)}{\Pr(b_{[i+6k:i+6k+5]} \mid \mathcal{M}_N)}\right)$$

$$= \sum_{k:0\le 6k\le j-i-5} \mathrm{lod}(b_{[i+6k:i+6k+5]});$$

this is just the sum of the scores of successive hexagrams beginning with the first six bases of the substring. On the other hand, as the underbraces in this diagram indicate,

$$\overbrace{\underbrace{\mathrm{bbbbbb}}\ \underbrace{\mathrm{bbbbbb}}\ \underbrace{\mathrm{bbbbbb}}\ \underbrace{\mathrm{bbbbbb}}\ \underbrace{\mathrm{bbbbbb}}\ \underbrace{\mathrm{bbb}}}$$

we can obtain a second estimate involving nearly independent events by shifting right three positions:

$$\mathrm{lod}(b_{[i:j]}) \approx \sum_{k:0\le 6k\le j-i-8} \mathrm{lod}(b_{[i+6k+3:i+6k+8]}).$$

We can obtain a presumably more accurate estimate by averaging the two:

$$\mathrm{lod}(b_{[i:j]})$$

$$\approx \frac{1}{2}\left(\sum_{k:0\le 6k\le j-i-5} \mathrm{lod}(b_{[i+6k:i+6k+5]}) + \sum_{k:0\le 6k\le j-i-8} \mathrm{lod}(b_{[i+6k+3:i+6k+8]})\right)$$

$$= \frac{1}{2}\sum_{k:0\le 3k\le j-i-5} \mathrm{lod}(b_{[i+3k:i+3k+5]}).$$

Since we will be mainly interested in the *relative* scores of the subsequences, we can ignore the factor of $1/2$ and assign the score

$$s(i, j) = \sum_{k:0\le 3k\le j-i-5} \mathrm{lod}(b_{i+3k} b_{i+3k+1} b_{i+3k+2} b_{i+3k+3} b_{i+3k+4} b_{i+3k+5}) \qquad (14.1)$$

to subsequence $b_i b_{i+1} b_{i+2} \ldots b_j$ and then look for the subsequence maximizing $s(\cdot, \cdot)$.

14.3   Predicting All Genes

We can now apply the statistical ideas we have developed to the task of finding all genes present in a sequence, on both strands and in any reading frame.

First, we address the reading-frame issue. There are three reading frames on "Watson" and three on "Crick" (as the two strands are sometimes called). To find the best prediction in any of Watson's reading frames, we must simply compute

$$\mathrm{lod}(b_0 b_1 b_2 b_3 b_4 b_5), \ \mathrm{lod}(b_1 b_2 b_3 b_4 b_5 b_6), \ \mathrm{lod}(b_2 b_3 b_4 b_5 b_6 b_7), \ldots,$$

and again apply (14.1) to find which of the roughly $n^2/6$ subsequences with length divisible by 3 has the highest score. We can do this either in three left-to-right passes or in a single careful pass that maintains the last score for each of the three reading frames. To search Crick's reading frames, we generate a representation of Crick and repeat the process.

Predicting multiple genes in a sequence is a little more complicated, especially in the eukaryotic case. We will discuss prokaryotes first.

Prokaryotic genes lack introns, and they tend to be densely packed with little DNA between genes. A greedy sort of divide-and-conquer strategy works well: Identify the most likely gene; remove the gene, splitting the sequence into two; and repeat the process on the two pieces. The process should be ended on any sequence for which the score of the predicted gene falls below some threshold, such as the lowest score of any real gene found in the training set.

The coding sequences of eukaryotic genes are interrupted by introns. Although the rough extent of a coding sequence can usually be detected by a simple dicodon model, intron–exon boundaries are identified only imprecisely unless the dicodon model is supplemented with statistical models of such boundaries.[4] Still, the greedy strategy outlined for prokaryotes may be adequate for some eukaryotic species. The Perl program described in the next section performed respectably on sequences from the eukaryote *Ashbya gossypii*, for example. This filamentous fungus, a minor pest to cotton growers, has very few multi-intron genes.

14.4   Gene Finding in Perl

This section describes a Perl module implementing the gene-finding ideas of the previous sections. The approach adopted is object-oriented; CSPredictor objects provide three methods to the external world as follows.

1. $csp = new CSPredictor $fileName;
   will create a new coding sequence predictor and assign a reference to $csp. The

---

[4] See Section 14.7 (p. 244).

new predictor will be based on the statistical model described in the named file. A typical file begins

```
45
aaaaaa 0.0003095
aaaaac 0.0004724
. . .
```

The first line gives a threshold score below which a subsequence will not be considered as a gene prediction. The remaining 4096 lines give the dicodon frequencies observed in the training set.

2. $predictions = $csp−>predictCodingSequences($seq);

will return a reference to a list of gene prediction for the DNA sequence $seq. The format of each item of this list is (*start, end, score*), where *start* and *end* are the indices of the first and last bases of the predicted coding sequence and *score* is the lod score of the predicted sequence as defined earlier in this chapter. If *start* is greater than *end*, this indicates that the predicted coding sequence lies on the complementary strand.

3. $fileName = $csp−>modelFile();

returns the name of the model file upon which predictor $csp is based.

Thus, a very simple program based on this module might be:

```
use SeqReader;
use CSPredictor;
my $predictor = new CSPredictor "E.coli.freqs";
my $reader = new SeqReader "E.coli.seqs";
while (my $seq = $reader−>getSeq()) {
    my $predictions = $predictor−>predictCodingSequences($seq);
    foreach (@$predictions) { displayPrediction($_); }
}
```

where we leave the precise effect of displayPrediction unspecified.

A CSPredictor object consists of a blessed reference to a hash containing three key–value pairs.

1. The value associated with key modelFile is a string giving the name of the model file from which the object was constructed.

2. Associated with key lodArr is a reference to an array giving the lod score for each of the 4096 possible hexagrams. Each hexagram is mapped to an integer in the set $\{0, 1, 2, \ldots, 4095\}$ by a method to be described shortly.

3. Associated with key threshold is the numerical threshold for accepting a subsequence lod score as a gene prediction.

We now turn to subroutine new, which reads the probabilities of model $\mathcal{M}_C$ from the model file, converts them to lod scores, creates the three-element hash, and returns the blessed reference. To process a line from the model file, it must first be split into a hexagram and a floating-point number. The loop control **while** (<FREQS>) assigns the next line in the model file to $_. The following line matches $_ against a complicated pattern containing two parenthesized subpatterns: one for the hexagram and another, $fpPat, designed to match a floating-point number. The $fpPat pattern is able to match numbers in scientific form, such as $-2.584E-88 = -2.584 \times 10^{-88}$. In this pattern, optional elements (such as minus signs) are followed by ?; if a group of optional items must appear together (such as an exponent E-88), they can be enclosed in (?:)?.

The first line within the loop assigns the hexagram to $k and the probability to $kprob. Here, we see the pattern-matching operator for the first time in a list context; rather than returning just "success" or "failure", it returns the list ($1,$2,$3,...) when it succeeds.

We can save a lot of calls to **substr** and therefore a lot of time by encoding every hexagram in the input sequence as an integer. This allows our table of lod scores to be stored as a list instead of a hash, which is larger and slower to access. The hexagram is translated into an array index by changing the four bases to the four digits 0, 1, 2, 3 and regarding the result as a base-4 number. Thus,

acttgc $= (013321)_4$

$$= (0 \times 4^5) + (1 \times 4^4) + (3 \times 4^3) + (3 \times 4^2) + (2 \times 4^1) + (1 \times 4^0)$$

$$= 0 + 256 + 192 + 48 + 8 + 1$$

$$= 505.$$

Once the lod scores are computed, the constituents of the new object are packaged into a hash and a blessed reference is returned.

---

sub new
{
    my ($this,   *## class name CSPredictor or an instance*
          $modelFile,   *## ref to file of dicodon frequencies*
) $= @_;$
    printf STDERR "Creating CSPredictor: $modelFile\n";

    my @lods $= ((0)$ **x** 4096);
    open FREQS, "<$modelFile" **or die** "can't open $modelFile\n";
    my $threshold $=$ <FREQS>;
    $threshold $=\tilde{}$ s/[^0-9]//;   *## remove any comments, etc.*
    my $fpPat $= '-?\d*(?:\.\d*)?(?:[Ee]-?\d+)?';$

```
    while (<FREQS>) {
        my ($k,$kprob) = m/([acgt]{6})\s*($fpPat)/;
        $k =~ tr/acgt/0123/;
        my $hexinx = 0;
        foreach (split //,$k) {
            $hexinx = $hexinx*4+int($_);
        }
        $lods[$hexinx] = ($kprob==0.0) ? -1E99 : (12+lg($kprob));
    }
    close FREQS;
    return bless {modelFile=>$modelFile,
                 lodArr=>\@lods,
                 threshold=>int($threshold)};
}
```

The method predictCodingSequences prepares the input sequence $watson for greedyPCS, which actually oversees the prediction process. The complement strand $crick must be created, and both strands must be preprocessed by calls to prepStrand.

```
sub predictCodingSequences
{
    my ($this,$watson) = @_;
    (my $crick = reverse $watson) =~ tr/acgt/tgca/;
    my @watson = $this->prepStrand(\$watson);
    my @crick = $this->prepStrand(\$crick);
    my $seqlen = length $watson;
    my @pcs = $this->greedyPCS(1,$seqlen,$seqlen,\@watson,\@crick);
    return \@pcs;
}
```

Subroutine prepStrand is responsible for initializing the dynamic programming arrays used by greedyPCS. It is called twice, once for $watson and once for $crick. To avoid repeated recomputation of hexamer lod scores, the array @lod is initialized so that entry $lod[$i] contains the lod score of the subsequence substr($$seq,$i−1,6). This is done in several stages. First, $lod[$i] is assigned the base-4 digit of the $ith base in the sequence. Next, it is updated to contain the numerical index of the trigram beginning at the $ith base, and then the index of the hexamer there. Finally, the hexamer index is replaced by the lod score drawn from the object's lodArr.

Concurrently, the array entries $best[$i] and $start[$i] are filled with score and starting point of the highest-scoring subsequence ending at position $i.

Eventually, prepStrand returns a list of references to these three arrays, the sequence, and its length.

```perl
sub prepStrand
{
    my ($this,      ## reference to this CSPredictor
        $seq        ## DNA sequence to be analyzed
       ) = @_;

    $$seq =~ tr/acgt/0123/;
    my @lod= (-1, (split //,$$seq),0,0,0,0,0,0);     ## pad
    my @best; $#best = $#lod-6;        ## allocate in one shot for speed
    my @start; $#start = $#lod-6;      ## allocate in one shot for speed
    my $lodArr = $this->{lodArr};      ## take out of the loop for speed

    for (my $i=1; $i<=3; $i++) {       ## initialize three frames
        $lod[$i] = 16*$lod[$i]+4*$lod[$i+1]+$lod[$i+2];
        $lod[$i+3] = 16*$lod[$i+3]+4*$lod[$i+4]+$lod[$i+5];
        $lod[$i] = $$lodArr[64*$lod[$i]+$lod[$i+3]];
        my $t = $lod[$i];
        ($best[$i],$start[$i]) = $t>0 ? ($t,$i) : (0,0);
    }
    for (my $i=4; $i<@start; $i++) {       ## the "real"loop
        $lod[$i+3] = 16*$lod[$i+3]+4*$lod[$i+4]+$lod[$i+5];
        $lod[$i] = $$lodArr[64*$lod[$i]+$lod[$i+3]];
        my $t = $best[$i-3]+$lod[$i];
        ($best[$i],$start[$i]) = $t>0 ? ($t, ($start[$i-3] || $i)) : (0,0);
    }
    $#lod = $#start;       ## remove padding from @lod
    return (\@lod, \@best, \@start, $seq, length $$seq);
}
```

The recursive subroutine greedyPCS calls reviewStrand to find the highest-scoring subsequence in either strand. If its score exceeds the threshold, then subroutine retrieveBest is called to tinker with the subsequence's boundaries so that they coincide with start and stop codons. The boundaries are added to the prediction list, and the subsequence is removed from further consideration. Then greedyPCS calls itself twice to make predictions to the left and right of the new prediction.

```perl
sub greedyPCS
{
    my ($this,$lo,$hi,$size,$watson,$crick) = @_;
```

```
   return ( ) if $lo>$hi−15;     ## too short for a coding sequence
   my ($winx,$wval) = reviewStrand($watson,$lo,$hi);
   my ($clo,$chi) = ($size+1−$hi,$size+1−$lo);
   my ($cinx,$cval) = reviewStrand($crick,$clo,$chi);
   my $bestv = ($wval>$cval) ? $wval : $cval;
   return( ) if $bestv < $this−>{threshold};
   my ($beststart,$bestend,$bestlo,$besthi);
   if ($wval>=$cval) {
       ($bestlo,$besthi) =
           ($beststart,$bestend) = retrieveBest($watson,$winx,$lo,$hi);
   } else {
       ($beststart,$bestend) = retrieveBest($crick,$cinx,$clo,$chi);
       ($besthi,$bestlo) =
           ($beststart,$bestend) = ($size+1−$beststart,$size+1−$bestend);
   }
   return ($this−>greedyPCS($lo,$bestlo−1,$size,$watson,$crick),
       [$beststart,$bestend,$bestv],
       $this−>greedyPCS($besthi+1,$hi,$size,$watson,$crick));
}
```

Subroutine reviewStrand searches a single strand's arrays in the current range of interest to find the highest-scoring subsequence. However, before this is possible, it is necessary to correct scores of any subsequences that end inside the current range but start outside it. The reverse situation is not problematic because we do not examine ending points outside the range of interest. The correction must be carried out separately for each reading frame. It commences at the lower end of the range of interest and terminates as soon as it detects a starting point within the current range. The subsequent search for highest score is a simple linear search.

```
sub reviewStrand
{
   my ($arr,      ## reference to the list of dynamic programming arrays
                  ## constructed by prepStrand
       $lo,$hi    ## range of current interest in the sequence
       ) = @_;
   my ($lod,$best,$start,$seq,$size) = @$arr;
   $hi = min($hi,$size−2);
   foreach my $i ($lo..$lo+2) {    ## for each reading frame
       my $t = $$lod[$i];
       ($$best[$i],$$start[$i]) = ($t>0) ? ($t,$i) : (0,0);
```

```perl
        for (my $j=$i+3; $j<=$hi; $j+=3) {
            ## stop as soon as entire sequence lies within range of interest
            last unless $$start[$j] && $$start[$j]<$lo;
            my $t = $$best[$j-3]+$$lod[$j];
            ($$best[$j],$$start[$j]) = $t>0 ? ($t, ($$start[$j-3] || $j)) : (0,0);
        }
    }
    ## find the best prediction's value and its location
    my ($besti,$bestv) = ($hi,$$best[$hi]);
    for (my $i=$hi-1; $i>=$lo; $i--) {
        ($besti,$bestv) = ($i,$$best[$i]) if $$best[$i] > $bestv;
    }
    return ($besti,$bestv);
}
```

The task of subroutine retrieveBest is to tinker with the boundaries of the highest-scoring subsequence to make them coincide with start and stop codons, if possible. This consists of forward and backward searches in the relevant reading frame.

```perl
sub retrieveBest
{
    my ($arr,       ## list of dynamic programming arrays
        $besti,     ## the index of the last codon of the best prediction
        $lo,$hi     ## current range of interest (don't exceed it!)
        ) = @_;
    my ($lod,$best,$start,$seq,$size) = @$arr;

    ### feel forward for stop codon
    my ($begin,$end) = ($$start[$besti]+3,$besti+2);
    $hi = min($hi,$besti+500,$size-2);
    for (my $i=$besti; $i<=$hi; $i+=3) {
        ($end=$i-1), last if $$lod[$i]==-1E99;
    }

    ### feel backward for start (M) codon
    $lo = max($lo,$begin-500);
    for (my $i=$begin-1; $i>$lo; $i-=3) {
        my $codon = substr($$seq,$i,3);
        last if $codon =~ /(302|320|300)/;     ## don't back up over "stop"
        return ($i+1,$end) if $codon eq "032";     ## atg – Methionine
    };
```

```
### feel forward for start (M) codon
$hi = min($begin+500,$end);
for (my $i=$begin−1; $i<$hi; $i+=3) {
    my $codon = substr($$seq,$i,3);
    return ($i+1,$end) if $codon eq "032";     ## atg – Methionine
};
return ($begin,$end);     ## no start nearby; return initial guess
}
```

14.5 Exercises

1. Write a Perl program to extract coding sequences from a file of sequences in Gen-Bank format by analyzing sequence annotations. This program could be used to develop a training set for the trigram or hexagram model.

2. In practice, it may be productive to establish background probabilities $p_N(A)$, $p_N(C)$, $p_N(G)$, $p_N(T)$ empirically and define $p_N(b_1 b_2 b_3) = p_N(b_1) \cdot p_N(b_2) \cdot p_N(b_3)$. Modify the format of the model file to include the background probabilities, and modify CSPredictor.pm to read them and use them to compute lod scores. What sort of sequences should be analyzed to establish background probabilities: coding, noncoding, or a representative combination?

3. Do noncoding sequences really satisfy the "independent, uniformly distributed" models of trigrams and hexagrams to which the coding sequence model is compared in this chapter? Investigate this question for one or more species using real genomic data and, if necessary, modify the models and programs in this chapter.

14.6 Complete Program Listings

Directory chap14/ of the software distribution includes package file CSPredictor.pm and a short driver driver.pl.

14.7 Bibliographic Notes

The causes of codon bias are explored in Powell and Moriyama (1997) and Sharp et al. (1993). Its use for gene finding is explored in Fickett and Tung (1992) and Staden and McLachlan (1982).

Salzberg et al. (1998a) describe a highly sophisticated gene-finding program for prokaryotes. Stormo (2000) outlines approaches for eukaryotes. Galas (2001) discusses gene finding in the human genome.

## CHAPTER FIFTEEN

Satellite Identification

We have already learned that the process of DNA replication is not perfect and that, in fact, this is a source of mutations both beneficial and deleterious in sequences encoding amino acid chains and regulating their production. Intergenic ("junk") DNA is also subject to mutations, and since these mutations pose no impediment to survival, they are preserved at much greater rates than mutations to coding and regulatory sequences. In such DNA, it is common to find *tandem repeats* consisting of several contiguous repetitions of the same short sequence.

The repetitions themselves may vary slightly as a result of point mutations. Furthermore, individuals within a population often carry different numbers of repetitions as a result of deviations from normal DNA replication. For example, the feature known as HUMHPRTB, which consists of varying numbers of repetitions of AGAT, was found to exist in nine different forms in a group of 417 humans; 314 group members carried two different forms or *alleles*. The sites of such variation are collectively called *VNTR* (*variable number of tandem repeat*) loci. Since mutations at VNTR loci are so frequent – up to 1% per gamete per generation – and are inherited, they form the basis of the highly publicized "DNA fingerprinting" techniques used to resolve paternity disputes and to free the wrongfully incarcerated. VNTRs can also be powerful tools for reconstructing pedigrees and phylogenies.

VNTR sequences, or *satellites*,[1] are commonly subdivided into two categories, which originate by distinct processes.

- *Microsatellites* are very short – 2–5 nucleotides – and arise from unusual events around the replication fork that cause the same nucleotides to be replicated repeatedly. Microsatellites are distributed rather uniformly throughout the genome.[2]

---

[1] The term "satellite" was attached to tandem repeats as a result of early work in which restriction fragments of DNA were subjected to centrifugation. The density of the fragments was found to be distributed according to a more-or-less bell-shaped curve with a "bump". The bump was named a satellite, and the name later transferred to the tandem repeats that were found to populate the bump.

[2] Microsatellites are also called *SSRs* (*simple sequence repeats*).

- *Minisatellites* are somewhat longer – 20 or so nucleotides. They arise from re-combination events between maternally and paternally inherited chromosomes and are more common near the telomeres (ends of chromosomes).

In this chapter, we will be concerned with the problem of identifying tandemly repeated sequences within a long sequence. Once VNTRs are identified in a single sequence, they can be counted and analyzed in all members of a population and used for comparison.

15.1   Finding Satellites Efficiently

Our algorithmic task is to find, in an input sequence, all tandem repeats subject to re-strictions specified by the user. Specifically, the user specifies the lengths l_{min} and l_{max} of the shortest and longest acceptable satellite, the minimum acceptable number of consecutive repetitions r_{min}, and the maximum relative mismatch ε permitted be-tween the model satellite and any one of its repetitions. Mismatch is measured by *edit distance,* the minimum number of insertions, deletions, and substitutions required to mutate one string to another. Edit distance is simply similarity computed with no re-ward for a match and a penalty of -1 for a mismatch or a gap. For example, the edit distance $d(\texttt{accgtgggtt}, \texttt{acgtttgttt})$ is 4; an optimal series of edit operations is easily derived from the global alignment

$$\begin{bmatrix} \texttt{accgtgggtt-} \\ \texttt{a-cgtttgttt} \end{bmatrix}.$$

More formally, given a DNA sequence s and the parameters $l_{min}, l_{max}, r_{min},$ and ε, we seek satellites t satisfying

$$l_{min} \leq |t| \leq l_{max},$$
$$s = ut_1 t_2 t_3 \ldots t_{r_{min}} v \tag{15.1}$$

and

$$d(t_i, t) \leq \varepsilon |t| \quad \text{for } 1 \leq i \leq r_{min}.$$

Since there are four bases, there are

$$\sum_{i=l_{min}}^{l_{max}} 4^i \approx \frac{4^{l_{max}+1}}{3}$$

possible satellites in the range of interest, and it is important to develop a strategy that avoids explicit consideration of each string. Our strategy is to build strings from right to left, analyzing each string for its potential to form a *suffix* of a satellite. If a string can be shown to be unusable as a suffix, all longer strings based on it can be immediately eliminated from consideration.

15.1.1   Suffix Testing

What are some necessary conditions for a string z to be a possible suffix of a satellite? First of all, there must be some substrings z_i of s that match z approximately. Second, at least r_{min} of these approximate matches must appear roughly periodically at intervals comparable to the length of the satellites being sought. We will call such a set of approximate matches a *supporting chain* for suffix z. To be useful for programming, the conditions defining a supporting chain need to be refined and formalized.

To determine the required degree of matching, we recall that each repetition t_i of our satellite t must satisfy $d(t_i, t) \le \varepsilon |t|$. How closely must a substring z_i match candidate suffix z to be of interest? Suppose that

$$s = u y z_1 y z_2 y z_3 \ldots y z_{r_{min}} v$$

and

$$d(y z_i, y z) = d(z_i, z) \le \varepsilon \cdot |yz| \quad \text{for all } i.$$

Then yz is a satellite, and z is a suffix of a satellite. Conversely, if $d(z_i, z) > \varepsilon \cdot |yz|$ for some i, then $(z_1, z_2, \ldots, z_{r_{min}})$ is not a supporting chain for z. Since $|yz|$ could be as great as l_{max}, only substrings z' failing to satisfy $d(z', z) \le \varepsilon \cdot l_{max}$ match too weakly to be part of any possible supporting chain for z's candidacy for satellite suffix.

To determine permissible spacing between repetitions, we reconsider equation (15.1) and the possible lengths of the t_i. If one string is k letters longer than another then their edit distance is at least k, since k gaps must be included in the shorter string in any alignment. It follows that, if $d(t_i, t) \le \varepsilon |t|$, then $(1 - \varepsilon)|t| \le |t_i| \le (1 + \varepsilon)|t|$. Since $|t|$ is not precisely known, the strongest possible conclusion is that

$$l_{min} - \varepsilon l_{max} \le |t_i| \le l_{max} + \varepsilon l_{max}.$$

Thus, an approximate match z_i can lend support to candidate suffix z only if another match z_j is positioned so that the distance δ between the left ends of the two substrings satisfies

$$\delta_{min} \le \delta \le \delta_{max},$$

where

$$\delta_{min} = l_{min} - \varepsilon l_{max},$$

$$\delta_{max} = l_{max} + \varepsilon l_{max}.$$

Furthermore, z_i must be part of a chain of at least r_{min} approximate matches separated by such an interval. To find such chains, we define three functions on positions in the string s. The function $r^{\ge}(i)$ gives the length of the longest chain of appropriately spaced approximate suffix matches beginning at position i, where position i is the position of the *left* end of the first match. Similarly, $r^{\leq}(i)$ gives the length of the longest chain *ending* with the match at position i, where position i is the position of the left end of the *last* match. Finally, $r(i)$ gives the length of the longest chain that includes the match at position i. These functions can be defined recursively as follows:

$$r^{\leq}(i) = \begin{cases} 1 + \displaystyle\max_{\delta_{\min} \leq \delta \leq \delta_{\max}} r^{\leq}(i - \delta) & \text{if } i \text{ is the left end of some } z' \\ & \text{such that } d(z', z) \leq \varepsilon \cdot l_{\max}, \\ 0 & \text{otherwise;} \end{cases}$$

$$r^{\geq}(i) = \begin{cases} 1 + \displaystyle\max_{\delta_{\min} \leq \delta \leq \delta_{\max}} r^{\geq}(i + \delta) & \text{if } i \text{ is the left end of some } z_i \\ & \text{such that } d(z', z) \leq \varepsilon \cdot l_{\max}, \\ 0 & \text{otherwise;} \end{cases}$$

$$r(i) = r^{\leq}(i) + r^{\geq}(i) - 1.$$

If $r(i) < r_{\min}$, then the match beginning at position i can lend no support to the proposition that the candidate suffix z can be expanded to a satellite.

Once we know the positions of the approximate matches z_i, it is easy to compute values of $r^{\leq}$ by scanning these positions in left-to-right order. Similarly, values of $r^{\geq}$ and r can be computed simultaneously in a second scan, this time from right to left.

To find these positions, we could modify Chapter 3's dynamic programming algorithm for pairwise alignment to find, for each i and k, the edit distance of the best match of the last k letters of the suffix z with any substring beginning at position i of the original sequence. Specifically, if $E_{i,k}$ represents this quantity, we have

$$E_{i,0} = 0;$$

$$E_{|s|+1,k} = k;$$

$$E_{i,k} = \min \begin{cases} E_{i,k-1} + 1, \\ E_{i+1,k} + 1, \\ E_{i+1,k-1} + (\text{if } w_k = s_i \text{ then } 0 \text{ else } 1), \end{cases} \tag{15.2}$$

where w is z reversed, so that w_k is the kth letter from the right end of z. We have usable matches at positions i with $E_{i,|z|} \leq \varepsilon \cdot l_{\max}$.

A crucial fact for our algorithm is that $E_{i,k} \leq b$ only if $E_{i-1,k+1} \leq b$. For if $E_{i-1,k+1} > b$, then both $E_{i-1,k} > b - 1$ and $E_{i,k+1} > b - 1$. This is because adding a letter to one string can at best match a letter previously paired with a gap, lowering edit distance by 1. But this implies that the three quantities $E_{i-1,k} + 1$, $E_{i,k+1} + 1$, and E_{ik} all exceed b.

Like its original, this dynamic programming algorithm requires time proportional to $|s| \cdot |z|$, the product of the lengths of the suffix and the input sequence. On the other hand, since we are building suffixes by prepending one character at a time, all but the last row of the table for suffix $z = bz'$ will have previously been filled for suffix z'. Since we are directly interested only in $E_{i,|z|}$ and since these can be computed from only $E_{i,|z'|}$, we can reduce the time spent for each suffix from $O(|s| \cdot |z|)$ to just $O(|s|)$ by saving (with each suffix considered) the values of $E_{i,|z|}$ for $1 \leq i \leq |s|$. In

```
0000000001111111111222222222233333333334444444444555555555
123456789012345678901234567890123456789012345678901234567 8
-----------------------------------------------------------
aggtcagtcaaggtcaggtaaggtcaagggtcaggcaggtagtccggtcacgtcggat
1   11  101    11  101    101      11  11  11       11    11 /aa/ matches
    1  101    1   11    101      1   1           1          /caa/ init matches
    7  666    5   44    333      2   1           1          /caa/ r>=
    1  122    3   44    555      6   7           1          /caa/ r<=
    7  677    7   77    777      7   7           1          /caa/ r
    1  101    1   11    101      1   1                      /caa/ final matches
```

Figure 15.1: Suffix testing with $l_{min} = 5$, $l_{max} = 10$, $r_{min} = 5$, and $\varepsilon = 0.1$.

fact, in order to grow longer suffixes from z efficiently, it is necessary only to remember values for positions that actually support z's candidacy, the *match set*

$$m(z) = \{(i, E_{i,|z|}) \mid E_{i,|z|} \le \varepsilon \cdot l_{max} \text{ and } r(i) \ge r_{min}\},$$

since no other positions can contribute to the construction of a usable match for a longer suffix. Typically, the size of match sets can be expected to decrease rapidly as characters are added to a suffix, unless the strings involved are actually suffixes of a genuine satellite.

The process of extending a suffix is illustrated by Figure 15.1. The search parameters were set to $l_{min} = 5$, $l_{max} = 10$, $r_{min} = 5$, and $\varepsilon = 0.1$; the edit distance between a suffix and its match must therefore be no more than $0.1 \cdot 10 = 1$. The first line below the string shows, for suffix $z = \text{aa}$, the best edit distance $E_{i,|z|}$ in each position i at which $E_{i,|z|} \le 1$. The next line shows approximate matches to the extended suffix caa computed according to (15.2). For example, the best matches at positions 24, 25, and 26 are reflected in the alignments

$$\begin{bmatrix} \text{tcaa} \\ \text{-caa} \end{bmatrix}, \quad \begin{bmatrix} \text{caa} \\ \text{caa} \end{bmatrix}, \quad \text{and} \quad \begin{bmatrix} \text{-aa} \\ \text{caa} \end{bmatrix}.$$

The next line illustrates the computation of $r^{\ge}(i)$. We have $r^{\ge}(36) = 1$ because the match at 49 is too far away (farther than 11) to form a chain. On the other hand, position 19 can chain with 24, 25, and 26, and each of these can chain with 32, and 32 can chain with 36. So $r^{\ge}(19) = 4$. (Positions 25 and 26 can also chain directly with 36.) For $r^{\le}$ we have, for example, $r^{\le}(8) = 1$ because the match at 5 is too *close* (i.e., <4) to form a chain. When computing r, we see that the match at position 49 can never be part of a sufficiently long chain to be of interest. Hence we can omit this position from the match set for caa. The final match set is

$$m(\text{"caa"}) = \{(5, 1), (8, 1), (9, 0), (10, 1), (15, 1), (19, 1),$$
$$(20, 1), (24, 1), (25, 0), (26, 1), (32, 1), (36, 1)\}.$$

15.1.2 Satellite Testing

Once a string has passed the suffix test, we must consider whether it is a satellite in its own right. To be absolutely certain that a suffix z is a satellite, we must consider

```
00000000001111111111222222222233333333334444444444555555555
12345678901234567890123456789012345678901234567890123456 78
-----------------------------------------------------------
aggtcagtcaaggtcaggtaaggtcaagggtcaggcaggtagtccggtcacgtcggat
      1  101    1    1    101    1                          /tcaa/ matches
      1  101    1    1    101    1                          /gtcaa/ init matches
      6  555    4    3    222    1                          /gtcaa/ r>=
         1                 1                                 /gtcaa/ R>=
```

Figure 15.2: Satellite testing with $l_{min} = 5$, $l_{max} = 10$, $r_{min} = 5$, and $\varepsilon = 0.1$.

individually all approximate matches starting at a given position i and determine for each, based on its individual length, which other positions offer opportunities for chaining. In this case, the matches to be chained must fit together exactly, with no intervening bases, as described by (15.1). We need to know

$$E'_{i,j} = d(z, s_i s_{i+1} s_{i+2} \cdots s_{j-1}),$$

but only where $E'_{i,j} \leq \varepsilon|z|$.[3] We can compute all values for a single i by aligning z to $s_i s_{i+1} s_{i+2} \cdots s_{i+\lfloor(1+\varepsilon)|z|\rfloor}$. Although this might appear to require $O((1 + \varepsilon)|z|^2)$ time per value of i, we can limit ourselves to $O(\varepsilon|z|^2)$ time by observing that entries farther than $\varepsilon|z|$ from the main diagonal of the dynamic programming table will always exceed $\varepsilon|z|$.

We use these values to determine chain lengths as follows:

$$R^{\geq}(i) = \begin{cases} 1 + \max\{R^{\geq}(i) \mid E'_{i,j} \leq \varepsilon|z|\} & \text{if } E_{i,|z|} \leq \varepsilon|z|, \\ 0 & \text{otherwise.} \end{cases}$$

As already noted, we can have $E'_{i,j} \leq \varepsilon|z|$ only if

$$i + |z| - \varepsilon|z| \leq j \leq i + |z| + \varepsilon|z|.$$

If $R^{\geq}(i) \geq r_{min}$ for any i, then the string z can finally be confirmed as a satellite.

The difference between $r^{\geq}$ and $R^{\geq}$ can be seen in the example of suffix and satellite testing of `gtcaa` in Figure 15.2. With search parameters $l_{min} = 5$, $l_{max} = 10$, $r_{min} = 5$, and $\varepsilon = 0.1$, we allow match error of $(0.1)10 = 1$ and have $5 \leq \delta \leq 11$ as possible distances between repeats during suffix testing. Thus, $r^{\geq}(22) = r^{\geq}(23) = r^{\geq}(24) = 2$, since $\delta = 8, 9, 10$ lie within the permissible range and allow the repeat at position 30 to be reached. But during satellite testing we allow only $\lfloor(0.1)5\rfloor = 0$ errors in approximate matches, a criterion met only at positions 7 and 23. We allow

---

[3] From this it follows that

$$E_{i,|z|} = \min_{k \geq 0} d(z, s_i s_{i+1} s_{i+2} \cdots s_{i+k-1})$$

$$= \min_{j \geq i-1} d(z, s_i s_{i+1} s_{i+2} \cdots s_{j-1})$$

$$= \min_{j \geq i-1} E'_{i,j}.$$

intervals of $(0.9)5 \leq \delta \leq (1.1)5$, or $\delta = 5$ exactly, making it impossible for these positions to chain. Even if we still had a qualifying match at position 13, it could no longer chain with position 7 under the tighter criterion for intervals. This suffix passes the suffix test and will be expanded further, but it fails the satellite test.

Since the definitive satellite test just described requires $O(|m(z)| \cdot \varepsilon |z|^2)$ time, it makes sense to seek faster tests that can eliminate many suffixes as satellites and/or reduce the size of the match set. Most obviously, we can reject immediately any suffix shorter than l_{min}. Another test – a variation of the one we have already applied to validate suffixes – requires only $O(|m(z)|)$ time and improves the overall speed of our program significantly.

Most of the inequalities we developed for suffix testing depended directly or indirectly on the inequality

$$l_{min} \leq |t| \leq l_{max}.$$

Now that we are dealing with a fixed, known satellite candidate z, we can replace l_{min} and l_{max} by $|z|$ in all of those inequalities. Specifically, we can reduce the match set to

$$\mu(z) = \{(i, E_{i,|z|}) \mid E_{i,|z|} \leq \varepsilon \cdot |z| \text{ and } \rho(i) \geq r_{min}\},$$

where

$$\rho^{\leq}(i) = \begin{cases} 1 + \max_{(1-\varepsilon)|z| \leq \delta \leq (1+\varepsilon)|z|} \rho^{\leq}(i - \delta) & \text{if } i \text{ is the left end of some } z' \\ & \text{such that } d(z', z) \leq \varepsilon \cdot l_{max}, \\ 0 & \text{otherwise;} \end{cases}$$

$$\rho^{\geq}(i) = \begin{cases} 1 + \max_{(1-\varepsilon)|z| \leq \delta \leq (1+\varepsilon)|z|} \rho^{\geq}(i + \delta) & \text{if } i \text{ is the left end of some } z' \\ & \text{such that } d(z', z) \leq \varepsilon \cdot l_{max}, \\ 0 & \text{otherwise;} \end{cases}$$

$$\rho(i) = \rho^{\leq}(i) + \rho^{\geq}(i) - 1.$$

Even when this test does not wipe out the match set completely, it often reduces it and thereby the number of times the costly alignment algorithm must be run.

15.2 Finding Satellites in Perl

This section describes the Perl implementation of our algorithm. We consider the five elements of the implementation in this order:

1. the input format and how to process it;
2. the representation of match sets;
3. subroutine filterForRepeats, which pares down a match set by computing either the $r(\cdot)$ or the $\rho(\cdot)$ function on its elements;
4. subroutine suffixCheck, which extends a candidate suffix by one base and tests the extended suffix by invoking filterForRepeats; and

5. subroutine satelliteCheck, which invokes filterForRepeats to perform final checks on a candidate satellite.

The input files for our program are organized as follows. The first line is used to set the options l_{min}, l_{max}, ε, and r_{min}. The remaining lines are concatenated to form the DNA sequence in which satellites are to be sought. To make the input files somewhat self-documenting, the first line contains Perl statements to set any or all of the global variables $maxSatLength, $minSatLength, $minSatRepeats, and $epsilon. The program sets each of these to a default value and then executes the user's statements with the lines

```
my $s = <STDIN>;     ## first line gives options
eval $s;             ## sets options maxSatLength, etc.
```

in subroutine main. In general, the **eval** operator accepts a string argument and evaluates it as if it were written as part of the Perl program.

In this program, match sets are represented by lists. The first element is the suffix being matched. The odd-indexed entries contain positions of the target sequence where the suffix matches approximately. These positions are given in *descending* order. For each odd-indexed entry i, the following even-indexed entry gives the edit distance $E_{|z|,i}$ between the suffix z and the best-matching substring beginning at that position in the target sequence. The match set $m(\texttt{"caa"})$ given in Figure 15.1 is represented by the list

```
("caa",36,1,32,1,26,1,25,0,24,1,20,1,19,1,15,1,10,1,9,0,8,1,5,1)
```

The top-level subroutine, findSatellites, begins the expansion process by constructing a match list for the empty string, which matches without error at every position, and pushing this list onto a stack named suffixes. From then on, the procedure is to (i) pop a suffix from the stack, (ii) attempt to extend it by prepending each of the four nucleotides in turn, and (iii) push back onto the stack the suffixes resulting from any successful extensions. If any suffixes are found to be satellites in their own right, they are saved in another list named @satellites. Potential suffixes and satellites are tested by calls to suffixCheck and satelliteCheck.

Subroutines suffixCheck and satelliteCheck use subroutine filterForRepeats to compute the functions r and ρ, respectively, by providing different arguments for parameters $maxErr and $jumps. Parameter $pattern is a reference to a match list in the format described previously. Parameter $maxErr gives the permissible error between the string given in $pattern and its approximate matches; $jumps is a reference to a list of permissible intervals between matches in the same chain. This subroutine implements the formulas for $r^{\leq}$, $r^{\geq}$, $\rho^{\leq}$, and $\rho^{\geq}$ rather directly, storing

the results in the arrays @repeatsBefore and @repeatsAfter. It uses this information to build and return a new match list with disqualified matches removed from
further consideration.

```perl
sub filterForRepeats {
    my ($pattern,$maxErr,$jumps) = @_;
    my (@repeatsBefore,@repeatsAfter);
    $#repeatsBefore = $#repeatsAfter = $$pattern[1];     ## fast allocate
    for (my $p=$#$pattern−1; $p>0; $p−=2) {
        next if $$pattern[$p+1] > $maxErr;
        my $i = $$pattern[$p];
        $repeatsBefore[$i] = 1+max(map($repeatsBefore[$i−$_],@$jumps));
    }
    my @filteredPattern = ($$pattern[0]);
    for (my $p=1; $p<$#$pattern; $p+=2) {
        next if $$pattern[$p+1] > $maxErr;
        my $i = $$pattern[$p];
        $repeatsAfter[$i] = 1+max(map($repeatsAfter[$i+$_],@$jumps));
        push @filteredPattern,$i,$$pattern[$p+1]
            if $repeatsBefore[$i]+$repeatsAfter[$i] > $minSatRepeats;
    }
    return $#filteredPattern ? \@filteredPattern : "";
}
```

Before calling filterForRepeats, suffixCheck must compute a preliminary match
set for the extended suffix using formula (15.2). It traverses the match list of the original suffix from left to right, so that the input sequence is scanned right to left.

Since $E_{i,k} \leq b$ only if $E_{i-1,k+1} \leq b$, we don't need to consider any position k for
the match list of the extended suffix unless $k + 1$ is in the match list of the original.
Furthermore, suppose that position $k + 1$ is found in array entry p of the match list of
the original, with $E_{i-1,k+1}$ in entry $p + 1$. Then position k, if present, must be in the
very next location, entry $p + 2$, with $E_{i-1,k}$ in entry $p + 3$. Likewise, when computing $E_{i,k}$, we will find $E_{i,k+1}$ (if present at all) to be the most recent entry in the match
list of the extended suffix. In the following code, the match lists of the original and
extended suffixes are called @$suffix and @eSuffix, respectively.

```perl
sub suffixCheck {
    my ($base,$suffix,$sref) = @_;
    my @eSuffix = ($base.$$suffix[0]);      ## string for extended suffix
    my $eLen = length $eSuffix[0];
    my $maxErr = int($epsilon * $maxSatLength);      ## permissible error
```

```
    for (my $p=1; $p<@$suffix; $p+=2) {
        my $k = $$suffix[$p]−1;
        last if $k < 0;
        my $eq = (substr($$sref,$k,1) eq $base) ? 0 : 1;
        my $eErr = $$suffix[$p+1]+$eq;      ## pair base to base
        $eErr = $$suffix[$p+3]+1            ## put gap in sequence
            if ($$suffix[$p+2]==$k) && ($$suffix[$p+3]+1<$eErr);
        $eErr = $eSuffix[$#eSuffix]+1       ## put gap in satellite
            if ($eSuffix[$#eSuffix−1]==$k+1)
                && ($eSuffix[$#eSuffix]+1<$eErr);
        push @eSuffix,$k,$eErr if ($eErr<=$maxErr);
    }

    my @jump = (max($eLen,$minSatLength)−$maxErr
                  .. $maxSatLength+$maxErr);
    my $filteredESuffix = filterForRepeats(\@eSuffix,$maxErr,\@jump);
#   print "/$eSuffix[0]/ ",($filteredESuffix ? "accepted\n" : "rejected\n");
    return $filteredESuffix;
}
```

For each satellite found, the program displays one or more ranges of the input sequence where its tandem repeats can be found. This information is collected by subroutine satelliteCheck while verifying the existence of sufficient numbers of repetitions. Just as array entry $repeatAfter[$i] contains the number of repeats in the longest chain of repeats beginning at position $i, $stringEnd[$i] contains the position where this chain ends, and $stringErr[$i] contains the chain's cumulative error. Since many values of $stringEnd[$i] will be identical and since we wish to identify maximal runs of tandem repeats, we use hash %span to associate with each identified right end of a string in @stringEnd the leftmost possible starting point. At the end, the maximal runs are exactly the intervals contained in %span. We return a list of these intervals together with their number of repetitions and total error.

```
sub satelliteCheck {
    my ($sat,$sref) = @_;
    my $satString = $$sat[0];
    ...
    for (my $p=1; $p<@$sat; $p+=2) {
        my $i = $$sat[$p];
        my $goodJ = goodEndings($satString,$sref,$i,$maxErr);
        for (my $q=0; $q<@$goodJ; $q+=2) {
            my $j = $$goodJ[$q];
```

```
        if ($repeatsAfter[$j]>=$repeatsAfter[$i]) {
            $repeatsAfter[$i] = 1+$repeatsAfter[$j];
            $stringEnd[$i] = $stringEnd[$j] || $j;
            $stringErr[$i] = $stringErr[$j]+$$goodJ[$q+1];
        }
    }
    next if ($repeatsAfter[$i]<$minSatRepeats);
    my $end = $stringEnd[$i];
    $span{$end} = $i
        if !defined($span{$stringEnd[$i]})
            || ($end-$i-$stringErr[$i]      ## length minus error
                >$end-$span{$end}-$stringErr[$span{$end}]);
}
## we return a list of the maximal chains for the satellite
return [map([$satString,$span{$_},$_,
            $stringErr[$span{$_}],$repeatsAfter[$span{$_}]],
        (keys %span)
        )];
}
```

Subroutine satelliteCheck uses subroutine goodEndings to determine edit distances between the suffix and all substrings beginning at a position $i of the DNA sequence. This is carried out by the usual Needleman–Wunsch algorithm as described in Chapter 3. The last row of the dynamic programming table contains the score of the alignment of each of the relevant substrings with the entirety of the suffix. Using this information, subroutine goodEndings assembles a list rather like a match list containing positions of right ends of substrings in even-numbered entries and the corresponding edit distances in the following odd-numbered positions. Because our interest is restricted to distance less than $maxErr, we need only fill entries within this distance of the main diagonal of the dynamic programming table.

15.3 Exercises

1. Suppose that the line

```
    my $suffix = pop @suffixes;
```

in subroutine findSatellites is replaced with

```
    my $suffix = shift @suffixes;
```

How will this change the program's output? Running time? Memory requirements?

2. Develop a program that finds *exactly* repeated satellites by first constructing a suffix tree of the sequence.

3. Often, the program described in this chapter detects many satellites that differ only slightly and that span the same (more or less) range of positions in the input sequence. Design and implement a procedure to decide whether one satellite (with range) supersedes another – based on length, number of repetitions, cumulative error (edit distance), et cetera. Use this procedure to eliminate superseded satellites from the output, leaving only the best models of the tandem repeats.

15.4   Complete Program Listings

This chapter's program is found in file `chap15/satellite.pl` in the software distribution.

15.5   Bibliographic Notes

Some general properties of satellites are given in the work of Jeffreys et al. (1985, 1988, 1994). The algorithm described in this chapter is a simplification of one described by Sagot and Myers (1998).

## CHAPTER SIXTEEN

Restriction Mapping

One of the ways biologists begin to analyze a long sequence of DNA is to develop a *restriction site map*. Restriction sites are the locations at which the sequence is cut by enzymes known as *restriction enzymes* – or, more precisely, *restriction endonucleases*. Restriction enzymes are found in bacteria, where they provide some protection against viral invasion by destroying viral DNA. Each bears a name such as *EcoRI* and *HindIV,* derived from the bacterium in which it was discovered. Each restriction enzyme can cut DNA at any location containing a specific short sequence. Examples of some commonly used restriction enzymes and the sequences they cut are given in Figure 16.1. The sequence is typically a palindrome of even length. In most cases, the enzyme cuts the double strand unevenly so that a small group of unpaired nucleotides – a *sticky end* – remains on both sides of the cut. In nature, this feature may facilitate further degradation of the invading DNA by other enzymes; in the laboratory, it is often used to "cut and paste" new sequences into DNA at known locations.[1]

A restriction map of a sequence is simply a list of the locations at which one or more restriction enzymes are known to cut the sequence. Such a map can be used to pinpoint the origin of a gene or other subsequence within a larger sequence.

Restriction maps are generally constructed by performing *digestion experiments*: The sequence to be mapped is replicated by PCR[2] and then exposed to one or more restriction enzymes, together and/or separately. Then the lengths of the resulting fragments are measured by electrophoretic techniques such as *pulsed-field gel electrophoresis* (PFGE). Finally, these lengths, representing distances between restriction sites, are used to construct a map. One common approach is a *double digest,* in which the sequence is exposed to each of two enzymes separately and then again to both at once. Unfortunately, probabilistic analysis of the double digest technique shows

[1] You may be wondering how the bacterium avoids cutting up its own DNA. The answer is that the bacterium also produces specialized *DNA methylases,* which catalyze the addition of methyl groups to certain bases of the bacterial DNA as shields against the activity of its own endonucleases.
[2] See Chapter 3.

```
AluI                                          BstNI
         5'-A-G|C-T-3'                                 5'-C-C|A-G-G-3'
         3'-T-C|G-A-5'                                 3'-G-G-T|C-C-5'

HindIII                                       EcoRI
         5'-A|A-G-C-T-T-3'                             5'-G|A-A-T-T-C-3'
         3'-T-T-C-G-A|A-5'                             3'-C-T-T-A-A|G-5'
```

Figure 16.1: Some restriction enzymes and the sequences they cut.

a high probability that the resulting distances will be consistent with more than one restriction map unless the total number of restriction sites is rather small.

An alternative approach is a *partial digest*. This technique exposes many copies of the sequence to a single enzyme, but not for so long a time that every restriction site of every copy is cut. The goal is to obtain and measure a set of fragments representing each of the distances between any restriction site and every other. It should then be possible, at least in some cases, to determine the relative positions of the n sites from these $n(n-1)/2$ distances.

The partial digest approach is not without problems. If the sequence is digested for too short a time then some of the shorter distances may be absent from the set of resulting fragments, whereas excessively long digestion may cut all of the fragments representing the longer distances into short pieces. This problem can be mitigated by performing two or more digests for different lengths of time, or by varying the concentration of the enzyme. Still, there is never any guarantee that all $n(n-1)/2$ distances are represented by the fragments from the digest. These difficulties increase with the number of restriction sites. Nonetheless, the backtracking technique we will consider in the following section, in an enhanced version, has been able to deal with both missing fragments and imprecise measurements with reasonable success.

We will look at two versions of the technique: one that assumes the data set to be complete and perfectly accurate, and one allowing imprecise data.

16.1   A Backtracking Algorithm for Partial Digests

To begin our study of the partial digest problem, we will assume that we have been presented with a "perfect" input set containing exact measurements of the lengths of every one of the $m = n(n-1)/2$ distances between pairs of a set of n restriction sites at unknown locations. Once we have worked out the details of an algorithm for this idealized case, we will be in a position to consider modifications to handle defective data sets.

Now suppose we are presented with the set of distances[3]

$$D_0 = \{100, 85, 70, 65, 55, 55, 45, 35, 35, 30, 30, 25, 20, 15, 10\};$$

---

[3] These distances are distinctly unrealistic; fragments ranging from a few hundred to tens of thousands of base pairs are more typical.

strictly speaking, this is a *multiset* because some distances are repeated. Obviously, the largest distance (100) is the distance between the leftmost and rightmost sites. We can assign them locations $s_0 = 0$ and $s_1 = 100$ and then eliminate 100 from further consideration, giving

$$D_1 = \{85, 70, 65, 55, 55, 45, 35, 35, 30, 30, 25, 20, 15, 10\} \quad \text{and} \quad S_1 = \{0, 100\}.$$

Now the largest remaining distance, 85, must be the distance between some site s_2 and either s_0 or s_1. For if 85 is the distance between s_2 some s_j and $s_0 < s_j < s_2$, then the distance from s_2 to s_0 is greater than 85, which is inconsistent with the input set. A similar inconsistency arises if $s_2 < s_j < s_1$. Therefore, either $s_2 = 0 + 85 = 85$ or $s_2 = 100 - 85 = 15$, or the data set is inconsistent. Since $100 - 85 = 15$ is in the set, we can place s_2 at either 85 or 15, remove both 85 and 15 from the set, and proceed with

$$D_2 = \{70, 65, 55, 55, 45, 35, 35, 30, 30, 25, 20, 10\} \quad \text{and} \quad S_2 = \{0, 85, 100\}.$$

We will return to the possibility $s_2 = 15$ later. To place a site s_3, we again note that either $s_3 - s_0 = 70$ or $s_1 - s_3 = 70$ or the input is inconsistent. If $s_3 = 70$, then $|s_2 - s_3| = 15$. However, this is inconsistent since no 15 remains in D_2. If $s_3 = s_1 - 70 = 30$ then $|s_2 - s_3| = 55$, and two 55s remain in D_2. Therefore, we set $s_3 = 30$ and remove 30, one 55, and 70 from D_2, giving

$$D_3 = \{65, 55, 45, 35, 35, 30, 25, 20, 10\} \quad \text{and} \quad S_3 = \{0, 30, 85, 100\}.$$

Site s_4 must be either 65 or 35. If 35, then D_3 must contain all of $35 - 0$, $35 - 30$, $85 - 35$, and $100 - 35$. But $50 \notin D_3$ and $5 \notin D_3$. If $s_4 = 65$, then D_3 must contain $65 - 0$, $65 - 30$, $85 - 65$, and $100 - 65$. All are present, so we set $s_4 = 65$ and hence

$$D_4 = \{55, 45, 30, 25, 10\} \quad \text{and} \quad S_4 = \{0, 30, 65, 85, 100\}.$$

Candidate locations for s_5, our last site, are 55 and 45. If $s_5 = 45$ then the distances are 45, 15, 20, 40, and 55; if $s_5 = 55$, the distances to the other sites are exactly those remaining in D_4. We end up with sites

$$S = \{0, 30, 55, 65, 85, 100\}.$$

Now, when placing s_2, we found that $s_2 = 15$ was also consistent with the input. Normally, if both options are consistent then both must be pursued independently to avoid overlooking a solution to the problem. But it is not too hard to see that, when placing the third site, we always have either two options or none and that the two options lead to solutions that are simply reversals of each other. Pursuing $s_2 = 15$ in this case leads to the solution

$$S' = \{0, 15, 35, 45, 70, 100\}.$$

This example reflects our general approach, which is to place the leftmost and rightmost restriction sites first and then to repeatedly use the greatest remaining distance to place an additional site, always working inward toward the middle from the two ends. Let S_i be the set of the first $i + 1$ sites and let D_i be the set of distances remaining after their placement. If m_i is the largest element in D_i and if

$$\{|m_i - s| : s \in S_i\} \subseteq D_i,$$

then $s_i = m_i$ is consistent with the input data and the algorithm must continue with[4]

$$D_{i+1} = D_i \setminus \{|m_i - s| : s \in S_i\} \quad \text{and} \quad S_{i+1} = S_i \cup \{m_i\}.$$

Likewise, if

$$\{|(s_1 - m_i) - s| : s \in S_i\} \subseteq D_i$$

then $s_i = s_1 - m_i$ is consistent with the input data, and the algorithm must continue with

$$D_{i+1} = D_i \setminus \{|(s_1 - m_i) - s| : s \in S_i\} \quad \text{and} \quad S_{i+1} = S_i \cup \{s_1 - m_i\}.$$

If both $s_i = m_i$ and $s_i = s_1 - m_i$ are consistent, the two possibilities must be considered independently.

16.2 Partial Digests in Perl

Our Perl program for constructing restriction maps from perfect partial digest data is very short – just fifty lines.

We must make decisions about two data structures: one to represent the set of sites, and one to represent the set of distances. The site set will be represented by a list that will be filled in sorted order by working inward from its ends. The distance set will be concurrently represented by two structures: a sorted list @distances of all distances, and a hash %distAvail whose keys are distances and whose values give the number of copies of the key still unaccounted for by the current incomplete placement of sites. This combination makes it easy to find the largest remaining distance when it is time to place a new site: we can move forward through @distances from the position of the last site through progressively smaller distances, checking in %distAvail to see whether these distances are still available. These data structures are set up as global variables by our main program after reading from the input file a set of distances separated by whitespace (or by any other nondigit characters).[5]

---

[4] The symbol $\setminus$ represents set subtraction; $A \setminus B$ is set of all elements of A that are *not* also in B.

[5] Although good software engineering practice disparages the use of global variables, the brevity of our program and our variables' use in both of our subroutines seem to weigh against strict application of the "no globals" rule.

```
my %distAvail;
my @solutions;
foreach (map(split(/\D+/,$_),<STDIN>)) { $distAvail{$_}++; };
my @distances = sort {$a<=>$b} keys %distAvail;
my $numDists;
foreach (values %distAvail) { $numDists += $_; }
my @sites = (("-") x (1+int(sqrt(2*$numDists))));
$sites[0] = 0;
$sites[-1] = $distances[-1];
$distAvail{$distances[-1]}--;
placeSite(1,$#sites-1);
foreach (@solutions) { print "Solution: @$_\n"; }
exit( );
```

The main program also places the leftmost and rightmost sites and removes their separation from %distAvail. The remainder of the sites are placed by the subroutine placeSite.

Subroutine placeSite is a recursive subroutine, albeit indirectly – the recursive calls are hidden in verifyDistances. Each recursive call is responsible for placing a single restriction site, specifically, the one corresponding to the largest remaining distance. placeSite's first step is to set $i to the index in @distances of the largest remaining distance. If no distances remain unaccounted for, then the current contents of the @sites array represents a placement that is consistent with all of the data; in this case, @sites is copied and saved in @solutions.

```
sub placeSite {
    my ($l,$r) = @_;
    my $i = $#distances;
    while (($i>=0) && !$distAvail{$distances[$i]}) { $i--; }
    if ($i<0) {     ## all distances accounted for
        push @solutions, [@sites];     ## makes copy of @sites
        return;
    }
    my $dist = $distances[$i];
    my $newSite = $sites[-1]-$dist;
    verifyDistances($l,$r,"L",$newSite);
    verifyDistances($l,$r,"R",$dist);
}
```

Arguments $l and $r hold the left and right boundaries of the available entries in the @sites array as we fill it from its two ends. We must try putting a new site in each of these entries. If it is left of center, then the largest remaining distance is the distance between the new site and the rightmost site. If it is right of center, the largest distance is the distance between the new site and the leftmost site, 0. The subroutine verifyDistances is called for each of these possible placements in order to check the availability of the distances from the new site to all previously placed sites.

Subroutine verifyDistances operates by forming the logical conjunction ("and") of the counts of the new distances found in %distAvail and simultaneously post-decrementing the counts. If all counts are positive, then $allAvail will also be positive. If one or more 0 counts are encountered then $allAvail will be set to 0, signaling that the proposed location of the new site is inconsistent with the distance set. If the current partial placement is consistent, verifyDistances calls placeSite recursively to attempt to extend the partial placement. In either case, the new site must eventually be removed from the placement and the corresponding distances must be restored to %distAvail so that the program can explore other possible placements.

```
sub verifyDistances {
    my ($l,$r,$LR,$newSite) = @_;
    my $n = ($LR eq "L") ? $l : $r;
    my $allAvail = 1;
    foreach my $j ((0..$l−1),($r+1..$#sites)) {
        $allAvail = $distAvail{abs($newSite−$sites[$j])}−− && $allAvail;
    }
    if ($allAvail) {    ## we have found all the distances we need!
        $sites[$n] = $newSite;
        if ($LR eq "L") { placeSite($l+1, $r); }
        else { placeSite($l, $r−1); }
        $sites[$n] = "-";
    }
    foreach my $j ((0..$l−1),($r+1..$#sites)) {
        $distAvail{abs($newSite−$sites[$j])}++;
    }
}
```

16.3   Uncertain Measurement and Interval Arithmetic

The backtracking algorithm of the previous section has one serious shortcoming: it requires the fragment lengths to be given *exactly* in the input set. In reality, partial digest experiments give mere *estimates* of the actual fragment lengths, generally accurate to within 3–5%.

One method used to cope with inexact measurements in a computational setting is to represent estimated quantities as ranges of numbers – that is, as *intervals* of the real line. To carry out computations with such estimates, methods are required for *interval arithmetic.* For example, if two quantities represented by intervals are added together, the result must be another interval representing all possible results of adding any two numbers in the given intervals. Rather than asking whether two quantities are exactly equal, we must be content to know whether it is *possible* that they are equal – if their intervals overlap – or whether their equality is inconsistent with what is known about the quantities because their intervals are disjoint.

For example, if we perform a partial digest experiment, measure three fragments of length 20, 31, 55, and believe our measurements are accurate within 10%, then our input consists of the three intervals [18, 22], [27.9, 34.1], and [49.5, 60.5]. In this case, the data are consistent with a map of three restriction sites, because [18, 22] + [27.9, 34.1] = [45.9, 56.1] and [49.5, 60.5] overlap. If the first site is pegged at 0, then the third can lie anywhere in [49.5, 56.1], the intersection of the two intervals. On the other hand, if we believe our measurements are accurate to within 1% then we have a problem, since [19.8, 20.2] + [30.69, 31.31] = [50.49, 51.51] and [54.45, 55.55] do not overlap. We would conclude, perhaps, that some fragments were lost and never measured, or that some of the measurements were of contaminants.

Typically, intervals are represented by pairs of floating-point numbers corresponding to the endpoints of the interval. In our case, since our fragments consist of discrete units (nucleotides), we will restrict our endpoints to be integers.

16.3.1 Backtracking with Intervals

Now let's consider how our algorithm for analyzing partial digests will change if we assume that there are measurement errors in the distances presented in the input. We will still assume that every one of the $m = n(n - 1)/2$ distances between restriction sites is represented by a measurement in the input set; techniques for dealing with missing measurements are explored in Section 16.4.

For concreteness, suppose we are presented with the distances

$$\{100, 85, 70, 65, 55, 55, 45, 35, 35, 30, 30, 25, 20, 15, 10\}.$$

If these distances are in terms of base pairs and are accurate to within 5%, then the corresponding multiset of intervals is

$$D_0 = \{[95, 105], [81, 89], [67, 73], [62, 68], \ldots, [24, 26], [19, 21], [15, 15], [10, 10]\}.$$

Interval [95, 105] must estimate the distance between the leftmost and rightmost sites, so we assign locations $s_0 = [0, 0]$ and $s_1 = [95, 105]$, leaving

$$D_1 = \{[81, 89], [67, 73], [62, 68], \ldots, [29, 31], [24, 26], [19, 21], [15, 15], [10, 10]\}.$$

Now the largest remaining distance, [81, 89], must be the distance between s_2 and either s_0 or s_1. The interval [95, 105] − [81, 89] = [6, 24] intersects with *four* different intervals in D_1:

[24, 26], [19, 21], [15, 15], and [10, 10].

To be sure we do not miss a valid solution, we must therefore explore eight different possibilities:

- We can place s_2 in $[6, 24] \cap [24, 26] = [24, 24]$ and continue the process with

$$D_2 = \{[67, 73], [62, 68], \ldots, [29, 31], [19, 21], [15, 15], [10, 10]\}.$$

- We can place s_2 in $[6, 24] \cap [19, 21] = [19, 21]$ and continue the process with

$$D_2 = \{[67, 73], [62, 68], \ldots, [29, 31], [24, 26], [15, 15], [10, 10]\}.$$

- We can place s_2 in $[6, 24] \cap [15, 15] = [15, 15]$ and continue the process with

$$D_2 = \{[67, 73], [62, 68], \ldots, [29, 31], [24, 26], [19, 21], [10, 10]\}.$$

- We can place s_2 in $[6, 24] \cap [10, 10] = [10, 10]$ and continue the process with

$$D_2 = \{[67, 73], [62, 68], \ldots, [29, 31], [24, 26], [19, 21], [15, 15]\}.$$

Finally, we can also use the same four intervals as distances to s_1, giving four more placements for s_2:

$$([95, 105] - [24, 26]) \cap [81, 89] = [81, 81],$$
$$([95, 105] - [19, 21]) \cap [81, 89] = [81, 86],$$
$$([95, 105] - [15, 15]) \cap [81, 89] = [81, 89],$$
$$([95, 105] - [10, 10]) \cap [81, 89] = [85, 89].$$

When measurements were exact, we never had to deal with more than two different placements of a new site relative to the previously placed sites (and, more often than not, one of the two could be eliminated immediately). That was because a required measurement was either in the set or not in the set. Now that the measurements are represented by intervals, there may be many distinct – even disjoint – intervals in the set that can account for a needed measurement. Thus, we must now backtrack among different choices of intervals to account for distances in addition to backtracking for the two choices of left or right placement of the sites.

To see this, we continue with the second of the eight possible placements of s_2:

$$s_0 \in [0, 0], \quad s_1 \in [95, 105], \quad s_2 \in [19, 21],$$
$$D_2 = \{[67, 73], [62, 68], \ldots, [29, 31], [24, 26], [15, 15], [10, 10]\}.$$

The longest remaining distance is $[67, 73]$; if we place s_3 on the left, we require that $s_3 \in [95, 105] - [67, 73] = [22, 38]$. The candidates in S_2 for the distance from s_0 to s_3 are $[34, 36]$, $[29, 31]$, and $[24, 26]$; all three are subsets of $[22, 38]$. We must also account for the distance from s_3 to s_2.

- If $s_3 \in [34, 36]$, distance $|s_2, s_3|$ must intersect with $[34, 36] - [19, 21] = [13, 17]$. The interval $[15, 15]$ is the only one that can account for this distance. In this case, s_3 can lie anywhere in $[19, 21] + [15, 15] = [34, 36]$, and we continue by exploring

$$s_0 \in [0, 0], \quad s_1 \in [95, 105], \quad s_2 \in [19, 21], \quad s_3 \in [34, 36],$$

$$D_3 = \{[62, 68], \dots, [29, 31], [24, 26], [10, 10]\}.$$

- If $s_3 \in [29, 31]$, distance $|s_2, s_3|$ must intersect with $[29, 31] - [19, 21] = [8, 12]$, and only $[10, 10]$ can account for this. In this case, s_3 must lie in $[19, 21] + [10, 10] = [29, 31]$, and we continue with

$$s_0 \in [0, 0], \quad s_1 \in [95, 105], \quad s_2 \in [19, 21], \quad s_3 \in [29, 31],$$

$$D_3 = \{[62, 68], \dots, [29, 31], [24, 26], [15, 15]\}.$$

- If $s_3 \in [24, 26]$, distance $|s_2, s_3|$ must intersect with $[24, 26] - [19, 21] = [3, 7]$. But no interval in D_2 intersects with $[3, 7]$, so we can eliminate this partial solution from further consideration.

On the other hand, we could place s_3 on the right, in the interval $[67, 73]$. Then we must also account for distances

$$|s_1, s_3| \in [95, 105] - [67, 73] = [22, 38],$$

$$|s_2, s_3| \in [67, 73] - [19, 21] = [46, 52].$$

The only measurement that can account for $[46, 52]$ is $[43, 47]$, so $|s_2, s_3| \in [46, 47]$. From this, we can narrow s_3 down to

$$s_3 \in ([43, 47] + [19, 21]) \cup [67, 73] = [62, 68] \cup [67, 73] = [67, 68].$$

From this, we deduce that

$$|s_1, s_3| \in [95, 105] - [67, 68] = [27, 38],$$

and there are two satisfactory measurements: $[29, 31]$ and $[34, 36]$.

16.3.2   Interval Arithmetic in Perl

Integer interval arithmetic is not hard to implement. Our intervals are references to lists of two integers, blessed into the package IntegerInterval.

We create new intervals with IntegerInterval−>new. This method's two arguments are the lower and upper limits of the interval. Floating-point arguments are allowed; however, the arguments are converted to integers so that the resulting closed integer interval contains exactly the same integers as the real interval described by the arguments. Although our application does not use them, negative limits are correctly

implemented. This package differs from our usual practice in that we bless a reference to a *list* rather than a reference to a hash.

```
sub new {
    my ($this,$lo,$hi) = @_;
    my $iLo = int($lo);
    $iLo++ if $iLo < $lo;
    my $iHi = int($hi);
    $iHi-- if $iHi > $hi;
    bless [$iLo,$iHi];
}
```

To describe the other methods of the package, it is helpful to assume that $x and
$y are references to two intervals representing (respectively) the possible values of
the quantities x and y.

To determine whether it is possible that $x = y$, we need to compute the intersection (overlap) of $x and $y. We can obtain the intersection by either of the method
calls $x->intersect($y) or $y->intersect($x). These calls return an interval if $x
and $y intersect; in this case, $x = y$ could possibly hold if both x and y belong to
the interval returned. If $x and $y are disjoint intervals, then **undef** is returned and
$x \neq y$ is certain.

The logic of the intersect method is rather simple. The lower limit of the intersection is the greater of the lower limits of the operands; its upper limit is the lesser of
the two upper limits. If the lower limit exceeds the upper limit, the interval is empty.

```
sub intersect {
    my ($this,$other) = @_;
    my ($a,$b) = @$this;
    my ($c,$d) = @$other;
    $a>$c or $a=$c;
    $b<$d or $b=$d;
    $b>=$a or return undef;
    return $this->new($a,$b);
}
```

The line $b>=$a **or return undef**; illustrates one of the ways Perl programmers
use **and** and **or** as control structures. If $b>=$a is true then the **or** is true, so the
second disjunct is never evaluated and the interpreter continues on the next line. If
$b>=$a is false, however, the interpreter attempts to determine the value of the **or**
by evaluating **return undef**; this, of course, terminates the subroutine. The whole
statement has exactly the same effect as **return undef if** $b<$a.

Another method, $x->$minus($y), returns an interval representing the possible values of $x - y$. We find the largest possible value of the difference by subtracting the smallest possible value of y from the largest possible value of x.

```perl
sub minus {
    my ($this,$other) = @_;
    my ($a,$b) = @$this;
    my ($c,$d) = @$other;
    return $this->new($a-$d,$b-$c);
}
```

We may sometimes need a string representation of the interval. This is important not just for printing but also for use as a hash key. To save the time that could be consumed by repeatedly converting the integer limits to strings, we save the string representation as the third element of the interval array. To save space, we don't produce the string representation until the first time it is requested. Neither space nor time is expended for string representations of the many intervals that will never be printed or used as hash keys.

```perl
sub toString {
    my ($this) = @_;
    $$this[2] ||= "$$this[0]:$$this[1]";
}
```

The Perl code for the other methods in the package, $x->$plus($y) and $x->$print(@L), is generally self-explanatory. The optional arguments of the **print** method are a convenience worth mentioning; they are printed following the interval. This makes it possible, for example, to write a list of intervals, one per line, with a statement like

```perl
foreach (@intervals) { $_->print("\n"); }
```

Since our application does not require multiplication and division of intervals, we leave their implementation as Exercise 10.

16.3.3 Partial Digests with Uncertainty in Perl

Many of the program modifications necessary to work with intervals are straightforward. The contents of the @sites and @distances arrays are now intervals (references blessed to IntegerInterval); the keys of %distAvail are the string forms of the

intervals, as returned by the method IntegerInterval−>toString( ). Only the most
noteworthy changes – those to the subroutine verifyDistances – will be discussed
in detail.

With distances measured exactly, the distance between two sites was either available or unavailable as a measurement. If available, there were no choices as to which
measurement would represent the distance. With intervals, we must introduce a second phase of backtracking between the placement of any two sites, during which we
explore different possible assignments of measurement intervals to distances between
placed sites. To achieve this, we transform verifyDistances into a chain of recursive
calls to verifyDistance, each of which is responsible for assigning a measurement to
the distance between the newly placed site and a single one of the previously placed
sites. The previously placed site in question is indicated by the last argument to subroutine verifyDistance, which we name $j.

```
sub verifyDistances {
    verifyDistance(@_,0);
}
```

Successive recursive calls to verifyDistance will run the value of $j through the in-
dices in the list (0..$l−1,$r+1..$#sites), just as in verifyDistances in `exact.pl`.
The first part of verifyDistance terminates this "loop" of recursive calls when $j
passes beyond the last index. When this happens, the recursive calls have successfully assigned measurements for all of the new site's distances, and the program moves
on and tries to place another site.

```
sub verifyDistance {
    my ($l,$r,$LR,$newSite,$j) = @_;
    my $n = ($LR eq "L") ? $l : $r;
    if ($j==@sites) {    ## we have found all the distances we need!
        $sites[$n] = $newSite;
        if ($LR eq "L") { placeSite($l+1, $r); }
        else { placeSite($l, $r−1); }
        $sites[$n] = "-";
        return;
    }
    ...
}
```

If $j still points to a site in @sites, we must find a measurement to assign to the
distance between $sites[j] and the new site; the interval describing this distance is

assigned to $distSought. We then search through the list of measurements until we find a measurement that is both available and compatible with this distance. The assignment may be compatible with only a *portion* of the interval describing the new site; we restrict the new site to that portion by computing an intersection.

```perl
sub verifyDistance {
    ...
    my $distSought =
        ($j<$n) ?
            $newSite->minus($sites[$j]) : $sites[$j]->minus($newSite);
    my $i = @distances;
    while (--$i>=0) {
        next unless $distAvail{$distances[$i]->toString};
        next unless $distances[$i]->intersect($distSought);
        $distAvail{$distances[$i]->toString}--;
        $d[$j][$n] = $d[$n][$j] = $distances[$i];
        my $range = ($j<$n) ? $sites[$j]->plus($distances[$i])
                            : $sites[$j]->minus($distances[$i]);
        my $nextJ = ($j==$l-1) ? $r+1 : $j+1;
        verifyDistance($l,$r,$LR,$newSite->intersect($range),$nextJ);
        $distAvail{$distances[$i]->toString}++;
    }
}
```

Since the **while**-loop iterates through the entire array of distances, *every* available distance is considered as a candidate to represent $distSought, and a recursive call to verifyDistance is made for *every* viable candidate.

The two-dimensional global array @d saves the measurements that are assigned in verifyDistance. We will use these in the next section to make one last refinement to the placement of the restriction sites.

16.3.4   A Final Check for Interval Consistency

It turns out that, despite all our efforts, the site placement produced by placeSite and verifyDistances may still be inconsistent with the input. The reason for this is an asymmetry in the way sites are placed: the earlier placement of the outer sites affects the later placement of the inner sites, but not vice versa.

As an example, consider the input set (accurate to $\pm 5\%$)

$$\{10, 21, 21, 30, 43, 52, 52, 74, 94, 100\}.$$

One of the placements generated by placeSite and verifyDistances is

$[0, 0]$, $[50, 54]$, $[71, 75]$, $[90, 95]$, $[95, 105]$,

with measurements assigned as follows:

	0	1	2	3
1	[50, 54]			
2	[71, 77]	[20, 22]		
3	[90, 98]	[41, 45]	[20, 22]	
4	[95, 105]	[50, 54]	[29, 31]	[10, 10]

It is not hard to verify that the intervals in the tables are compatible with the differences of the intervals into which the sites have been placed; for example, $[95, 105] - [90, 95] = [0, 15]$ is compatible with $[10, 10]$. However, if s_3 is restricted to $[90, 95]$ and if $[10, 10]$ represents the distance to s_4, then s_4 must lie in the smaller interval $[100, 105]$.

Suppose that our placement process successively assigns intervals to restriction sites $s'_0 = s_0$, $s'_1 = s_n$, $s'_2 \in \{s_1, s_{n-1}\}, \ldots, s'_n$ while assigning distances d'_{ik} for the distance from s'_i to s'_k — so that $s_0, s_1, s_2, \ldots, s_n$ is the left-to-right order of the sites and $s'_0, s'_1, s'_2, \ldots, s'_n$ is the "before-to-after" order of the same sites. A little thought shows that we would like our final placement to satisfy

$$s'_0 = [0, 0]$$

and that, for all i,

$$s_i = \bigcap_{k:k \neq i} ((s_k - d_{ik}) \cup (s_k + d_{ik})) \qquad (16.1)$$

or (equivalently)

$$s'_i = \bigcap_{k:k \neq i} ((s'_k - d'_{ik}) \cup (s'_k + d'_{ik})).$$

The placement process, however, guarantees only that

$$s'_i = \bigcap_{k:k < i} ((s'_k - d'_{ik}) \cup (s'_k + d'_{ik})).$$

To narrow the placement of sites to satisfy the stricter set of constraints, we can work from left to right in n stages, so that, at the end of the jth stage,

$$s_i = \bigcap_{k:k \leq j} ((s_k - d_{ik}) \cup (s_k + d_{ik})) \quad \text{for } i \leq j.$$

We accomplish this by first narrowing s_j's interval so that

$$s_j = \bigcap_{k:k \leq j} ((s_k - d_{kk}) \cup (s_k + d_{kk}))$$

and then narrowing $s_0, s_1, \ldots, s_{j-1}$ to be consistent with s_j. The whole process takes time proportional to the third power of the number of restriction sites and is implemented in the subroutine checkGlobalConstraints. If an empty interval is assigned to any s_i by equation (16.1), the subroutine returns **undef**. If the global constraints can be satisfied, then a reference to a list of intervals is returned.

```perl
sub checkGlobalConstraints {
    my @revisedSites = @sites;
    foreach my $j (1..$#sites) {
        foreach my $i (0..$j-1) {
            $revisedSites[$j] =
                $revisedSites[$j] ->
                    intersect($revisedSites[$i]->plus($d[$i][$j]));
            $revisedSites[$j] or return undef;
        }
        foreach my $i (reverse(0..$j-1)) {
            foreach my $k ($i+1..$j) {
                $revisedSites[$i] =
                    $revisedSites[$i] ->
                        intersect($revisedSites[$k]->minus($d[$i][$k]));
                $revisedSites[$i] or return undef;
            }
        }
    }
    return \@revisedSites;
}
```

In the case of our example, the placements of two of the sites can be narrowed, resulting in the placements

$$[0, 0], \quad [50, 54], \quad [71, 75], \quad [91, 95], \quad [101, 105].$$

16.4 Exercises

1. In the first program, could we safely change the line

```perl
$allAvail = $distAvail{abs($sites[$l]-$sites[$j])}-- && $allAvail;
```

to

$allAvail &&= $distAvail{**abs**($sites[$l]−$sites[$j])}−−;

---

and, if so, would other modifications be required? *Hint:* The suggested modification is equivalent to

---

$allAvail = $allAvail && $distAvail{**abs**($sites[$l]−$sites[$j])}−−;

---

2. When presented with the distance set
$$\{2, 2, 2, 2, 2, 3, 3, 4, 4, 4, 5, 5, 6, 6, 6, 7, 8, 8, 8, 8, 9,$$
$$10, 10, 10, 11, 12, 12, 13, 14, 14, 16, 16, 16, 18, 18, 20\},$$
   the program for exact input prints eight copies each of two solutions,
$$\{0, 2, 4, 7, 10, 12, 16, 18, 20\} \quad \text{and} \quad \{0, 2, 4, 8, 10, 13, 16, 18, 20\}.$$
   Under what circumstances can such multiple copies arise? Is it possible to speed up the program by avoiding the work required to find the same solution more than once?

3. Suppose that verifyDistances and verifyDistance have been called to assign (inexact) measurements to distances from site s_3 (i.e., $sites[3]) to sites s_0, s_1, s_2, s_6, s_7, and s_8. Suppose further that there are many possible assignments for s_0, s_1, s_2 but absolutely none for s_7. As written, the program will spend a lot of time unproductively generating partial assignments to s_0, s_1, and s_2.

   Modify verifyDistances to check quickly that there is at least one possible assignment for each distance before it calls verifyDistance to explore combinations of individual assignments. Conduct some experiments to determine how effectively this change speeds up the program.

4. The distances in @distances are sorted from longest to shortest. Modify the **while**-loop in verifyDistance to terminate as soon as the first distance too short to represent $distSought is detected. Does this change speed up the program significantly?

5. Will the program run faster if global constraints are checked prior to the placement of the last site? Implement more frequent checking and conduct some experiments.

6. One approach to speeding up the program for inexact data is to reduce the number of distance intervals that must be considered for each entry in @d by making barely distinguishable intervals totally indistinguishable. In other words, two distance intervals that overlap by more than a certain fraction can be transformed into two copies of one and the same interval. Let $[(1 − \varepsilon)a, (1 + \varepsilon)a]$ and $[(1 − \varepsilon)b, (1 + \varepsilon)b]$ be two overlapping intervals. They could be replaced by two

copies of their union, $[(1 - \varepsilon)a, (1 + \varepsilon)b]$ or by two copies of their "average", $[(1 - \varepsilon)(a + b)/2, (1 + \varepsilon)(a + b)/2]$. This approach can be extended to groups of three or more overlapping intervals in a straightforward fashion.

Work out the details of this approach and implement them in Perl. Compare the speed of the program with and without your modifications. Do your changes make it possible for the program to miss any solutions? If so, can you evaluate how likely missed solutions are?

7. Often, small fragments are lost during electrophoresis – they move all the way off the end of the gel. Modify the programs to assume that measurements of all fragments shorter than a certain threshold are absent. A placement of the restriction sites is deemed to be acceptable if every pairwise distance greater than the threshold is in the input set and if every measurement in the input set (greater than the threshold) is accounted for by some pairwise distance in the placement. (Absent other information, a logical choice for the threshold is the smallest measurement in the input set.)

8. The *multiplicities* of fragment lengths may also be inaccurate or totally missing from input sets; it may be possible to say only, for example, that there was "at least one" fragment of length $\approx 27{,}400$, or "4 ± 1" fragments of length ≈ 3280. How can the program be modified to handle these situations?

9. Research the advantages and disadvantages of blessing list references instead of hash references to represent objects. Can you imagine circumstances under which one might wish to bless scalar references – for example, references to strings?

10. Add multiplication and division methods to `Interval.pm`. Investigate and apply the overload package, which will allow the methods of `Interval.pm` to be invoked by the normal arithmetic operators $+$, $-$, $*$, and $/$.

11. *(Difficult)* Another approach to handling measurement errors is to regard each measurement in the input set as defining a probability distribution on the actual fragment length. For example, for an input value of a, we could assume a normal (Gaussian) distribution with mean a and standard deviation εa for some fixed value (like 0.01) of ε. For simplicity, we could then assume that the individual measurement errors are independent and thus seek to place the restriction sites and assign measurements to pairs of sites that maximize the product of the probabilities of all of the intersite distances. Implement this maximum likelihood method.

16.5 Complete Program Listings

Directory `chap16/` of the software distribution contains our program for perfect data in file `exact.pl`. Our package for interval arithmetic is contained in `Integer-Interval.pm`, and our program applying intervals to the partial digest problem is found in `interval.pl`.

16.6 Bibliographic Notes

The biology behind partial digest mapping can be found in Dowling et al. (1996).

This chapter's partial digest program is based on the work of Skiena and Sundaraman (1994). An algorithm for attacking the *complete* digest problem can be found in Fasulo et al. (1999). Alternative approaches can be found in Parida (1998), Chuan (1993), Wright et al. (1994), and Lee, Dancik, and Waterman (1998).

Rearranging Genomes: Gates and Hurdles

So far, we have been concerned with sequences corresponding roughly to one gene or protein – a few hundred residues or a thousand or so base pairs. As the various on-going whole-genome projects approach completion and as others are proposed and carried out, it is becoming possible to compare species not just on the basis of differences in sequence for homologous proteins but also on the basis of the arrangement of whole genes on the chromosomes.

Two genes on the same chromosome are said to be *syntenic,* and synteny often is preserved among closely related species. However, the arrangement of genes on chromosomes can be altered by the following sorts of mutations.

- *Translocations,* in which two chromosomes "trade ends". More formally, if the sequences of two chromosomes are $s = s_1 s_2$ (with complementary strand $s' = s_2' s_1'$) and $t = t_1 t_2$, then after a translocation the two new sequences might be either $s_1 t_2$ and $t_1 s_2$ or $s_1 t_1'$ and $t_2' s_2$ (with complements). This sort of event may occur when the two chromosomes are physically close during DNA replication and the replication machinery accidentally "changes tracks" by jumping from one to the other.
- *Reversals,* in which a stretch of DNA internal to a single chromosome reverses its orientation relative to its chromosome. More formally, $s = s_1 s_2 s_3$ may change to $s_1 s_2' s_3$. Reversals may occur when the chromosome forms a loop and the replication machinery changes tracks where the DNA crosses itself.

Studies of mammalian species indicate that many genes can be grouped into "blocks" or "chromosomal segments" that have been conserved through most of mammalian evolution. Studies of different species of the genus *Brassica* (which includes the cole vegetables, turnips, and mustards) have shown that their mitochondrial genes differ little in sequence but radically in arrangement.

When sequence offers few hints about evolutionary distance, the number of reversals and translocations required to transform one species's gene order to another's can provide valuable information for reconstructing evolutionary pedigrees. Figure 17.1 illustrates the sequence of reversals required to transform the cabbage's mitochondrial

(cabbage)

$$\begin{bmatrix} 5\text{'}-a-d-b-3\text{'} \\ 3\text{'}--e-c--5\text{'} \end{bmatrix} \quad \text{reverse b}$$

$$\begin{bmatrix} 5\text{'}-a-d---3\text{'} \\ 3\text{'}--e-cb-5\text{'} \end{bmatrix} \quad \text{reverse d}$$

$$\begin{bmatrix} 5\text{'}-abcde-3\text{'} \\ 3\text{'}-------5\text{'} \end{bmatrix} \quad \text{reverse edcb}$$

(turnip)

Figure 17.1: Transforming cabbage to turnip by reversals.

DNA into the turnip's. The focus of this chapter will be an algorithm for finding the shortest sequence of reversals that transforms one given arrangement of genes into another. This process is commonly called *sorting by reversals.*

17.1 Sorting by Reversals

Although genome rearrangement may not seem to be a sorting problem, rearrangement within a single chromosome can in fact be turned into one quite easily. Suppose our two species have the following chromosomes (each letter representing one gene):

$$\begin{bmatrix} 5\text{'}-p-gka-b-d--r-3\text{'} \\ 3\text{'}--c---h-e-zx--5\text{'} \end{bmatrix} \quad \text{and} \quad \begin{bmatrix} 5\text{'}-pz-e-h--x-gr-3\text{'} \\ 3\text{'}---d-b-ak-c---5\text{'} \end{bmatrix}.$$

We can number the genes left to right in the first species and then use the same numbers in the second species, adding a minus sign if the gene appears on the opposite strand in the second species. Then the two chromosomes become

$$(1, 2, 3, \ldots, 12) \quad \text{and} \quad (1, -10, -9, -8, -7, -6, -5, -4, -11, 2, 3, 12).$$

For example, k, the fourth gene on the left, is represented by 4. On the right, k is the eighth gene and appears on the opposite strand, so -4 appears in the eighth position on the right. Since the first list is always in ascending order under this numbering scheme, the task is to sort the second list into ascending order by a series of reversal operations. Each reversal operation both reverses and negates some contiguous sublist. The goal of *sorting by reversals* is to find the *shortest* series of operations that sorts the second list. For this example, two reversals suffice: one of positions 9 through 11, then another of positions 2 through 10 of the list.

In this chapter, we will study an algorithm proven to produce a shortest sorting sequence of reversals. Because the correctness proof itself is rather technical, we will omit its many details. Instead, we will focus on imparting an intuitive understanding of the steps of the algorithm and provide examples that illustrate the plausibility of their correctness. Mathematically sophisticated readers should be in a good position to tackle the original sources for this algorithm after reading the chapter.

The mathematical literature refers to our problem more specifically as "sorting signed permutations by reversals". Since the unsigned version (in which the numbers in the permutation are at all times positive) had already been shown to be NP-complete, many researchers were surprised by the publication of a polynomial-time solution to the signed version in 1995. Since then, the algorithm has been both simplified and sped up, and the method we will study requires time proportional to the product of the number of genes and the number of reversals required.[1] The more general problem involving translocations among a number of chromosomes has also been solved in polynomial time.

Oddly enough, we will begin to solve the problem by transforming to an unsigned version. Since the general unsigned version is NP-complete, this may seem counter-productive. However, the unsigned problem will include a restriction on the reversals permitted, and the restricted version will be tractable.

We will transform a signed permutation of n items into an unsigned permutation of $2n$ numbers by replacing each signed number by two unsigned numbers. Specifically, we will replace a positive number i with $(2i - 1, 2i)$, in this order. We will replace a negative number i with $(-2i, -2i - 1)$. For example, we replace 3 by $(5, 6)$ and -3 by $(6, 5)$. If we transform

$$(1, -10, -9, -8, -7, -6, -5, -4, -11, 2, 3, 12)$$

into an unsigned permutation, the result is

$$(1, 2, 20, 19, 18, 17, 16, 15, 14, 13, 12, 11, 10, 9, 8, 7, 22, 21, 3, 4, 5, 6, 23, 24).$$

We will restrict the reversals considered to *even reversals,* which are precisely those reversals that do not split up the unsigned pairs like $5, 6$ or $6, 5$ that together represent a single signed integer. An even reversal of the pair $5, 6$, representing 3, turns it into $6, 5$, representing -3. So the encoding of the signs in the pairs is accurately maintained by even reversals. For readability, we will write semicolons in our unsigned permutations at the positions that can be separated by even reversals:

$$(1, 2; 20, 19; 18, 17; 16, 15; 14, 13; 12, 11; 10, 9; 8, 7; 22, 21; 3, 4; 5, 6; 23, 24).$$

We will use *twist* as a synonym for even reversal. It will also be convenient to add two fixed elements, 0 and $2n + 1$, to the ends of the lists at positions 0 and $2n + 1$, so that we will write

$$(0; 1, 2; 20, 19; 18, 17; 16, 15; 14, 13; 12, 11; 10, 9; 8, 7; 22, 21; 3, 4; 5, 6; 23, 24; 25)$$

for our example permutation.

---

[1] If the number of reversals is very small then the time is dominated by a small term involving the inverse of Ackermann's function, an extremely slow-growing function.

17.2   Making a Wish List

The first step in solving the problem of sorting by reversals is to identify the positions of pairs of numbers that are next to each other in a sorted list but are not next to each other in the current list. In the list

(0; 1, 2; 20, 19; 18, 17; 16, 15; 14, 13; 12, 11; 10, 9; 8, 7; 22, 21; 3, 4; 5, 6; 23, 24; 25),

for example, 2 and 3 should be adjacent but are separated; their positions are 2 and 19. We will call the pair (2, 19) a *wish*, since we wish for the numbers in these positions to be next to each other. The other wishes in this list are (16, 22), (3, 18), and (17, 23). Wishes are *simple* if they can be *granted* by a single twist, that is, if the numbers in the two positions can be made adjacent by a single twist. The wish (17, 23) is a simple wish because reversing positions 17 through 22 will bring the two numbers 22 (in postition 17) and 23 (in position 23) together; (16, 22) is simple for the same reason. Wishes (3, 18) and (2, 19) are *twisted* wishes; no single twist will put 20 next to 21 or 2 next to 3. Simple wishes are easy to recognize, because the two positions have the same *parity*: either both are even or both are odd. Conversely, the positions of a twisted wish have opposite parity: one even and one odd.

If we grant wish (17, 23) by carrying out the single twist, we obtain

(0; 1, 2; 20, 19; 18, 17; 16, 15; 14, 13; 12, 11; 10, 9; 8, 7; 6, 5; 4, 3; 21, 22; 23, 24; 25).

The numbers 2 and 3 are still separated, but the corresponding wish has changed to (2, 20), a simple wish. The new wish for 20 and 21 is also simple: (3, 21). We can finish sorting with just one more twist.

The interesting point here is that a simple wish and a twisted wish *overlapped*, so that granting the simple wish "untwisted" the twisted wish. It is not hard to see that this is generally true, since granting the simple wish inverts the parity of the index of one of the endpoints of the twisted wish. (We say that parity is inverted when an odd index becomes even or vice versa.) The key to finding the shortest sequence of twists that grants all wishes will be to tabulate all the wishes, classify them as simple or twisted, and then analyze their overlaps. The result of this process will be a choice of which twist to perform next to make best progress toward the goal.

Let's examine further the interaction of overlapping wishes. In the permutation

(0; 5, 6; 2, 1; 8, 7; 3, 4; 9, 10; 13, 14; 15, 16; 12, 11; 17),

the wishes are (0, 4), (3, 7), (1, 8), (2, 6), and (5, 9) on the left and (10, 16), (11, 15), and (14, 17) on the right. Since none of the intervals on the left intersects with any interval on the right, we can really handle the disentanglement of the two halves as subproblems to be solved independently.

But it turns out that subproblems may be independent even when intervals *do* intersect. For example, consider the list

(0; 1, 2; 12, 11; 8, 7; 6, 5; 9, 10; 4, 3; 13, 14; 15).

Three of its wishes – $(8, 11)$, $(5, 9)$, and $(4, 10)$ – interact, but $(2, 12)$ and $(3, 13)$ are independent of this group even though they intersect. Closer inspection shows that $(2, 12)$ and $(3, 13)$ *contain* the other intervals. The twist that grants these two wishes inverts parity of *both* ends of each of the other three intervals, which leaves simple wishes simple and twisted wishes twisted.

The sort of "overlapping" of intervals that interests us is clearly more than simple intersection. We define it precisely as follows. If (a, b) and (c, d) are two wishes, then (a, b) and (c, d) *overlap* if and only if either $a < c < b < d$ or $c < a < d < b$.

We can now define "interacting" wishes: If (a, b) and (c, d) are two wishes, then (a, b) and (c, d) *interact* if and only if either (i) (a, b) and (c, d) overlap or (ii) some wish (g, h) overlaps (a, b) and interacts with (c, d).

17.3 Analyzing the Interaction Relation

Interaction divides the wishes into closed blocks in which each wish interacts with every other in its own block, but none outside its own block. After listing the wishes, the next step in the algorithm is to identify these blocks. This can be done rather easily by sorting the wishes by left endpoint and then scanning them from left to right; we will not discuss this process further until we implement it in Perl. Once the blocks are identified, we must examine each block and consider the specific interactions of simple and twisted wishes.

Although we have spoken of finding "the" shortest sequence of twists that sorts the input permutation, there will usually be many shortest sequences. For example, if there are two independent blocks of permutations then we may choose to sort out the leftmost or the rightmost block first or to alternate between the two blocks, twist by twist. We will call a twist *safe* if it is the next twist in *some* shortest sequence. Obviously, if we repeatedly choose a safe twist then we will find a shortest sorting sequence.

Any block of wishes that contains at least one simple wish will be called a *simple block*. Blocks consisting entirely of twisted wishes will be called *twisted blocks*. Simple blocks are relatively easy to sort. Usually, granting a simple wish in a simple block will untwist a twisted wish in the same block, so that it is possible to proceed by granting simple wish after simple wish. We must exercise some caution that granting a certain wish does not split a simple block into, say, a simple block and a twisted block. Safe twists never create new twisted blocks. Fortunately, others have shown rigorously that we never have to search far for a safe twist in a simple block: If w is any simple wish and $w_1, w_2, \ldots, w_k$ are the simple wishes overlapping w, then at least one among these $k + 1$ wishes is safe.

Our strategy for twisted blocks will be to use twists to transform them into simple blocks, a process we call *simplification*. Although we could simplify by performing exactly one twist in each twisted block, this turns out to be inefficient. By twisting these blocks together in pairs, we can simplify two, three, or more blocks with a single twist.

Section 17.4 discusses the best way to transform all the twisted blocks to simple blocks with the minimum number of twists. Section 17.5 describes how to grant the wishes of each simple block with as few twists as possible.

17.4 Clearing the Hurdles

This section continues our discussion of simplification of twisted blocks of wishes. First, let's look at a few examples of possible ways to simplify.

Example 1. Consider the permutation

$$(0; 3, 4; 1, 2; 7, 8; 5, 6; 9)$$

with a single block of wishes, all twisted:

$$(0, 3)\ (1, 4)\ (2, 7)\ (5, 8)\ (6, 9).$$

We can diagram the wishes like this, with T marking the endpoints of twisted wishes, S of simple wishes, and G of wishes just granted.

```
0  1  2 3 4 5 6 7 8 9
T-----T    T-----T
   T-----T    T-----T
      T---------T
```

There is no alternative but to perform a twist to make some of the wishes simple. Twisting the leftmost wish in the block is satisfactory; it gives

$$(0; 4, 3; 1, 2; 7, 8; 5, 6; 9)$$

with wish list

```
0  1  2 3 4 5 6 7 8 9
T-----T    T-----T
      S---S    T-----T
      S-----------S
```

We can represent this transformation schematically by

[AAA]AAA -> aaaaaa.   (17.1)

Here the As represent a sequence of positions that fall into the same twisted overlap block, the brackets [ ] represent the region to be twisted, and the as represent the resulting simple block.

Example 2. Another possibility is illustrated by the permutation

$$(0; 3, 4; 1, 2; 5, 6; 7, 8; 11, 12; 9, 10; 15, 16; 13, 14; 17),$$

with a wish list that falls into two twisted blocks,

A: (0, 3)(1, 4)(2, 5),

B: (8, 11)(9, 12)(10, 15)(13, 16)(14, 17),

with diagram

```
                          1 1 1 1 1 1 1 1
        0 1 2 3 4 5 6 7 8 9 0 1 2 3 4 5 6 7
        T-----T             T-----T   T-----T
          T-----T             T-----T   T-----T
            T-----T             T---------T
```

We could simplify each block independently by a single twist within each; but we can also twist the two blocks together into a *single simple block* with just one twist, yielding

(0; 3, 4; 1, 2; [8, 7; 6, 5]; 11, 12; 9, 10; 15, 16; 13, 14; 17),

with wish list

(0, 3) (1, 4) (2, 8) (5, 11) (9, 12) (10, 15) (13, 16) (14, 17)

or, diagramatically,

```
                          1 1 1 1 1 1 1 1
        0 1 2 3 4 5 6 7 8 9 0 1 2 3 4 5 6 7
        T-----T   S-----------S   T-----T
          T-----T             T-----T   T-----T
          S-----------S   T---------T
```

Schematically, we can write

        AAAAA[A B]BBBBB -> abababababab.   (17.2)

The right-hand side intersperses as and bs to indicate that the two blocks have been merged. Lower case indicates a simple block.

In this notation, simple blocks may exist between or be contained within the twisted blocks on the left. Since the presence or absence of simple blocks on the left does not change the efficacy of the simplifying twists, they are always omitted from the left side of these schematics.

Example 3. It might be surprising that a single twist can simplify three, four, or more twisted blocks. The wish list of the permutation

(0; 1, 2; 25, 26; 21, 22; 3, 4; 27, 28; 15, 16; 5, 6; 9, 10; 7, 8; 11, 12; 19, 20;

17, 18; 13, 14; 29, 30; 23, 24; 31, 32; 37, 38; 33, 34; 35, 36; 39, 40; 41)

has three twisted blocks,

A: (14, 17)(15, 18)(16, 19),

B: (32, 35)(33, 38)(34, 39),

C: (2, 7) (3, 30) (4, 9) (5, 22) (6, 29) (8, 13) (10, 27)

    (11, 26) (12, 23) (20, 25) (21, 24) (28, 31),

with diagram

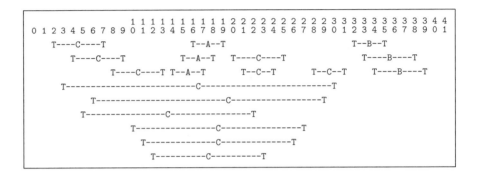

Again, twisting the right end of A and the left end of B creates a single, large simple block:

```
CCCCCC AAAAA[A CCCCCC B]BBBBB
-> abcabcabcabcabcabcabc.                                    (17.3)
```

The twist creates at least one simple edge joining A to B. But this edge overlaps the edge(s) joining the two "pieces" of C. (These edges, too, have become simple, since the parity of their right endpoints has been inverted.) The result is wish list

(0; 1, 2; 25, 26; 21, 22; 3, 4; 27, 28; 15, 16; 5, 6; 9, 10; 7, 8;

[32, 31; 24, 23; 30, 29; 14, 13; 18, 17; 20, 19; 12, 11];

37, 38; 33, 34; 35, 36; 39, 40; 41),

with diagram

```
                  1 1 1 1 1 1 1 1 1 1 2 2 2 2 2 2 2 2 2 2 3 3 3 3 3 3 3 3 3 3 4 4
  0 1 2 3 4 5 6 7 8 9 0 1 2 3 4 5 6 7 8 9 0 1 2 3 4 5 6 7 8 9 0 1 2 3 4 5 6 7 8 9 0 1
    T---------T                   S-----------------------------S
                                    S-----------------------------S
        T---------T         T-----T                 T---------T T---------T
            T---------T T-----T         T-----T         T-----T         T---------T
      S-----------------------------S
         S-----------------------------S
        S-------------------------------------S
             S---------------------------S
            S-----------------------------S
             S-----------------------------S
```

The same technique extends to larger numbers of intervening blocks:

CCCCCC AAAAA[A CCCCCC EEEEE DDDDD B]BBBBB DDDDD EEEEE

  -> abcdeabcdeabcdeabcdeabcdeabcdeabcdeabcdeabcde.   (17.4)

Of course, if any one of the intervening blocks consists of a single piece then it is not simplified, since *both* endpoints of its wishes undergo parity change:

CCCCCC AAAAA[A CCCCCC EEEEE DDDDD B]BBBBB EEEEE

  -> abceabceabceabceabceabce DDDDD abceabceabce.   (17.5)

We are now ready to define *hurdle* and *superhurdle.*

- A *hurdle* is a twisted block that has only one "piece". More formally, a hurdle is a twisted block lacking any single interval that strictly contains all the positions of another twisted block. The twisted block of Example 1 is a hurdle.
- A *straddle* is a twisted block that has exactly two pieces. More formally, a straddle is a twisted block having intervals that contain all the positions of exactly one other twisted block. Block C of Example 3 is a straddle.

  The twisted block contained by a straddle must be a hurdle; for if not, the block it contained would also be contained by the straddle.
- A *superhurdle* is a twisted block contained by a straddle. Block A of Example 3 is a superhurdle.

The following observations explain why hurdles, straddles, and superhurdles are interesting.

- If there are no hurdles then there are no twisted blocks.
- A twisted block can be simplified only by a twist that *cuts* it by twisting some but not all of its elements. If the block has several pieces then the cut can come between the pieces, but a hurdle *must* be cut *within* its single piece.
- If a superhurdle is simplified by scheme (17.1), then the straddle that contains it becomes a hurdle. It is better to simplify a superhurdle by twisting it with another hurdle; that simplifies both the hurdles and also the straddle. Or, if there are no other hurdles, it should be twisted with its straddle:

  BBBBBB AAAAA[A B]BBBBB -> ababababababab.   (17.6)

- If two adjacent hurdles are twisted together as in scheme (17.2), a twisted block that "straddles" both may become a hurdle:

  CCCCCC AAAAA[A B]BBBBB CCCCCC

    -> CCCCCC ababababababab CCCCCC = CCCCCCCCCCCCCC.

It is preferable to twist *nonadjacent* hurdles as in scheme (17.4).

Tying these observations together, we can articulate our priorities for simplifying twisted blocks as follows.

1. If possible, twist together two superhurdles. This will simplify both superhurdles and their straddles.
2. Otherwise, if possible, twist together a superhurdle and a regular hurdle. This simplifies at least the two hurdles and one straddle.
3. Otherwise, if possible, twist together two nonadjacent hurdles. This simplifies the hurdles and at least one twisted block between.
4. Otherwise, if possible, twist together two adjacent hurdles. This simplifies the two hurdles.
5. Otherwise, if possible, twist together a superhurdle and its straddle. This simplifies the superhurdle and its straddle.
6. Otherwise, if possible, twist a single hurdle. This simplifies the only hurdle.
7. Otherwise, there are no hurdles and simplification is complete.

17.5 Happy Cliques

To sort out simple blocks of wishes, we proceed by identifying a *happy clique* of wishes within the block. A happy clique is a subset of the block having the following properties.

1. Every wish in a happy clique is simple.
2. Every wish in a happy clique overlaps every other wish in the happy clique.
3. If a simple wish w_1 outside the clique overlaps with any wish w_2 inside the clique, then w_1 must also overlap with some other simple wish w_3 that is outside the clique and does not overlap with w_2.

Once a happy clique is identified, we will choose the wish from the happy clique that overlaps the largest number of twisted wishes. Although we will not prove rigorously that this is the correct choice, the following example will help illustrate what a happy clique is and why its properties help us sort simple blocks.

The signed permutation

$$(-12, -3, -4, -6, -9, 1, 11, 2, 5, 10, 8, 13, 7, 14)$$

corresponds to the unsigned permutation

$$(0; 24, 23; 6, 5; 9, 8; 12, 11; 18, 17; 1, 2; 21, 22;$$
$$3, 4; 9, 10; 15, 16; 25, 26; 13, 14; 27, 28; 29).$$

Its simple wishes are

$$(1, 23) \ (2, 14) \ (4, 16) \ (5, 17) \ (7, 25) \ (8, 18) \ (9, 19) \ (10, 22)$$

and its twisted wishes are

$$(0, 11) \ (3, 6) \ (12, 15) \ (13, 20) \ (21, 26) \ (24, 27).$$

We can diagram the wishes as before:

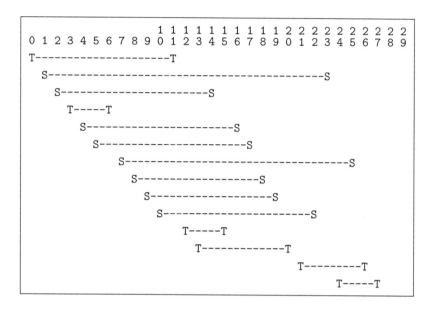

It is not hard to see that every wish in this block interacts with every other. For example, $(0, 11)$ interacts with $(24, 27)$ because $(0, 11)$ overlaps $(1, 23)$, which overlaps $(21, 26)$, which overlaps $(24, 27)$.

The wishes

$$(2, 14)\ (4, 16)\ (5, 17)\ (8, 18)\ (9, 19)\ (10, 22)$$

form a happy clique; each is simple, and each overlaps with every other, so that Properties 1 and 2 are satisfied. As for Property 3, the wish $(1, 23)$ is outside the clique and doesn't overlap with any clique wishes. The wish $(7, 25)$ overlaps $(2, 14)$, $(4, 16)$, and $(5, 17)$. But $(7, 25)$ also overlaps $(1, 23)$, and $(1, 23)$ does not overlap any of $(2, 14)$, $(4, 16)$, or $(5, 17)$.

Every reversal in the shortest sequence that sorts a simple block will grant a simple wish. Generally speaking, we prefer simple wishes to twisted wishes and try to avoid creating new twisted wishes. More importantly, we need to avoid creating new twisted *blocks* at all costs. A theorem stated and proved in Section 17.8 asserts that our choice of wish satisfies this need.

There are three factors we might want to consider in choosing which wish to grant.

- How many simple wishes in the clique become twisted?
- How many simple wishes outside the clique become twisted?
- How many twisted wishes outside the clique become simple?

If we choose to grant any single wish in a happy clique, then the others (in the case of our example, five) will inevitably become twisted wishes. But we must at some point begin to handle these wishes. By finding subsets that satisfy Properties 1 and 2, we are identifying wishes that are equal in terms of this inevitable negative aspect. We can then apply the other criteria to find the best wish within the clique.

At first, it appears that we should avoid granting a wish that twists a simple wish outside the clique. But Property 3 says that we don't need to be concerned about that, because each such wish can be untwisted later by granting a simple wish (w_3) outside the clique. In our example, granting (2, 14), (4, 16), or (5, 17) will twist (7, 25). But it really doesn't hurt us to do this, since (7, 25) can be untwisted later by granting (1, 23).

So, after identifying a happy clique, we can choose the best wish to grant by concentrating on the third factor. We grant the wish that untwists the largest number of twisted wishes. Granting (2, 14) untwists (0, 11), (12, 15), and (13, 20); no other wish in the clique overlaps more than two twisted wishes. Reversing positions 3–14 changes the wish list from

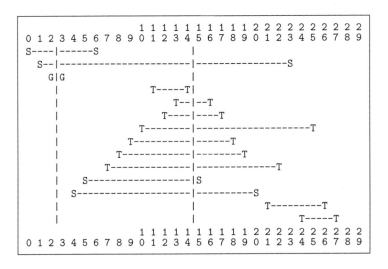

to

At the next step, we begin with the following wishes:

```
                    1 1 1 1 1 1 1 1 1 1 2 2 2 2 2 2 2 2 2 2
  0 1 2 3 4 5 6 7 8 9 0 1 2 3 4 5 6 7 8 9 0 1 2 3 4 5 6 7 8 9
S--------|--S                      |
  S------|-------------------|----------------S
      S|-------------------|---------S
      |S-----------------|S
      |       T-------------|-------------T
      |         T-----------|--------T
      |         T---------|------T
      |           T-------|--------------------T
      |             T-----T|
      |               T----|----T
      |                 T--|--T
      |                    |              T---------T
      |                    |                  T-----T
```

The wish (5, 15) by itself forms a happy clique; it overlaps (0, 6), but so does (1, 23). Granting (5, 15) gives us

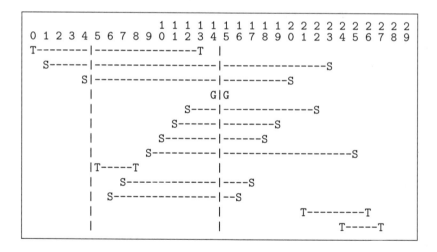

17.6 Sorting by Reversals in Perl

Our main program for sorting by reversals consists of two successive loops. The first loop repeatedly twists together hurdles until none remain. The second loop repeatedly grants simple wishes selected from happy cliques.

```
my @spi = readSignedPerm( );
my @pi = signedToUnsigned(@spi);
printPerm(\@pi);
```

```perl
my $count;
while (1) {
    my @wishes = makeWishList(\@pi);
    my $ov = findOverlapBlocks(\@wishes);
    last unless clearHurdles(\@wishes,$ov);
    $count++;
}

while (1) {
    my @wishes = makeWishList(\@pi);
    my @happy = findHappyClique(@wishes);
    last unless @happy;
    my @interval = @{findMaxTwistedDegree(\@happy,\@wishes)};
    twist(@interval,\@pi," >grant wish>");
    $count++;
}
print "$count reversals required.\n";
```

During the hurdle-eliminating phase, the first step to finding the next reversal is to build the wish list, which is carried out by subroutine makeWishList. The next step – grouping interacting wishes together in blocks – is assigned to findOverlapBlocks. Subroutine clearHurdles calls subroutine classifyBlocks to classify the blocks as hurdles, superhurdles, and so on. Subsequently, subroutine clearHurdles uses the classifications to search for the actual positions to twist.

Once all hurdles are removed, the program begins to locate wishes to grant in simple blocks. There are two steps to this process: finding a happy clique (subroutine findHappyClique) and finding the wish in the happy clique that overlaps the largest number of twisted wishes.

We will represent a wish as a reference to a list with two elements, the left and right positions. The wish list will be a list of these references. Subroutines L and R take a single argument – a wish (reference) – and return its left or right endpoint.

```perl
sub L { $_[0][0]; }
sub R { $_[0][1]; }
```

The unsigned permutation is represented in the obvious way by a list, @pi. To construct the wish list, it is more convenient to work with the inverse permutation @piInv. Each position of @piInv tells where the corresponding number can be found in @pi. For example, if the number 23 is found at list position 7 in @pi, then position 23 of @piInv contains 7. With this information, it is easy to construct the wish list. For example, the numbers 22 and 23 "wish" to be next to each other. Their actual positions can be found in @piInv[22] and @piInv[23]. If they are not already adjacent, then [@piInv[22],@piInv[23]] (or [@piInv[23],@piInv[22]]) is a wish. We simply step through @piInv two entries at a time, constructing wishes and collecting them in the list @wishes. As a final step, we sort them by left endpoint. All this happens in the subroutine makeWishList.

```perl
sub makeWishList {
    my ($pi) = @_;      ## a reference to the list containing the permutation
    my @piInv = (1) x @$pi;
    foreach my $i (0..$#pi) { $piInv[$$pi[$i]] = $i; }
    my @wishes = ();
    for (my $i=0; $i<$#piInv; $i+=2) {
        my ($j,$j1) = ($piInv[$i],$piInv[$i+1]);
        ($j,$j1) = ($j1,$j) if ($j1<$j);
        next if ($j==$j1−1);
        push @wishes, [$j,$j1];
    }
    return(sort { L($a)<=>L($b) } @wishes);
}
```

After constructing the wish list, the next step is to group interacting wishes together in blocks. The blocks form disjoint subsets of the set of wishes. Before we compare any wishes to determine overlap, each wish will be alone in a subset of one. Discovering that two wishes overlap amounts to discovering that their sets are subsets of the same block. When this happens, we will form the union of the two subsets. Solving this problem efficiently is a classic topic in the theory of data structures, where is often referred to as the *union-find problem*. We will implement a Perl module called UnionFind.pm with three subroutines.

- $uf = new UnionFind
  creates a new union-find structure and assigns a reference to $uf.
- $uf−>find($key)
  returns the representative element of the (unique) subset containing $key in union-find structure $uf.
- $uf−>union($key1,$key2)
  merges the two subsets containing $key1 and $key2 in union-find structure $uf. It returns the representative of the new subset.

The users of this module need not be concerned about which element is the representative of a particular subset. The important thing is that each subset has a single representative that is returned consistently by find. This makes it possible to determine whether two elements are in the same subset by the test ($uf−>find($key1) **eq** $uf−>find($key2)). Our particular implementation is an extremely fast one known as "union by rank with path compression"; it is described in detail in Appendix C.

Subroutine findOverlapBlocks accepts a reference to a wish list as its only argument, and it returns a reference to a union-find object in which interacting wishes are assigned to the same subset and noninteracting wishes are assigned to different subsets. Actually, the keys of the union-find structure are not wishes but rather *endpoints* of wishes, which is a bit more convenient for later processing. The "handle" of each subset is the wish that extends furthest to the right.

Subroutine findOverlapBlocks works by scanning the number line from left to right and maintaining a list of wish intervals that contain the current scan point. This list must be updated by adding each wish at its left endpoint and removing it at its right endpoint. Therefore, the subroutine begins by constructing and sorting a list of scan "events" consisting of an endpoint and either "add" or "remove". Concurrently, the @handle list and the union-find structure are initialized by placing the two endpoints of each wish in the same subset with the wish itself as handle.

```perl
sub findOverlapBlocks {
    my ($wishes) = @_;
    my $ov = new UnionFind( );
    my @handle;
    my @events;
    foreach (@$wishes) {
        $ov->union(L($_),R($_));
        $handle[L($_)] = $handle[R($_)] = [L($_),R($_)];
        push @events, ([L($_),"add"], [R($_),"remove"]);
    }
    @events = sort { $$a[0]<=>$$b[0] } @events;
    ...
}
```

Once constructed, the events are processed left to right. When the scan reaches the left endpoint of a wish, the endpoint is simply pushed onto @stack, to remain there until either (a) the scan reaches its own right endpoint or (b) an overlapping wish is discovered.

When the right endpoint of a wish is reached, it is possible to decide which wishes it overlaps by examining the stack. Every other wish pushed onto the stack (represented by its left endpoint) after the current wish clearly has a later (i.e., further-to-the-right) left endpoint, because wishes are pushed in order by left endpoint. It must also have a later right endpoint, because right endpoints are also processed in order and – as we will shortly see – the processing removes the wish. Therefore, the current wish overlaps each such wish, and we merge their subsets together.

Now, among the wishes pushed onto the stack *before* the current wish and still remaining, we can be sure that they have earlier endpoints than both the current wish and all the wishes just merged with it. They should be joined to the new subset if they also have an earlier right endpoint than *at least one* of the newly merged wishes.

Therefore, it suffices to retain on the stack the one wish among the newly merged subset that has the *latest* right endpoint – the new subset's handle. To process a right endpoint, we will pop off the stack the current wish and all others above it on the stack, merge these together, and replace them on the stack with the handle of the newly formed subset.

Thinking ahead to wishes lying entirely to the right of the current scan point, we see that each has a later left endpoint than *all* of the wishes in the newly formed subset. Such a wish will overlap a wish in the new subset if and only if it has a later right endpoint than *some* wish in the new subset. But we do not need to retain any information on the stack in order to detect these overlaps; they will be detected when the right end of the overlapping wish is processed.

```perl
sub findOverlapBlocks {
    ...
    my @stack = (−1);
    foreach my $e (@events) {
        my ($endpt,$action) = @$e;
        if ($action eq "add") {
            push @stack, $endpt;
        } elsif ($action eq "remove") {
            my $block = $ov−>find($endpt);
            my $Lh = L($handle[$block]);
            while ($Lh<=$stack[$#stack]) {
                my $block1 = $ov−>find(pop @stack);
                $block = $ov−>union($block,$block1);
                $handle[$block] = $handle[$block1]
                    if R($handle[$block1]) > R($handle[$block]);
            }
            push @stack, L($handle[$block]) if R($handle[$block]) > $endpt;
        }
    }
    return $ov;
}
```

After constructing the blocks, the next step is to classify them. First, we will classify them as "simple" or "twisted" by examining each wish and updating its block's entry in the list @blocktype. If a wish is simple then its block's type is updated to simple; if twisted, type is updated to twisted only if undefined.

```perl
sub twisted { odd(L($_[0])+R($_[0])) }
sub simple { !odd(L($_[0])+R($_[0])) }
sub odd { (1 & shift) }
```

```
sub classifyBlocks {
    my ($wishes,$ov) = @_;
    my @blockType;
    foreach my $e (@$wishes) {
        my $block = $ov->find(L($e));
        $blockType[$block] = (simple($e) && "simple")
            || $blockType[$block] || "twisted";
    }
    ...
```

In the next step, hurdles and superhurdles are identified among the twisted blocks. The underlying process can be described as follows. A list of the same length as the unsigned permutation is constructed, in which the ith entry gives the block containing the wish (if any) with an endpoint at position i. Next, all but twisted blocks are removed from this list. Then, runs of consecutive block numbers are compressed to a single entry. In practice, we can conflate these three stages into a single pass through the positions of the permutation.

```
sub classifyBlocks {
    ...
    my @runs = (-1);      ## condense twisted runs to a single entry
    my @numRuns;          ## count number of runs for each block
    foreach (map { $ov->find($_) } (0..$#pi)) {
        push @runs, $_
            if ($blockType[$_] eq "twisted") && ($runs[$#runs] != $_);
    }
    shift @runs;      ## remove the -1
    ...
```

If we discover that the entire list is compressed to a single number, then we have a single simple hurdle and can return immediately. The return value is a list consisting of the numbers of superhurdles and simple hurdles followed by a reference to the array of runs and a reference to the list of block classifications.

```
sub classifyBlocks {
    ...
    if (@runs==1) {      ## special case for one giant twisted block ...
        $blockType[$runs[0]] = "simple hurdle";
        return (0,1,\@runs,\@blockType);
    }
    ...
```

Otherwise, we begin to count the number of runs for each of the blocks present in the @runs list. Some special handling is required if the first and last runs are of the same block; this must be treated as a single "wraparound" run. Next, we identify hurdles by processing the runs one by one. If a run is the only one for its block then the block is a hurdle. If, in addition, the previous and next runs form a straddle, then the block is a superhurdle.

```perl
sub classifyBlocks {
    ...
    foreach (@runs) { $numRuns[$_]++; }
    if ($runs[0]==$runs[$#runs]) { $numRuns[0]--; }
    else { push @runs, $runs[0] }
    push @runs, $runs[1];     ## extra entries facilitate superhurdle detection

    my ($superhurdles,$simplehurdles);
    for (my $i=1; $i<$#runs; $i++) {
        if ($numRuns[$runs[$i]]==1) {     ## consecutive...a hurdle
            if (($runs[$i-1]==$runs[$i+1])
                && ($numRuns[$runs[$i-1]]==2)) {
                $blockType[$runs[$i]] = "superhurdle";
                $superhurdles++;
            } else {
                $blockType[$runs[$i]] = "simple hurdle";
                $simplehurdles++;
            }
        }
    }
    $#runs -= 2;
    return ($superhurdles,$simplehurdles,\@runs,\@blockType);
}
```

Subroutine classifyBlocks returns the information it gathers to the subroutine clearHurdles, which is responsible for using it to choose a twist consistent with the priorities enumerated in Section 17.4. clearHurdles returns 1 if it finds one or more hurdles to clear or 0 if no more hurdles exist. First, clearHurdles identifies which blocks will be twisted together using subroutine blockSearch, which searches for left-to-right occurrences of blocks of the type specified by its arguments. The easiest cases are when there is no superhurdle and at most one simple hurdle:

```perl
sub clearHurdles {
    my ($wishes,$ov) = @_;
    my ($superhurdles,$simplehurdles,$runs,$blockType)
        = classifyBlocks($wishes,$ov);
```

```
   my ($note,$h1,$h2);
   if ($superhurdles+$simplehurdles==0) { return 0; }
   if ($superhurdles==0 && $simplehurdles==1) {
      ($note,$h1) = blockSearch($runs,$blockType,"simple hurdle");
      ## find and untwist wish with leftmost endpoint
      my ($i,$j);
      for ($i=0; $i<@pi; $i++) { last if $ov->find($i)==$h1};
      foreach (@$wishes) { $j=R($_) if (L($_)==$i) }
      twist($i, $j, \@pi, " >twist simple hurdle>");
      return 1;
   }
   ...
```

Otherwise, two blocks must be twisted together; they must be chosen according to the priorities of Section 17.4

```
sub clearHurdles {
   ...
   ## choose two blocks to twist together
   if ($superhurdles>=2) {
      ($note,$h1,$h2) =
         blockSearch($runs,$blockType,"superhurdle","superhurdle");
   } elsif (($superhurdles==1) && ($simplehurdles>=1)) {
      ($note,$h1,$h2) =
         blockSearch($runs,$blockType,"superhurdle","simple hurdle");
      ($note,$h1,$h2) =
         blockSearch($runs,$blockType,"simple hurdle","superhurdle")
            unless $h2;
   } elsif ($superhurdles==1) {
      ($note,$h1,$h2) =
         blockSearch($runs,$blockType,"superhurdle","twisted");
      ($note,$h1,$h2) =
         blockSearch($runs,$blockType,"twisted","superhurdle")
            unless $h2;
   } elsif ($simplehurdles>=2) {
      ($note,$h1,$h2) =
         blockSearch($runs,$blockType,
                     "simple hurdle","simple hurdle","simple hurdle");
   }
   ...
```

After subroutine clearHurdles has identified blocks to twist together, it searches for the actual positions to twist. Normally, these are the last position of the first block

and the first position of the last block; however, a wraparound block can complicate
matters slightly.

```perl
sub clearHurdles {
    ...
    my $end1, my $start2;
    for (my $i=0; $i<@pi; $i++) { $end1=$i if $ov->find($i)==$h1 };
    for (my $i=@pi−1; $i>=0; $i−−) { $start2=$i if $ov->find($i)==$h2 };
    if ($h1==$$runs[0]) {     ## first hurdle could wrap around
        for ($end1=$start2; $end1>=0; $end1−−) {
            last if $ov->find($end1)==$h1;
        }
    }
    if ($h2==$$runs[0]) {     ## second hurdle could wrap around
        for ($start2=$end1; $start2<@pi; $start2++) {
            last if $ov->find($start2)==$h2;
        }
    }
    ## twist together
    twist($end1, $start2, \@pi, " >twist $note>");
    return 1;
}
```

Once all hurdles are removed, the program begins to locate wishes to grant in simple blocks. There are two steps, implemented in the subroutines findHappyClique
and findMaxTwistedDegree. Both operate by scanning the wish list from left to
right. findHappyClique accumulates a happy clique in list @C. The set of wishes
in this list forms a happy clique of all the wishes processed so far. When processing
a new wish, we may discover that it should be added to the current clique, that it in-
validates and should replace the current clique, or that the search can be terminated
because the new wish (and thus all remaining wishes) is too far to the right to invalidate the current clique. In addition to the clique @C, the algorithm also maintains a
simple wish $t that strictly contains the current clique.

```perl
sub findHappyClique {
    my @wishes = @_;
    my @C = ();
    my $t = undef;
    foreach my $w (@wishes) {
        if (twisted($w)) { next; }
        elsif (@C==0) { push @C, $w; }
        elsif (R(@C[@C−1])<L($w)) { return @C; }
```

```
        elsif ($t && R($w)>R($t)) { next; }
        elsif (R($w)<R($C[@C−1])) { $t = pop @C; @C = ($w); }
        elsif (L($w)<=R($C[0])) { push @C, $w; }
        elsif (L($w)>R($C[0])) { @C = ($w); }
    }
    return @C;
}
```

The seven cases of findHappyClique can be explained as follows.

1. **if** (twisted($w)) { **next**; }
   Happy cliques contain only simple wishes, and their properties deal only with simple wishes.

2. **elsif** (@C==0) { **push** @C, $w; **next**; }
   The first simple wish by itself satisfies the happy-clique properties.

3. **elsif** (R(@C[@C−1]) < L($w)) { **last**; }

```
        $C[0]:        |-----------|
     $C[@C-1]:              |-----------|
          $w:                                  |-----------|
```

If the new wish lies completely to the right of the happy clique, then so do all the remaining wishes. Nothing remaining to be processed can invalidate the happy clique, so we are done.

4. **elsif** ($t && R($w)>R($t)) { **next**; }

The new simple wish intersects with the last wish in the clique, but it also overlaps with simple wish $t, which is not part of the clique. Hence it satisfies Property 3 with respect to the current clique in @C, and we can ignore it.

5. **elsif** (R($w)<R($C[@C−1])) { $t = **pop** @C; @C = ($w); }

```
        $C[0]:        |-----------|
     $C[@C-1]:                |--------------------|
          $w:                  |-----------|
```

The new wish is contained in at least the last wish of the old clique, and it may overlap with earlier wishes in the current clique. We can make the new wish into a new singleton happy clique because (a) all the previous wishes in the current clique overlap the last wish and (b) the last wish does not overlap the new wish; Property 3 is satisfied. The last wish of the old clique becomes the new $t.

6. **elsif** (L($w)<=R($C[0])) { **push** @C, $w; }

```
      $t: |- - - - - - - - - - - - - - - - - - -|
   $C[0]:      |-----------|
$C[@C-1]:             |-----------|
      $w:               |-----------|
```

The new simple wish overlaps every wish in the current clique, and it is contained
in $t (if defined). If it overlaps any earlier simple wishes outside the clique, they
have already been found to satisfy Property 3. So the new wish can be added to
the clique.

7. **elsif** (L($w)>R($C[0])) { @C = ($w); }

```
      $t: |- - - - - - - - - - - - - - - - - - -|
   $C[0]:      |--------|
$C[@C-1]:             |-----------|
      $w:                |-----------|
```

The new wish overlaps the last wish but not the first wish of the current clique. It
does not overlap $t; in fact, $t (if defined) contains it. Perhaps it satisfies Property 3 with some wish further to the right; we don't know. Since our goal is to
find *any* happy clique quickly, we assume the worst about the current clique. We
replace the current clique with a singleton set consisting only of the new wish.
The old $C[0] intersects every member of the old clique and is not a member of
the new clique. Therefore, every member of the old clique satisfies Property 3.

It is worth reiterating that the assigned task is simply to find *some* happy clique –
not the largest, the smallest, the leftmost, or the rightmost. Since @C always contains a valid happy clique for the wishes processed so far and is never empty after the
first simple wish is processed, we are guaranteed to have a happy clique at termination.

Once a happy clique is found, the subroutine findMaxTwistedDegree finds the
wish in the happy clique that overlaps with the most twisted wishes. Although this
task seems straightforward, sophisticated programming is required to accomplish the
task in time proportional to the length of the permutation. Interested readers should
consult the primary sources for an explanation of our subroutine.

17.7 Exercise

1. The priorities in Section 17.4 are actually overelaborated. Argue that the following priorities suffice and then modify the program to implement these priorities.

   (a) If possible, twist together two nonadjacent hurdles (simple or super). This
   simplifies the hurdles and at least one twisted block between.
   (b) Otherwise, if possible, twist together two adjacent hurdles. This simplifies
   the two hurdles.

(c) Otherwise, if possible, twist together a superhurdle and its straddle. This simplifies the superhurdle and its straddle.

(d) Otherwise, if possible, twist a single hurdle. This simplifies the only hurdle.

(e) Otherwise, there are no hurdles and simplification is complete.

17.8   Appendix: Correctness of Choice of Wish from Happy Clique

Theorem. *Let C be a happy clique and let w be a wish in C that overlaps at least as many twisted wishes as any other. Then granting w does not create new twisted blocks.*

Proof. It suffices to show that every wish w' overlapping w before granting w either lies in a simple block after granting w or is granted by the same twist that grants w. There are four cases.

Case (i): w' was twisted before. Then it becomes simple, so it clearly lies in a simple block.

Case (ii): w' was simple before and was not in the clique C. By Property 2, it overlaps a simple wish outside the clique before – and after – w is granted. Hence it lies in a simple block.

Case (iii): w' was a simple wish in the clique C before, and some wish t was twisted before and overlapped w but not w'. Then t becomes simple and overlaps w' after, so w' lies in a simple block.

Case (iv): w' was a simple wish in the clique C before, and every wish t that was twisted before and overlapped w also overlapped w'. But since we chose w to overlap at least as many twisted wishes as any other wish in the clique, we must conclude that w and w' overlapped exactly the same twisted and simple wishes before w was granted. This means that, after w is granted, w' overlaps no other wishes whatsoever. This is possible only if granting wish w also grants w', since (as can be easily verified) no wish can form a block all by itself.   QED

17.9   Complete Program Listings

This chapter's program is found in `chap17/kst.pl` in the software distribution.

17.10   Bibliographic Notes

One researcher who published work in 1979 on the unsigned version of sorting by reversals was W. H. Gates of Microsoft – proof that, though writing mathematical papers may not make you the richest man in the world, it won't preclude it, either.

Hannenhalli and Pevzner (1995a) gave the first polynomial-time algorithm for sorting signed reversals. The algorithm described in this chapter is a faster and simpler one due to Kaplan, Shamir, and Tarjan (1997). Our terminology is much different, however: our wishes are their *gray edges,* our blocks are their *cycles,* and our simple (or twisted) wishes are their *oriented* (or *unoriented*) gray edges.

Hannenhalli and Pevzner (1995b) eventually solved a more difficult version of the problem in which several chromosomes are present and genes can be transposed from one chromosome to another.

Drawing RNA Cloverleaves

This appendix presents the program used to draw the RNA secondary structure diagrams of Chapter 2. It builds on the recursive evalRna subroutine of that chapter.

We assume the existence of the following graphics routines for generating simple pictures in the Cartesian plane.

- beginPicture( ) performs any initialization required by the graphics package before any lines or text are displayed.
- endPicture( ) performs any shutdown required by the graphics package to end the picture.
- drawLine($x1,$y1,$x2,$y2,$thickness) draws a line segment from ($x1,$y1) to ($x2,$y2). Its thickness is determined by $thickness, with 0 being the minimum visible thickness.
- drawLetter($x,$y,$letter) centers the letter $letter at position ($x,$y).

The complete listing of our program includes working graphics routines for Adobe Systems' PostScript graphics language. This language is understood by many common printers as well as by previewers such as ghostview and gv. (The Linux operating system includes filters that translate PostScript into commands understood by many printers that do not directly support PostScript.) We will focus on the algorithmic aspects of determining where to put the lines and letters rather than on the details of the PostScript to be generated.

Our subroutines inherit their recursive structure from our bond-counting subroutines. drawRnaStructure differs from evalRnaStructure only in its calls to beginPicture and endPicture:

```
sub drawRnaStructure {
    my ($basestring,$structurestring) = @_;
    @bases = split(//,5.$basestring.3);
    @structure = split(//,"($structurestring)");
    beginPicture();
```

```
    drawRna(0, $#structure, 0,0, 1,0);
    endPicture( );
}
```

As we shall see, drawRna is substantially more complicated than evalRna.

Our program works by decomposing the RNA structure into a set of connected rings. Adjacent rings share a single H-bonded pair. The signal recognition particle RNA depicted in Figure A.1 has a total of 22 rings: seventeen 4-base rings, two 9-base rings, and one ring each of 8, 13, and 16 bases. The first ring to be drawn in this case is the one at the right of the diagram, containing the $3'$ and $5'$ markers, which are treated like bases. Note that $5'$ and $3'$ are always drawn at positions $(0, 0)$ and $(1, 0)$, respectively.

Subroutine evalRna works by drawing the single ring containing the bases at positions \$l and \$r and then making recursive calls to draw neighboring rings. Since we want adjacent bases to be separated by a fixed distance, we must first determine how many bases lie on the ring being drawn. Returning to our idea of nesting level, we see that the bases lying on the ring are:

- the bases at positions \$l and \$r;
- the bases corresponding to dots at level 0; and
- the bases corresponding to parentheses (left or right) at level 1.

We must make two left-to-right scans. The first scan will count the number of bases on the ring, and the second will actually draw the ring. The first scan is even simpler than evalRna:

```
sub drawRna {
    my ($l,$r,$lx,$ly,$rx,$ry) = @_;          ## 1
    my $level = 0;                            ## 2
    my $count = 2;                            ## 3
    for (my $i=$l+1; $i<$r; $i++) {           ## 4
        $level-- if ($structure[$i] eq ")");  ## 5
        $count++ if $level==0;                ## 6
        $level++ if ($structure[$i] eq "(");  ## 7
    }
        ...
}
```

Next, we must use \$count to determine the center ($cx,$cy) and radius \$rad of the ring. We must also determine the angle \$theta (in radians) subtended by adjacent bases as well as the orientation \$alpha of the first base with respect to the center of the ring. This requires a little simple trigonometry, illustrated in the following figure.

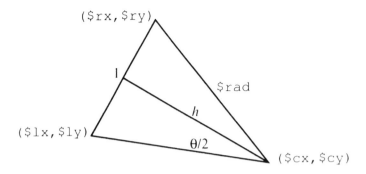

We want the bases on the ring to form a regular $count-gon with sides of length 1, so the sides subtend an angle of $\theta = 2\pi/\$count$. The radius r is the hypotenuse of a right triangle with a side of length $1/2$ and opposite angle of $\theta/2$; hence $(1/2)/r = \sin(\theta/2)$, or $r = 1/(2\sin(\theta/2))$. The length h of the third side of this triangle satisfies $h/r = \cos(\theta/2)$, or $h = r\cos(\theta/2)$. These calculations are carried out in Lines 8 through 10.

We can find the center of the circle by finding the midpoint of the line segment joining ($lx,$ly) and ($rx,$ry) and then adding $h times the perpendicular unit vector ($ly−$ry,$rx−$lx), as in Lines 11 and 12.

Finally, the orientation of ($lx,$ly) with respect to the center is the angle with tangent ($ly−$cy)/($lx−$cx). To avoid problems with floating-point overflows and underflows, Perl supplies a two-argument arctangent function **atan2**, used in Line 13.

```
sub drawRna {

    ...
    my $theta = 2 * 3.14159 / $count;               ## 8
    my $rad = 1 / (2*sin($theta/2));                 ## 9
    my $h = $rad * cos($theta/2);                    ## 10
    my ($cx,$cy) = ((($lx+$rx)/2.0)+$h*($ly−$ry),    ## 11
                    (($ly+$ry)/2.0)+$h*($rx−$lx));   ## 12
    my $alpha = atan2($ly−$cy,$lx−$cx);              ## 13

    ...

}
```

We now have the information we need to draw the bases around the ring. We make another left-to-right scan of the structure (Lines 15–27), paying special attention to draw bases corresponding to dots at nesting level 0 and parentheses at level 1 (Lines 17–25). Since we must draw lines between pairs of bases, we keep track of the index and coordinates of the last base drawn in the variables $ii and ($xx,$yy) (Lines 14, 23, 24). For each new base to be drawn, we move clockwise around the ring by substracting $theta from $alpha, the orientation of the current base with respect to the center of the ring (Line 18). When a right parenthesis at level 1 is found, drawRna is called recursively (Line 21) to draw the structure lying between the right parenthesis and its left mate.

The location of the last base in the ring is known, so the lines joining it to its neighbors are handled outside the loop (Lines 29–32). The final step is to draw the line joining the first and last bases of the ring. The width of this line reflects the number of hydrogen bonds between these two bases.

```
sub drawRna {
    ...
    my ($ii,$xx,$yy) = ($l,$lx,$ly);                                        ## 14

    for (my $i=$l+1; $i<=$r; $i++) {                                        ## 15
        $level−− if ($structure[$i] eq ")");                               ## 16
        if ($level==0) {                                                    ## 17
            $alpha −= $theta;                                              ## 18
            my ($x,$y)
                = ($cx+$rad*cos($alpha),$cy+$rad*sin($alpha));             ## 19
            if ($structure[$i] eq ")") {                                   ## 20
                drawRna($ii,$i,$xx,$yy,$x,$y)                              ## 21
            } else { drawLine($xx,$yy,$x,$y,0); }                          ## 22
            drawBase($xx,$yy,$bases[$ii]);                                ## 23
            ($xx,$yy)=($x,$y);                                            ## 24
            $ii = $i;                                                     ## 25
        }                                                                  ## 26
        $level++ if ($structure[$i] eq "(");                              ## 27
    }                                                                      ## 28
    drawLine($xx,$yy,$rx,$ry,0);                                          ## 29
    drawBase($xx,$yy,$bases[$r−1]);                                       ## 30
    drawBase($rx,$ry,$bases[$r]);                                         ## 31
    my %bonds
        = (GU=>1,UG=>1,AU=>2,UA=>2,CG=>3,GC=>3);                          ## 32
    drawLine($lx,$ly,$rx,$ry,$bonds{$bases[$l].$bases[$r]}+0)             ## 33
        unless ($lx==0 || $ly==0);    ## 3−5 pair                         ## 34
}
```

The main program does little more than read the input strings and pass them to the subroutines. The only feature of interest is the use of the standard file handle STDERR. By default, output goes to the file handle STDOUT. Output to STDOUT goes to the terminal screen unless redirected by the command line when the program is invoked. Because PostScript must be printed or viewed with a special previewer, the output of this particular program would almost always be redirected to a file. Output to the STDERR handle also goes to the terminal but is not affected by redirection. This makes it useful for error messages, requests for user input, and debugging

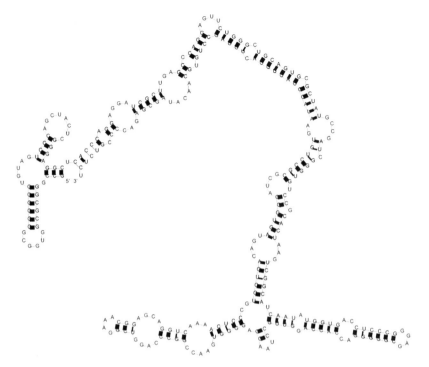

Figure A.1: A canine SRP RNA, as drawn by our program.

output. Our program sends the bond count to the screen even while the drawing is being saved in a file.

```
my $basestring = <STDIN>;
chomp($basestring);
my $parenstring = <STDIN>;
chomp($parenstring);
print STDERR evalRnaStructure($basestring,$parenstring),
    "hydrogen bonds in this structure.\n";
drawRnaStructure($basestring,$parenstring);
```

A.1 Exercises

1. Subroutines drawRnaStructure and drawRna are not really that efficient. The worst-case running time of this program grows as the square of the length of the RNA. The big inefficiency is that substrings are scanned repeatedly in successive recursive calls to determine the mates of parentheses. This problem can be solved by modifying drawRnaStructure to create and fill another array containing the indices of the mates of parentheses. For example, for the structure

$$((\ldots)\ldots(((\ldots))\ldots))()$$

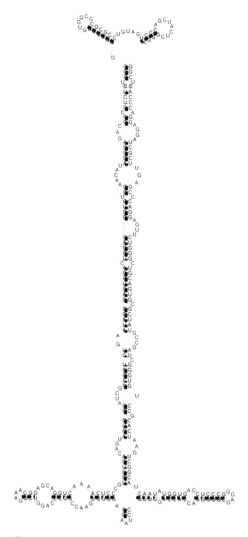

Figure A.2: A canine SRP RNA, drawn more conventionally.

the array (list) should contain

(19, 5, 2, 3, 4, 1, 6, 7, 18, 16, 15, 11, 12, 13, 14, 10, 9, 17, 8, 1, 20, 21).

Modify drawRnaStructure to create this list. Then modify drawRna to use the list to reduce the overall running time to a linear function of the length of the RNA.

2. The drawings produced by drawRnaStructure are also not quite up to publication standards. Normally, instead of placing all *bases* at equal intervals around a ring, the *H-bonded pairs* on the ring are spaced at equal intervals and the unbonded bases are placed in between. For example, Figures A.1 and A.2 (respectively) depict a canine SRP RNA drawn as our program would draw it and then drawn more conventionally. Modify our program to draw in the more conventional fashion.

A.2 Complete Program Listings

The complete program for drawing RNAs is found in `appA/drawRNA.pl` in the software distribution.

A.3 Bibliographic Notes

Adobe's PostScript language is described in detail in Adobe Systems (1999). The `ghostview` previewer for PostScript can be downloaded from `http://www.cs.wisc.edu/~ghost/`.

Space-Saving Strategies for Alignment

The pairwise global alignment algorithm of Chapter 3 (the Needleman–Wunsch algorithm) requires both time and space proportional to the product of the lengths of the two strings. It is not too hard to see that space to store two rows of the matrix M suffices if only the similarity score (and not the optimal alignment itself) is desired as output. After the ith row of matrix M is used to compute the $(i + 1)$th, it is never referred to again. In this appendix, we apply this idea with a bit of refinement and then show that linear space and quadratic time also suffice to produce the optimal global alignment itself. The exercises explore ways to apply the same method to other alignment problems.

B.1   Finding Similarity Scores Compactly

For simplicity, we will assume the $\pm 1/-2$ DNA scoring scheme. Specializing our algorithm from Chapter 3, we obtain

```perl
sub similarity {
    my($s,$t) = @_;
    foreach my $i (0..length($s)) { $M[$i][0] = −2*$i; }
    foreach my $j (0..length($t)) { $M[0][$j] = −2*$j; }
    foreach my $i (1..length($s)) {
        foreach my $j (1..length($t)) {
            my $m = (substr($s,$i−1,1) eq substr($t,$j−1,1)) ? +1 : −1;
            $M[$i][$j] =
                max($M[$i−1][$j]−2, $M[$i][$j−1]−2, $M[$i−1][$j−1]+$m);
        }
    }
    return ($M[length($s)][length($t)]);
}
```

Our goal is to avoid storing all the rows of @M. In fact, we can get by with storing just a single row @a with an extra entry. As a first attempt, we consider saving only the most recently computed row and then overwriting it left to right to compute the next row:

```
foreach my $i (1..length($s)) {
    $a[0] = -2*$i;
    foreach my $j (1..length($t)) {
        my $m = (substr($s,$i-1,1) eq substr($t,$j-1,1)) ? +1 : -1;
        $a[$j] = max($a[$j]-2, $a[$j-1]-2, $a[$j-1]+$m);     ## OOPS!!!
    }
}
```

The problem with this scheme, of course, is that $a[$j-1] doesn't simultaneously contain both $M_{i,j-1}$ and $M_{i-1,j-1}$; the latter has been overwritten with our attempt at computing the former by the time we try to compute $M_{i,j}$ for assignment to $a[$j].

The solution to this is to *move over* every entry of the array @a by one position to make room for the new value $M[$i][0]. This process establishes our new loop invariant,

$$\$a[\$k] = \begin{cases} M_{i-1,k-1} & \text{for } k \geq j, \\ M_{i,k} & \text{for } k < j. \end{cases}$$

From then on, we will find $M_{i-1,j-1}$, $M_{i,j-1}$, and $M_{i-1,j}$ in $a[j], $a[j-1], and $a[j+1], respectively.

The final subroutine, compactSimilarity, uses Perl's **unshift** operator to shoehorn $M_{i,0}$ into $a[0] to establish the loop invariant. At the end of the loop, $M_{i-1,|t|}$ is removed from the last position of the array with **pop**.

```
sub compactSimilarity {
    my ($s,$t) = @_;
    my ($sLen,$tLen) = (length($s),length($t));
    my @a;
    foreach my $j (0..$tLen) { $a[$j] = -2*$j; }
    foreach my $i (1..$sLen) {
        unshift @a, -2*$i;
        foreach my $j (1..$tLen) {
            my $m = (substr($s,$i-1,1) eq substr($t,$j-1,1)) ? +1 : -1;
            $a[$j] = max($a[$j]+$m, $a[$j-1]-2, $a[$j+1]-2);
        }
        pop @a;
```

```
    }
    return $a[$#a];
}
```

B.2   Finding Alignments Compactly

Our linear-space scoring algorithm gives correct similarity scores, but it doesn't leave us with the full table of values of M_{ij} for retracing the steps that led to the score. Constructing an optimal alignment in linear space requires a clever combination of our linear programming strategy and the divide-and-conquer algorithm paradigm.

To illustrate the strategy, let's use SASKATCHEWAN and SESQUICENTENNIAL as inputs and imagine that we already have the optimal alignment. By drawing a vertical line between the T and the C of SASKATCHEWAN, we divide SASKATCHEWAN into two equal substrings and SESQUICENTENNIAL into two possibly unequal substrings. The alignment of SASKAT and the left part of SESQUICENTENNIAL will be an optimal alignment of those two substrings, and the alignment of the two right-hand substrings will also be optimal. (Any possible change that could improve the left- or right-hand score would also improve the total score, contradicting the optimality of the overall alignment.) So our compact alignment subroutine compactAlignment first uses our compact scoring routine to figure out where to break SESQUICENTENNIAL in two. We need to know which of the following 17 possibilities gives the highest score:

$$\text{sim(SASKAT, } \Lambda) + \text{sim(CHEWAN, SESQUICENTENNIAL)},$$

$$\text{sim(SASKAT, S)} + \text{sim(CHEWAN, ESQUICENTENNIAL)},$$

$$\text{sim(SASKAT, SE)} + \text{sim(CHEWAN, SQUICENTENNIAL)},$$

$$\vdots$$

$$\text{sim(SASKAT, SESQUICENTENNI)} + \text{sim(CHEWAN, AL)},$$

$$\text{sim(SASKAT, SESQUICENTENNIA)} + \text{sim(CHEWAN, L)},$$

$$\text{sim(SASKAT, SESQUICENTENNIAL)} + \text{sim(CHEWAN, } \Lambda).$$

We could evaluate the 17 possibilities by calling compactSimilarity 34 times. But a closer look at compactSimilarity shows that the 17 quantities in the left-hand column are exactly the contents of the array @a at the end of the call

```
compactSimilarity('SASKAT', 'SESQUICENTENNIAL');
```

$a[$j] is the similarity of $s and the first $j letters of $t. We can get all 17 back by modifying the last line of compactSimilarity to read

---

```
    return @a;
```

The right-hand values are a little trickier, since they involve similarities with the *last* $j letters of $t. But we observe that similarity scores are not affected if both strings are reversed, so we can get all 17 quantities in the right-hand column by calling

    compactSimilarity('NAWEHC', 'LAINNETNECIUQSES');

In this particular case, it turns out that the best overall alignment comes from concatenating the best alignment of SASKAT and SESQUI with the best alignment of CHEWAN and CENTENNIAL. To determine the best alignment of these smaller strings, we simply call compactAlignment twice recursively:

    compactAlignment('SASKAT', 'SESQUI');

    compactAlignment('CHEWAN', 'CENTENNIAL');

The process of dividing the first argument in half, computing the best place to divide the second argument, aligning the left and right parts separately by recursive invocations, and finally joining the two alignments to form a single overall alignment is implemented by the following code.

```
sub compactAlignment {
    my($s,$t) = @_;
    my ($sLen,$tLen) = (length($s),length($t));
    if ($sLen>1) {
        my $mid = int($sLen/2);
        my $sBeg = substr($s,0,$mid);
        my $sEnd = substr($s,$mid,$sLen−$mid);
        my @aBeg = compactSimilarity($sBeg,$t);
        my @aEnd = reverse(compactSimilarity((scalar reverse $sEnd),
                                             (scalar reverse $t)));
        my ($kMax,$kMaxVal) = (0,$aBeg[0]+$aEnd[0]);
        foreach my $k (1..$tLen) {
            ($kMax,$kMaxVal) = ($k,$aBeg[$k]+$aEnd[$k])
                if ($aBeg[$k]+$aEnd[$k]) > $kMaxVal;
        }
        my ($ssBeg,$ttBeg,$scoreBeg) =
            split("\n", compactAlignment($sBeg, substr($t,0,$kMax)));
        my ($ssEnd,$ttEnd,$scoreEnd) =
            split("\n", compactAlignment($sEnd, substr($t,$kMax)));
        my $score = $scoreBeg+$scoreEnd;
        return "$ssBeg$ssEnd\n$ttBeg$ttEnd\n$score";
    }
    ...
}
```

The game ends when this subroutine's first argument consists of just a single letter. If this letter is present in the second argument then the matching letters must be put into the same column; otherwise, the single letter may be put in any column – we opt for the last.

```
sub compactAlignment {
    my($s,$t) = @_;
    my ($sLen,$tLen) = (length($s),length($t));
    if ($sLen>1) {
        ...
    } else {
        if ($t =~ /^(.*)$s(.*)$/) {
            my $ss = "-" x length($1) . $s . "-" x length($2);
            my $score = 3−2*$tLen;
            return "$ss\n$t\n$score";
        } else {
            my $score = 1−2*$tLen;
            return "-" x ($tLen−1) . "$s\n$t\n$score";
        }
    }
}
```

It should be clear that the amount of space used by this procedure is linear in the sum of the lengths of the two strings. On the other hand, the *time* required is less obvious.

It is clear enough that, if the two strings have length m and n, then subroutine compactSimilarity requires cmn time for some constant c. (In fact, the time should not be much different from that required for the algorithm in Chapter 3.) Now let $T(m, n)$ represent the time needed by compactAlignment for the same two strings. If $m = 1$, then by inspection we have

$$T(1, n) = c_1 n.$$

If $m > 1$ then for some $n_1 + n_2 = n$ we have

$$
\begin{aligned}
T(m, n) = {} & c\lfloor m/2 \rfloor n \quad \text{(for 1st call to compactSimilarity)} \\
& + c\lceil m/2 \rceil n \quad \text{(for 2nd call to compactSimilarity)} \\
& + T(m/2, n_1) \quad \text{(for 1st call to compactAlignment)} \\
& + T(m/2, n_2) \quad \text{(for 2nd call to compactAlignment)} \\
& + c_2(m + n) \quad \text{(for other operations).}
\end{aligned}
$$

From this recurrence relation, it is possible to derive that

$$T(m, n) \approx 2cmn.$$

Thus, reducing space by a factor of m costs us time, but only a small constant factor.

B.3   Exercises

1. Modify the program presented in this appendix to use protein substitution matrices. The most significant changes are to the section of compactAlignment dealing with single-letter $s.

2. Modify the program presented in this appendix to handle affine gap penalties, as described in Exercise 7 of Chapter 3.

3. Modify the program presented in direct.pl in Chapter 9 to compute the score (only) of the best three-sequence alignment in quadratic (rather than cubic) space. Use the resulting subroutine and a divide-and-conquer strategy to produce the optimal alignment itself using quadratic space and cubic time.

B.4   Complete Program Listings

The complete program for compact alignment is found in appB/compact.pl in the software distribution.

B.5   Bibliographic Note

See Myers and Miller (1988) and Huang, Hardison, and Miller (1990).

A Data Structure for Disjoint Sets

This appendix describes a simple yet very efficient Perl solution to a problem known as the *disjoint sets problem,* the *dynamic equivalence relation problem,* or the *union-find problem.* This problem appears in applications with the following scenario.

1. Each one of a finite set of keys is assigned to exactly one of a number of classes. These classes are the "disjoint sets" or the partitions of an equivalence relation. Often, the set of keys is known in advance, but this is not necessary to use our Perl package.
2. Initially, each key is in a class by itself.
3. As the application progresses, classes are joined together to form larger classes; classes are never divided into smaller classes. (The operation of joining classes together is called *union* or *merge.*)
4. At any moment, it must be possible to determine whether two keys are in the same class or in different classes.

To solve this problem, we create a package named UnionFind. Objects of type UnionFind represent an entire collection of keys and classes. The three methods of this package are:

- $uf = UnionFind−>new( ), which creates a new collection of disjoint sets, each of which has only one element;
- $uf−>inSameSet($key1,$key2), which returns true if its two arguments are elements of the same disjoint set or false if not;
- $uf−>union($key1,$key2), which combines the sets to which its two arguments belong into a single set.

The UnionFind structure itself contains two hashes. One contains a *parent* for each key, and the other contains a numerical *rank.* Each of the classes contains a single representative element; it can be recognized by its lack of value in the parent hash. In fact, it stands at the root of an *inverted tree* representing the class. Every (nonrepresentative) key's parent value is another key in the same class, and the root can be found by repeating the statement $key = $parent{$key} until a key with no parent is found. This is the job of subroutine find. If two keys are in the same class, then find will return the same representative key for both; if not, the representatives will be different.

Under this scheme, it is easy to join two classes together. union first finds the representatives of the two classes. Then, the parent value of one representative key is set to the other representative key. The key receiving the new parent loses its representative status. After the change, calling find on one of the keys in either of the two original classes will follow a chain of parents to one and the same representative.

C.1 Union by Rank

So far, we have not made use of our structure's second hash. Its purpose is to keep the maximum length of any path from a key to its class's representative relatively short. Every representative has a positive integer *rank* that gives a rough indication of the size of the class. When two classes are joined, union examines the rank of the two representatives to decide which will become the representative of the new, larger class. By way of example, suppose we know that one class has 37 elements and the other has only 23. If we are interested in minimizing the average number of steps required for a find operation, we should have the representative of the class of 37 represent the new class of 50. That way, we lengthen 23 possible find operations by a single step; the other way around, we would lengthen 37 finds.

Rather than recording the total size of a class, the rank hash records the largest number of steps required by find for any element of the class. This number is recorded under the key of the class's representative. The rank of a class consisting of a single key is 0. When two single-key classes are joined, the rank of the result is 1. If a third single-key class is joined, the rank of the result is still 1. If possible, union will choose the representative with larger rank to represent the joined class; in this case, the rank of the new representative is unchanged. If the two representatives have the same rank then either may be chosen to represent the joined class, but its rank increases by 1. Applying this strategy, it is not too hard to prove the following fact:

- If a class's representative key has rank k, then the class itself contains at least 2^k keys.

Turning this inequality around, we can say:

- If a class contains n keys then the rank of its representative is no more than $\lg n$, so any find operation on any key in the class takes no more than $\lg n$ steps.

Because a single union or inSameSet operation can involve at most two different keys, it follows that:

- By applying union by rank, any series of n union and inSameSet operations beginning with a new UnionFind structure can be completed in a number of steps proportional to $n \lg n$.

C.2 Path Compression

A further speed-up comes from the observation that we are never interested in knowing the parent of a key per se; we are only interested in the parent as a link to more distant ancestors. Thus, when we find the representative of a key, it makes sense to reset the key's parent value to point *directly* to the representative. Obviously this will save time if we call find on the same key later on. Less obviously, it also saves time for calls to find for *descendents* of the key. In fact, since we may walk through several nodes on the way to the representative, it makes sense to update *each of them* to point directly to the representative. This technique is called *path compression*.

The analysis of path compression is notoriously complicated, but the outcome is impressive:

- By simultaneously applying union by rank and path compression, any series of n union and inSameSet operations beginning with a new UnionFind structure can be completed in a number of steps proportional to $n \lg^* n$.

The notation $\lg^*$ represents a *very* slowly growing function. It is the inverse of a very rapidly growing function involving exponential "stacks" of 2s; $\lg^* n \leq 5$ for

$$n < 2^{2^{2^{2^2}}} = 2^{2^{2^4}} = 2^{2^{16}} = 2^{65536} > 10^{19728}.$$

Although $\lg^* n \to \infty$ as $n \to \infty$, it approaches so slowly that it can be regarded as bounded above by a small constant for any practical purpose.

C.3 Complete Program Listings

A complete listing of our package for disjoint sets is found in file `appC/Union-Find.pm` and below.

```
#########################################
package UnionFind;
#########################################
## Implements disjoint sets.
## Uses inverted trees, union by rank, and path compression.
#########################################

use strict;
use Util;
```

```perl
#########################################
sub new
## Creates a new union/find object.
#########################################
{
    my ($this) = @_;
    return bless [{ },{ }], (ref($this) || $this);
}


#########################################
sub union
#########################################
{
    my ($this,
        $key1,$key2) = @_;
    my ($parent,$rank) = @$this;
    $key1 = $this->find($key1);
    $key2 = $this->find($key2);
    return $key1 if $key1 eq $key2;
    ($key1,$key2) = ($key2,$key1) if $$rank{$key1} > $$rank{$key2};
    $$rank{$key2} = max($$rank{$key2}, 1+$$rank{$key1});
    return ($$parent{$key1} = $key2);
}


#########################################
sub find
#########################################
{
    my ($this,
        $key) = @_;
    my ($parent,$rank) = @$this;
    my $parentKey = $$parent{$key};
    if (!defined($parentKey)) {
        return ($key);
    } else {
        return ($$parent{$key} = $this->find($parentKey));
    }
}


#########################################
sub inSameSet
#########################################
```

```
{
    my ($this,$key1,$key2) = @_;
    $this->find($key1) eq $this->find($key2);
}

1;
```

C.4 Bibliographic Note

Data structures for disjoint sets are described in Chapter 22 of the textbook by Cormen, Leiserson, and Rivest (1990).

Suggestions for Further Reading

The following books are especially recommended for further reading. The citations in each category are arranged roughly from most to least accessible.

Algorithms and Programming in General Bentley (1999); Garey and Johnson (1979); Cormen et al. (1990)

Computational Molecular Biology Mount (2001); Setubal and Meidanis (1997); Salzberg, Searls, and Kasif (1998b); Lander and Waterman (1995); Pevzner (2000)

General Genetics and Molecular Biology Brown (1999); Nicholl (1994); Lewin (1999); Snustad and Simmons (1999); Russell (1992)

Perl Wall et al. (1997); Vromans (2000); Srinivasan (1997)

Public-Domain Databases and Software Gibas and Jambeck (2001); Baxevanis and Ouellette (2001)

Protein Structure Goodsell (1996); Brandon and Tooze (1999)

Statistical Modeling of Sequences Durbin et al. (1998); Baldi and Brunak (1998)

Bibliography

Adobe Systems, Inc. (1999). *PostScript*© *Language Reference,* 3rd ed. Harlow, U.K.: Addison-Wesley.

D. Akst (2000). "The forgotten plague." *American Heritage* 51 (December).

S. Altschul, M. Boguski, W. Gish, and J. C. Wooten (1994). "Issues in searching molecular sequence databases." *Nature Genetics* 6: 119–29.

S. F. Altschul (1991). "Amino acid substitution matrices from an information theoretic perspective." *Journal of Molecular Biology* 219: 555–65.

S. F. Altschul, R. J. Carroll, and D. J. Lipman (1989). "Weights for data related by a tree." *Journal of Molecular Biology* 207: 647–53.

S. F. Altschul and W. Gish (1996). "Local alignment statistics." *Methods in Enzymology* 266: 460–80.

S. F. Altschul, W. Gish, W. Miller, E. W. Myers, and D. J. Lipman (1990). "A basic local alignment search tool." *Journal of Molecular Biology* 215: 403–10.

S. F. Altschul, T. L. Madden, A. A. Schäffer, J. Zhang, Z. Zhang, W. Miller, and D. J. Lipman (1997). "Gapped BLAST and PSI-BLAST: A new generation of protein database search programs." *Nucleic Acids Research* 25: 3389–3402.

S. Anderson (1981). "Shotgun DNA sequencing using cloned DNase I-generated fragments." *Nucleic Acids Research* 9: 3015–27.

A. Bairoch and R. Apweiler (1996). "The SWISS-PROT protein sequence data bank and its new supplement TREMBL." *Nucleic Acids Research* 24: 12–15.

P. Baldi and S. Brunak (1998). *Bioinformatics: The Machine Learning Approach.* Cambridge, MA: MIT Press.

A. D. Baxevanis and B. F. F. Ouellette (Eds.) (2001). *Bioinformatics: A Practical Guide to the Analysis of Genes and Proteins.* New York: Wiley.

D. A. Benson, M. Boguski, D. J. Lipman, and J. Ostell (1996). "Genbank." *Nucleic Acids Research* 24: 1–5.

J. L. Bentley (1999). *Programming Pearls,* 2nd ed. Reading, MA: Addison-Wesley.

M. J. Bishop (Ed.) (1998). *Guide to Human Genome Computing,* 2nd ed. San Diego: Academic Press.

M. J. Bishop and C. J. Rawlings (1988). *Nucleic Acid and Protein Sequence Analysis: A Practical Approach.* Oxford: IRL Press.

C. Brandon and J. Tooze (1999). *Introduction to Protein Structure,* 2nd ed. New York: Garland.

T. A. Brown (1999). *Genomes.* New York: Wiley-Liss.

H. Carillo and D. Lipman (1988). "The multiple sequence alignment problem in biology." *SIAM Journal on Applied Mathematics* 48: 1073–82.

R. Y. Chuan (1993). "A constraint logic programming framework for constructing DNA restriction maps." *Artificial Intelligence in Medicine* 5: 447–64.

T. Cormen, C. E. Leiserson, and R. L. Rivest (1990). *Introduction to Algorithms.* Cambridge, MA: MIT Press.

F. Crick (1966). "The genetic code." *Scientific American* 215: 55–60.

W. E. Day, D. S. Johnson, and D. Sankoff (1986). "Computational complexity of inferring rooted phylogenies by parsimony." *Mathematical Biosciences* 81: 33–42.

M. Dayhoff, R. M. Schwartz, and B. C. Orcutt (1978). *A Model of Evolutionary Change in Proteins,* vol. 5, suppl. 3. Silver Spring, MD: National Biomedical Research Foundation.

S. Dean and R. Staden (1991). "A sequence assembly and editing program for efficient management of large projects." *Nucleic Acids Research* 19: 3907–11.

A. Dembo, S. Karlin, and O. Zeitouni (1994). "Limit distribution of maximal non-aligned two-sequence segmental scores." *Annals of Probability* 22: 2022–39.

T. E. Dowling, C. Moritz, J. D. Palmer, and L. H. Rieseberg (1996). "Nucleic acids III: Analysis of fragments and restriction sites." In Hillis et al. (1996b), pp. 249–320.

R. Durbin, S. Eddy, A. Krogh, and G. Mitchison (1998). *Biological Sequence Analysis: Probabilistic Models of Proteins and Nucleic Acids.* Cambridge University Press.

B. Ewing and P. Green (1998). "Base-calling of automated sequence traces using *phred,* II: Error probabilities." *Genome Research* 8: 186–94.

D. Fasulo, T. Jiang, R. M. Karp, R. J. Settergren, and E. Thayer (1999). "An algorithmic approach to multiple complete digest mapping." *Journal of Computational Biology* 6: 187–207.

J. Felsenstein (1982). "Numerical methods for inferring evolutionary trees." *Quarterly Review of Biology* 57: 379–404.

J. Felsenstein (1989). "PHYLIP – phylogeny inference package (version 3.2)." *Cladistics* 5: 164–6.

D.-F. Feng and R. F. Doolittle (1987). "Progressive sequence alignment as a prerequisite to correct phylogentic trees." *Journal of Molecular Evolution* 25: 251–360.

J. W. Fickett and C.-S. Tung (1992). "Assessment of protein coding measures." *Nucleic Acids Research* 20: 6441–50.

K. A. Frenkel (1991). "The human genome project and informatics." *Communications of the ACM* 34.

D. J. Galas (2001). "Making sense of the sequence." *Science* 291: 1257+.

M. R. Garey and D. S. Johnson (1979). *Computers and Intractibility: A Guide to the Theory of NP-Completeness.* San Francisco: Freeman.

W. H. Gates and C. H. Papidimitriou (1979). "Bounds for sorting by prefix reversal." *Discrete Mathematics* 27: 47–57.

R. F. Gesteland, T. R. Cech, and J. F. Atkins (2000). *The RNA World.* Cold Spring Harbor, NY: Cold Spring Harbor Laboratory Press.

C. Gibas and P. Jambeck (2001). *Developing Bioinformatics Computer Skills.* Sebastopol, CA: O'Reilly.

L. Gonick and M. Wheelis (1991). *The Cartoon Guide to Genetics.* New York: Harper-Perennial.

D. S. Goodsell (1996). *Our Molecular Nature.* New York: Springer-Verlag.

O. Gotoh (1996). "Significant improvement in accuracy of multiple protein sequence alignmenst by iterative refinements as assessed by reference to structural alignments." *Journal of Molecular Biology* 264: 823–38.

S. K. Gupta, J. Kececioglu, and A. A. Schäffer (1995). "Improving the practical space and time efficiency of the shortest-paths approach to sum-of-pairs multiple sequence alignment." *Journal of Computational Biology* 2: 459–72.

D. Gusfield (1997). *Algorithms on Strings, Trees, and Sequences: Computer Science and Computational Biology.* Cambridge University Press.

S. Hannenhalli and P. A. Pevzner (1995a). "Transforming cabbage into turnip (polynomial algorithm for sorting signed reversals)." In *Proceedings of 27th Annual ACM Symposium on the Theory of Computing,* pp. 178–89. New York: Association for Computing Machinery.

S. Hannenhalli and P. A. Pevzner (1995b). "Transforming men into mice (polynomial algorithm for genomic distance problem)." In *Proceedings of IEEE 36th Annual Symposium on Foundations of Computer Science,* pp. 581–92. Los Alamitos, CA: IEEE Computer Society Press.

J. Hawkins (1996). *Gene Structure and Expression.* Cambridge University Press.

S. Henikoff and J. G. Henikoff (1992). "Amino acid substitution matrices from protein blocks." *Proceedings of the National Academy of Sciences U.S.A.* 89: 10915–19.

D. M. Hillis, B. K. Mable, A. Larson, S. K. Davis, and E. A. Zimmer (1996a). "Nucleic acids IV: Sequencing and cloning." In Hillis et al. (1996b), pp. 321–80.

D. M. Hillis, C. Moritz, and B. K. Mable (Eds.) (1996b). *Molecular Systematics,* 2nd ed. Sunderland, MA: Sinauer Associates.

K. Hoffmann, P. Bucher, L. Falquet, and A. Bairoch (1999). "The PROSITE database: Its status in 1999." *Nucleic Acids Research* 27: 215–19.

X. Huang (1996). "An improved sequence assembly program." *Genomics* 33: 21–31.

X. Huang, R. C. Hardison, and W. Miller (1990). "A space-efficient algorithm for local similarities." *Computer Applications in the Biosciences* 6: 373–81.

A. J. Jeffreys, N. J. Royle, V. Wilson, and Z. Wong (1988). "Spontaneous mutation rates to new length alleles at tandem-repetitive hypervariable loci in human DNA." *Nature* 332: 278–81.

A. J. Jeffreys, K. Tamaki, A. MacLeod, D. G. Monckton, D. L. Neil, and J. A. L. Armour (1994). "Complex gene conversion events in germline mutations in human minisatellites." *Nature Genetics* 6: 136–45.

A. J. Jeffreys, V. Wilson, and S. L. Thein (1985). "Hypervariable 'minisatellite' regions in human DNA." *Nature* 314: 67–73.

H. Kaplan, R. Shamir, and R. E. Tarjan (1997). "Faster and simpler algorithm for sorting signed permutations by reversals." In *Proceedings of Eighth Symposium on Discrete Algorithms,* pp. 344–51. Philadelphia: SIAM Press.

S. Karlin and S. F. Altschul (1990). "Methods for assessing the statistical significance of molecular sequence features by using general scoring schemes." *Proceedings of the National Academy of Sciences U.S.A.* 87: 2264–8.

S. Karlin and S. F. Altschul (1993). "Applications and statistics for high-scoring segments in molecular sequences." *Proceedings of the National Academy of Sciences U.S.A.* 90: 5873–7.

R. Kosaraju, J. Park, and C. Stein (1994). "Long tours and short superstrings." In *Proceedings of IEEE 35th Annual Symposium on Foundations of Computer Science,* pp. 166–77. Los Alamitos, CA: IEEE Computer Society Press.

J. A. Lake and J. M. Moore (1998). "Phylogenetic analysis and comparative genomics." In *Trends Guide to Bioinformatics.* Amsterdam: Elsevier.

E. S. Lander and M. S. Waterman (Eds.) (1995). *Calculating the Secrets of Life.* Washington, DC: National Academy Press.

J. K. Lee, V. Dancik, and M. S. Waterman (1998). "Estimation for restriction sites observed by optical mapping using reversible-jump Markov chain Monte Carlo." *Journal of Computational Biology* 5: 505–15.

B. Lewin (1999). *Genes VII.* Oxford University Press.

M. A. McClure, T. K. Vasi, and W. M. Fitch (1994). "Comparative analysis of multiple protein-sequence alignment methods." *Molecular Biology and Evolution* 11: 571–92.

W. Miller (1993). "Building multiple alignments from pairwise alignments." *Computer Applications in the Biological Sciences* 9: 169–76.

J. A. Morgan (1999). "Bioinformatic foundations: An analysis of the GenBank feature table and its suitability for automated processing." Master's thesis, North Carolina State University, Raleigh.

R. Mott and R. Tribe (1999). "Approximate statistics of gapped alignments." *Journal of Computational Biology* 6: 91–112.

D. W. Mount (2001). *Bioinformatics: Sequence and Genome Analysis.* Cold Spring Harbor, NY: Cold Spring Harbor Laboratory Press.

K. B. Mullis (1990). "The unusual origin of the polymerase chain reaction." *Scientific American* 262: 56–65.

E. W. Myers and W. Miller (1988). "Optimal alignments in linear space." *Computer Applications in the Biological Sciences* 4: 11–17.

S. B. Needleman and C. D. Wunsch (1970). "A general method applicable to the search for similarities in the amino acid sequence of two proteins." *Journal of Molecular Biology* 48: 443–53.

M. Nei and S. Kumar (2000). *Molecular Evolution and Phylogenetics.* Oxford University Press.

D. S. T. Nicholl (1994). *An Introduction to Genetic Engineering.* Cambridge University Press.

C. Notredame, D. Higgins, and J. Heringa (2000). "T-COFFEE: A novel method for fast and accurate multiple sequence alignment." *Journal of Molecular Biology* 302: 205–17.

C.-Y. Ou et al. (1992). "Molecular epidemiology of HIV transmission in a dental practice." *Science* 256: 1165–71.

S. R. Palumbi (1996). "Nucleic acids II: The polymerase chain reaction." In Hillis et al. (1996b), pp. 205–48.

L. Parida (1998). "A uniform framework for ordered restriction map problems." *Journal of Computational Biology* 5: 725–39.

J. Pearl (1984). *Heuristics: Intelligent Search Strategies for Computer Problem Solving.* Reading, MA: Addison-Wesley.

W. R. Pearson (1989). "Rapid and sensitive sequence comparisions with FASTP and FASTA." *Methods in Enzymology* 183: 63–98.

W. R. Pearson (1995). "Comparison of methods for searching protein sequence databases." *Protein Science* 4: 1145–60.

W. R. Pearson and D. J. Lipman (1988). "Improved tools for biological sequence comparison." *Proceedings of the National Academy of Sciences U.S.A.* 85: 2444–8.

E. Pennisi (2001). "The human genome." *Science* 291: 1177–80.

P. A. Pevzner (2000). *Computational Molecular Biology.* Cambridge, MA: MIT Press.

J. R. Powell and E. N. Moriyama (1997). "Evolution of codon usage bias in *Drosophila.*" *Proceedings of the National Academy of Sciences U.S.A.* 94: 7784–90.

P. J. Russell (1992). *Genetics.* New York: HarperCollins.

M.-F. Sagot and E. W. Myers (1998). "Identifying satellites and periodic repetitions in biological sequences." *Journal of Computational Biology* 5: 539–53.

S. Salzberg, A. Delcher, S. Kasif, and O. White (1998a). "Microbial gene identification using interpolated Markov models." *Nucleic Acids Research* 26: 544–8.

S. L. Salzberg, D. B. Searls, and S. Kasif (Eds.) (1998b). *Computational Methods in Molecular Biology.* Amsterdam: Elsevier.

J. Setubal and J. Meidanis (1997). *Introduction to Computational Molecular Biology.* Boston: PWS.

P. M. Sharp, M. Stenico, J. F. Peden, and A. T. Lloyd (1993). "Mutational bias, translational selection, or both?" *Biochemical Society Transactions* 21: 835–41.

S. S. Skiena and G. Sundaraman (1994). "A partial digest approach to restriction site mapping." *Bulletin of Mathematical Biology* 56: 275–94.

R. F. Smith (1996). "Perspectives: Sequence data base searching in the era of large-scale genomic sequencing." *Genome Research* 6: 653–60.

T. Smith and M. S. Waterman (1981). "Identification of common molecular subsequences." *Journal of Molecular Biology* 147: 195–7.

P. Sneath and R. Sokal (1973). *Numerical Taxonomy: The Principles and Practice of Numerical Classification.* San Francisco: Freeman.

D. Snustad and M. Simmons (1999). *Principles of Genetics,* 2nd ed. New York: Wiley.

S. Srinivasan (1997). *Advanced Perl Programming.* Sebastopol, CA: O'Reilly.

R. Staden and A. D. McLachlan (1982). "Codon preference and its use in identifying protein coding regions in long DNA sequences." *Nucleic Acids Research* 10: 141–56.

G. Stormo (2000). "Gene-finding approaches for eukaryotes." *Genetics Research* 10: 394–7.

L. Stryer (1981). *Biochemistry,* 2nd ed. San Francisco: Freeman.

D. L. Swofford (1990). *PAUP, Phylogenetic Analysis Using Parsimony,* 3.5 ed. Champaign, IL.

D. L. Swofford, G. J. Olsen, P. J. Waddell, and D. M. Hillis (1996). "Phylogenetic inference." In Hillis et al. (1996b), pp. 321–80.

J. D. Thompson, D. G. Higgins, and T. J. Gibson (1994). "CLUSTAL-W: Improving the sensitivity of progressive multiple sequence alignment through sequence weighting, position-specific gap penalties and weight matrix choice." *Nucleic Acids Research* 22: 4673–80.

J. D. Thompson, F. Plewniak, and O. Poch (1999). "A comprehensive comparison of multiple sequence alignment programs." *Nucleic Acids Research* 27: 2682–90.

J. Vromans (2000). *Perl 5 Pocket Reference,* 3rd ed. Sebastopol, CA: O'Reilly.

L. Wall, T. Christiansen, and J. Orwant (1997). *Programming Perl.* Sebastopol, CA: O'Reilly.

M. S. Waterman and T. F. Smith (1986). "Rapid dynamic programming algorithms for RNA secondary structure." *Advances in Applied Mathematics* 7: 455–64.

J. D. Watson (1998). *The Double Helix,* reprint ed. New York: Simon & Schuster.

L. W. Wright, J. B. Lichter, J. Reinitz, M. A. Shifman, K. K. Kidd, and P. L. Miller (1994). "Computer-assisted restriction mapping: An integrated approach to handling experimental uncertainty." *Computer Applications in the Biological Sciences* 10: 435–42.

M. Zuker (1989). *The Use of Dynamic Programming Algorithms in RNA Secondary Structure Prediction.* Boca Raton, FL: CRC Press.

M. Zuker, D. Mathews, and D. Turner (1999). "Algorithms and thermodynamics for RNA secondary structure prediction: A practical guide." In J. Barciszewski and B. F. C. Clark (Eds.), *RNA Biochemistry and Biotechnology.* Dordrecht: Kluwer.

Index

GENOMIC PERL CONSULTANCY, INC.
Raleigh, North Carolina

DISCLAIMER OF WARRANTY and License Terms and Conditions

The Genomic Perl software ("our Software") consists of the computer programs that appear in the book *Genomic Perl: From Bioinformatics Basics to Working Code* ("the book") published by Cambridge University Press, plus a few incidental utility programs. Our software is copyrighted, and you can use it only if you have a license from us. Unless you have an individually signed license that supersedes the terms on this page, your use of our software signifies your agreement with these terms. READ THIS LICENSE AGREEMENT CAREFULLY. IF YOU DO NOT ACCEPT ITS TERMS, DO NOT BREAK THE SEAL ON THE MEDIA ENVELOPE AND RETURN THE BOOK TO PLACE OF PURCHASE FOR REFUND.

If you are (i) an owner of the book, or (ii) a full-time student at a bona fide institution of higher education, you may copy and execute our software, modify our software, and incorporate our software into your own programs provided that: (i) copies of our software, whether in original, modified, or incorporated form, contain the original copyright notice of our software; (ii) our software is available for execution either (a) on no more than three (3) CPUs or (b) by no more than three (3) users, whether in original, modified, or incorporated form; (iii) our software is not redistributed for commercial gain, whether in original, modified, or incorporated form; and (iv) our software is not installed on a World Wide Web server or anonymous ftp directory from which it could be downloaded or copied by unlicensed users, whether in original, modified, or incorporated form.

Any other use of our software is a violation of this license. Inquiries regarding other licenses may be directed to GenomicPerl@nc.rr.com.

The distribution media also include DNA and protein sequence files provided to illustrate the operation of the Genomic Perl software. These sequences have been extracted from public databases, and your use and redistribution of them is in no way subject to our control.

In using our software, you acknowledge and accept this DISCLAIMER OF WARRANTY:

> OUR SOFTWARE IS FURNISHED "AS IS". WE MAKE NO WARRANTIES, EXPRESS OR IMPLIED, THAT OUR SOFTWARE IS FREE OF ERROR, OR IS CONSISTENT WITH ANY PARTICULAR STANDARD OF MERCHANTABILITY, OR THAT IT WILL MEET YOUR REQUIREMENTS FOR ANY PARTICULAR APPLICATION. WE ACCEPT NO RESPONSIBILITY FOR ANY MATHEMATICAL OR TECHNICAL LIMITATIONS OF THE PROCEDURES AND FUNCTIONS WHICH MAKE UP OUR SOFTWARE. OUR SOFTWARE SHOULD NOT BE RELIED ON FOR SOLVING A PROBLEM WHOSE INCORRECT SOLUTION COULD RESULT IN INJURY TO A PERSON OR LOSS OF PROPERTY. NEITHER GENOMIC PERL CONSULTANCY, INC., NOR THE AUTHOR OF THE BOOK "GENOMIC PERL: FROM BIOINFORMATICS BASICS TO WORKING CODE", NOR THE PUBLISHER OF THAT BOOK SHALL IN ANY EVENT BE LIABLE FOR ANY DAMAGES, WHETHER DIRECT OR INDIRECT, SPECIAL OR GENERAL, CONSEQUENTIAL OR INCIDENTAL, ARISING FROM USE OF OUR SOFTWARE. YOUR USE OF OUR SOFTWARE IS ENTIRELY AT YOUR OWN RISK.